D1669468

Lipoxygenase
and Lipoxygenase Pathway Enzymes

Lipoxygenase and Lipoxygenase Pathway Enzymes

Editor

George Piazza
U.S. Department of Agriculture
Agricultural Research Service
Philadelphia, Pennsylvania

Champaign, Illinois

AOCS Mission Statement

To be a forum for the exchange of ideas, information, and experience among those with a professional interest in the science and technology of fats, oils, and related substances in ways that promote personal excellence and provide high standards of quality.

The paper used in this book is acid-free and falls within the guidelines established to ensure permanence and durability.

Library of Congress Cataloging-in-Publication Data

Lipoxygenase and lipoxygenase pathway enzymes/editor, G. Piazza.
p. cm.
Includes bibliographical references and index.
ISBN 0-935315-67-5 (alk. paper)
1. Lipoxygenases. I. Piazza, George J.
QP603.L56L55 1996
574.19′253—dc20 96-13078
CIP

Printed in the United States of America with vegetable oil-based inks.
00 99 98 97 96 5 4 3 2 1

Preface

This book presents timely and accurate information on various research approaches designed to determine the structure, mechanism of action, and physiological significance of lipoxygenase. To a lesser degree information on those enzymes that metabolize fatty acid hydroperoxides is also included in this monograph. Two of the chapters deal with lipoxygenase genetics and genes. Three chapters present the results of studies designed to elucidate the mechanism of action of lipoxygenase. Two chapters are concerned with the structure of lipoxygenase—one through X-ray crystallography and the other through site-directed mutagenesis. The oxylipin and peroxygenase pathways are discussed in two chapters. Three additional topics that are covered are the role of lipoxygenase in the resistance of plants to infection, the use of lipoxygenase as an aid to organic synthesis and the use of lipoxygenase-null soybeans in the preparation of foods.

George Piazza

Contents

Chapter 1

Lipoxygenases: Structure and Function

Sean T. Prigge[a], Jeffrey C. Boyington[a], Betty J. Gaffney[b], and L. Mario Amzel[a]

[a]Johns Hopkins School of Medicine, Department of Biophysics and Biophysical Chemistry, Baltimore, MD 21205, USA; [b]Johns Hopkins University, Department of Chemistry, Baltimore, MD 21218, USA.

Introduction

Lipoxygenases are a class of nonheme iron dioxygenase that catalyzes the hydroperoxidation of polyunsaturated fatty acids in plants, animals, and microorganisms. In addition to the natural substrates—linoleic acid and arachidonic acid—a wide variety of polyunsaturated fatty acids can act as substrates of these enzymes. Plant lipoxygenases catalyze the hydroperoxidation of linoleic acid (18:2) as the first step in the biosynthesis of the growth-regulatory substance—jasmonic acid—and of factors involved in wound healing—traumatin and traumatic acid (1–4). Mammalian lipoxygenases catalyze the hydroperoxidation of arachidonic acid (20:4), initiating the synthesis of two families of potent physiological effectors: leukotrienes and lipoxins (5–8).

Lipoxygenases form a closely related family with no similarities to other known sequences. Pairwise sequence identity between plant and mammalian lipoxygenases is 21–27%, while identity among pairs of plant sequences is 43–86%, and identity among pairs of mammalian sequences is 39–93%. The sequence identity between plant and mammalian lipoxygenases is highest in the portions of the catalytic domain near the iron atom. The 61 residues from W479 to N539 in soybean lipoxygenase-1 (SBL-1) include 16 residues that are conserved in all lipoxygenases. The amino-terminal 200 residues from the plant lipoxygenases have little homology to the mammalian sequences ($\leq$ 15% pairwise sequence identity). Since the mammalian enzymes (662–674 residues in length, including the amino-terminal methionine) are 165–261 residues shorter than the plant enzymes (839–923 residues in length), it is likely that large structural differences exist between the amino-termini of plant and mammalian lipoxygenases.

The iron atom, which is essential for enzymatic activity, exists in two oxidation states: Fe^{2+} and Fe^{3+}; in both forms the iron is in the high spin state (spin 4/2 and spin 5/2, respectively [9]). Spectroscopic data show that the metal is bound to nitrogen- and oxygen-containing groups in the protein; no evidence for heme or for sulfur ligands has been found (10–12). Structural data confirm that the iron atom is coordinated

This work was supported by Grant GM 44692 (to LMA) and Grant GM 36232 (to BJG) from the National Institute of Health.

by three imidazole nitrogens (Nε H499, Nε H504, and Nε H690) and one carboxylate oxygen (carboxy-terminal OT2 I839) that are conserved in all lipoxygenases (13).

Lipoxygenases require fatty acids containing a 1,4-diene with at least one of the olefins in the Z geometry (14). The lipoxygenase reaction proceeds with proton abstraction from the methylene carbon (carbon 3) of substrates containing a 1,4-diene followed by addition of dioxygen to one of the olefinic carbons (either carbon 1 or 5). The major product of reactions with fatty acids containing a 1Z,4Z-diene is the (S)-hydroperoxy fatty acid, and abstraction of one of the methylene pentadienyl protons occurs antarafacial to addition of dioxygen. The nomenclature of mammalian lipoxygenases is based on positional specificity and includes the carbon atom (5th, 8th, 12th, or 15th carbon) to which dioxygen binds in products of (5Z,8Z,11Z,14Z)-eicosatetraenoic acid (arachidonic acid).

Description of the Structure

Overall Structure

Soybean lipoxygenase-1 is a two-domain, single-chain prolate ellipsoid of dimension 90 × 65 × 60 Å (Fig. 1.1). The three-dimensional structure of SBL-1 contains 38% α-helix and 14% β-sheet. Domain I is an eight-stranded antiparallel β-barrel with jelly roll–like topology that comprises the 146 amino-terminal residues (Figs. 1.1 and 1.2). The interior of the barrel is densely packed with hydrophobic side chains, many of which are aromatic. This domain is identical in connectivity to a similar β-barrel observed in the carboxy-terminal portion of human pancreatic lipase (involved in colipase binding), even though the two regions have no significant sequence similarity (15). Domain I is separate and makes only a loose contact with the rest of the molecule.

The major domain of the protein, domain II, comprises the 693 carboxy-terminal residues in a bundle of 23 helices; two antiparallel β-sheets lie flat on the surface on opposite sides of the domain (Fig. 1.1). Seventeen of the helices are approximately parallel or antiparallel to each other and surround a long central helix of 43 residues that extends along the longest dimension of the molecule (Figs. 1.1 and 1.2). Although the central helix is surrounded on all sides by other helices, regions on each end are exposed to the solvent; the middle portion of this helix, though buried, is not exclusively hydrophobic either. Although none of the helices are completely hydrophobic, there is an extensive network of hydrophobic clusters throughout the domain, including five tryptophans that contact each other on one side of Helix 17.

Domain II contains two distinct sections of π-helix. A π-helix, with hydrogen bonding between residues i and $i + 5$ instead of i and $i + 4$, is rare. The π-helices lie in the middle of the two largest α-helices in the molecule, and each contains residues that serve as ligands to the iron. The first π-helix, located in Helix 9, is the longest reported in a globular protein so far. It consists of 13 residues from H494 to A506 (16). Two of these, H499 and H504, are iron ligands. The second π-helix (6 residues; I685–H690) occurs in Helix 18 and contains H690, the third iron ligand.

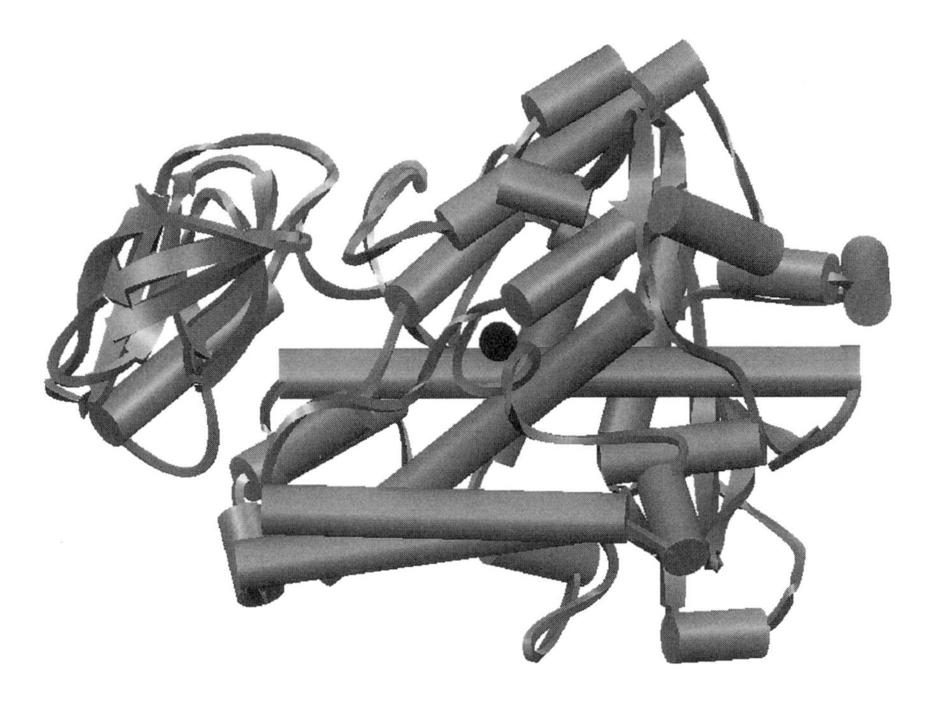

Figure 1.1. Schematic diagram of the three-dimensional structure of SBL-1. The α-helices are represented by cylinders, the strands in the β-sheets by arrows, the coils by narrow rods, and the iron by a sphere. Domain I, consisting of a single β-barrel, is on the left. Domain II contains the iron, all the helices in the structure, and two small β-sheet structures that lie on the surface of the enzyme.

Coordination of the Iron

The iron in SBL-1 is coordinated to four ligands at the center of domain II. The coordination can be best described as a highly distorted octahedron with two adjacent, unoccupied positions (Fig. 1.3 and Table 1.1). An alternative description is a tetrahedron in which one ligand position is significantly distorted from tetrahedral geometry.

Three of the coordination positions are occupied by the Nε of the three histidine residues described previously (H499, H504, and H690), and the fourth position is occupied by an oxygen of the carboxy-terminus (residue I839). The carboxylate provides the only negative charge in the immediate vicinity of the iron. The oxygen is about 1.5 Å away from the position expected for a tetrahedral arrangement, giving rise to an O–Fe–Nε angle with the Nε of H499 of 150.8° (164.6° in monomer 2), close to the 180° angle expected for octahedral coordination (Table 1.1). The three Fe–Nε and the Fe–O distances displayed in Table 1.1 are similar to other distances observed in small molecule iron complexes (17).

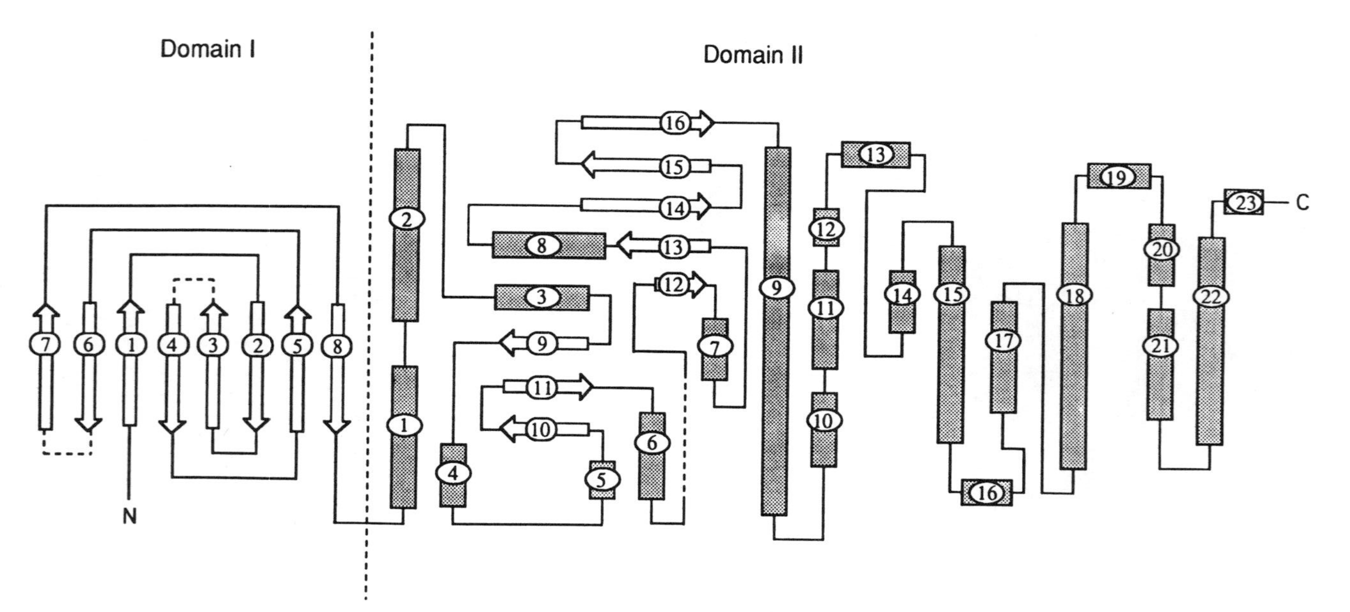

Figure 1.2. Schematic diagram of the secondary structure of SBL-1. The helices are represented by shaded rectangles, the β-sheet strands by arrows, and the connecting loops by continuous lines. Domain I is to the left of the vertical dotted line and domain II is to the right. The strands (1–16) and the helices (1–23) are numbered in the order in which they appear in the structure. Helices, represented as vertical rectangles, are almost parallel (average angle 28°) to the major helix, number 9, and the horizontal ones are almost perpendicular (average angle 73°) to it. The lengths of the rectangles are approximately proportional to the length of the helices. Dotted lines represent regions that were not built into the model. *Source:* Kabsch and Sander (16).

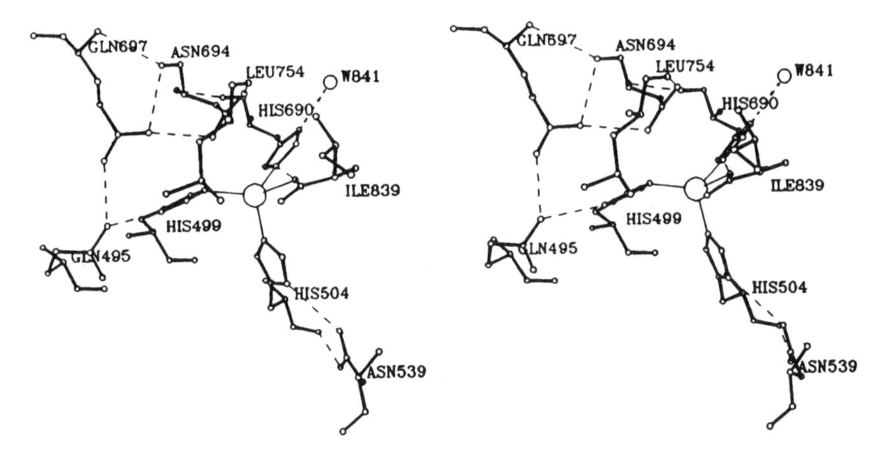

Figure 1.3. Stereo view of the iron center in SBL-1. The residues **H499, H504, H690**, and the terminal carboxylate of **I839** are directly coordinated to the iron. Dotted lines are hydrogen bonds.

TABLE 1.1 Coordination Geometry at the Iron Coordination Center[a]

Bonds	Metal–ligand distances (Å) Monomer 1	Monomer 2
Fe–H499 Nε	2.29	2.27
Fe–H504 Nε	2.13	2.24
Fe–H690 Nε	2.19	2.19
Fe–I839 OT2	2.07	2.08

Angles	Ligand–metal–ligand angles (degrees) Monomer 1	Monomer 2
H499 Nε–Fe–H504 Nε[b]	102.6	102.2
H499 Nε–Fe–H690 Nε[b]	100.9	109.6
H504 Nε–Fe–H690 Nε[b]	104.0	101.4
H690 Nε–Fe–I839 OT2[b]	78.1	73.5
H504 Nε–Fe–I839 OT2[b]	106.0	91.7
H499 Nε–Fe–I839 OT2[c]	150.8	164.6

[a]The geometry for both molecules in the asymmetric unit (monomer 1 and monomer 2) is displayed. OT2 is the carboxy-terminal oxygen number 2.
[b]109.5° in a tetrahedron and 90.0° in an octahedron.
[c]109.5° in a tetrahedron and 180.0° in an octahedron.

The plane defined by the Nε atoms of the three ligand imidazoles can be considered to be the base of the distorted tetrahedron. The planes of all three of the imidazole rings are perpendicular to this base. All of the Cδ–Nε–Fe and Cε–Nε–Fe angles formed by the ligands are close to 120°. As expected, two of the Fe–Nε bonds are in the plane of their imidazole rings. In contrast, the third Fe–Nε bond (H499) makes a

33.0° angle with the plane of the imidazole ring, suggesting the possibility of *sp*³ hybridization for this nitrogen (Fig. 1.3). Although it is possible to build the side chain at the position of H499 in a conformation similar to that of the other two histidine ligands, electron-density maps calculated with this residue omitted clearly indicate that the imidazole ring occupies the position described previously.

The orientations of the three histidine ligands are determined by their location in regions of π-helix and by their participation in a network of hydrogen bonds mediated by both main- and side-chain atoms (Fig. 1.3). The Nδ of H499 is hydrogen bonded to the side chain of Q495 at the beginning of a series of hydrogen bonds that connect Q697, N694, and the main-chain carbonyl of L754. The Nδ of H504 is hydrogen bonded to the side chain of N539. Similarly, H690 appears to be hydrogen bonded through a water (Wat841) to the main-chain carbonyls of S836 and S687. The Nδ of N694, a part of this network, is only 3.3 Å from the iron and is very close to the unoccupied coordination position opposite to H504. In six mammalian lipoxygenase sequences, N694 is replaced by histidine.

Spectroscopic studies to determine the coordination of the iron in ferrous lipoxygenases have favored octahedral coordination of the metal. On the basis of magnetic circular dichroism data, the effective site symmetry of ferrous SBL-1 was assigned as distorted octahedral (18). Extended X-ray absorption fine structure studies were best interpreted as indicating the presence of 6 ± 1 nitrogen or oxygen atoms (in any combination) at distances of 2.05–2.09 Å from the iron (19). The quadrupole-splitting parameters from Mössbauer measurements is also consistent with distorted octahedral ligand symmetry (10).

The coordination of the iron with only four ligands in a distorted octahedral arrangement leaves two unoccupied ligand positions. Because the spectroscopic evidence seems to favor more than four iron ligands, the possible presence of the fifth and sixth ligands, such as a water or a hydroxyl group, was investigated in detail. Difference (F_o - F_c) maps and $2F_o$ - F_c maps, calculated using data from 50 to 2.6 Å in the region close to the iron, showed no significant density in the position expected for these ligands. It is possible, however, that these positions become occupied during the catalytic cycle of the enzyme.

Tetrahedral iron is not common in small molecular complexes of either Fe^{2+} or Fe^{3+}; both oxidation states appear to favor the formation of six-coordinate octahedral complexes. However, tetrahedral coordination similar to that of lipoxygenase has been observed in iron superoxide dismutase (20). In both proteins, the iron is coordinated by three histidines and one carboxylate, the terminal carboxylate in lipoxygenase and a glutamate in superoxide dismutase. Both proteins have similar Fe–O and Fe–N distances, but the coordination is distorted from the tetrahedron in a different manner in the two structures, being closer to a trigonal bipyrimid in superoxide dismutase. The similarity of the iron centers is not a consequence of an overall similarity between the two proteins, because they have no sequence similarity and have completely different folding patterns. Even the structural elements that provide the side chains for iron coordination differ in the two proteins.

Internal Cavities

The coordinated iron faces two large internal cavities that can connect the metal to the exterior of the molecule. Cavity I is conical and forms a tunnel that connects the position opposite the Nε of H504 to the surface of the protein (Fig. 1.4). The tunnel is 18 Å in length. It is 8 Å in diameter for most of its length, but widens to 11 Å at the surface and narrows to 2.5 Å close to the iron center. The sides of the tunnel are lined with the side chains of 29 residues, most of which are hydrophobic. Cavity I residues are C357, V358, I359, R360, D408, Y409, I412, Y493, M497, S498, H499, L501, N502, T503, V570, N573, W574, V575, D578, Q579, L581, D584, K587, R588, Y610, W684, L689, H690, and V693. Some of these residues are conserved (or have only conservative substitutions) among all sequenced plant and mammalian lipoxygenases. This tunnel presents an ideal path for the movement of molecular oxygen from the outside into one of the two unoccupied coordination sites of the iron.

Cavity II faces the terminal carboxylate and two of the histidine ligands (H499 and H504). This 40 Å long, narrow cavity (less than 3.5 Å wide in some places) changes direction by more than 90° at two places (Fig. 1.4). One bend is adjacent to the iron center, close to the end of the cavity. Arachidonic acid (20:4) or even a slightly larger fatty acid can fit into the end of this bent region, in the innermost portion of the cavity. Arachidonic acid bound in this manner would come close to the iron with its 11,14-diene opposite to H690. Sequence comparisons of the different lipoxygenases reveal a high degree of conservation for the 46 residues lining this cavity. Eleven of the cavity II residues are conserved across the 22 known lipoxygenase sequences: W340, F346, G353, N355, K483, D490, H499, H504, Q697, N706, and L754. Most of the side chains in cavity II are either hydrophobic or neutral, with the exception of

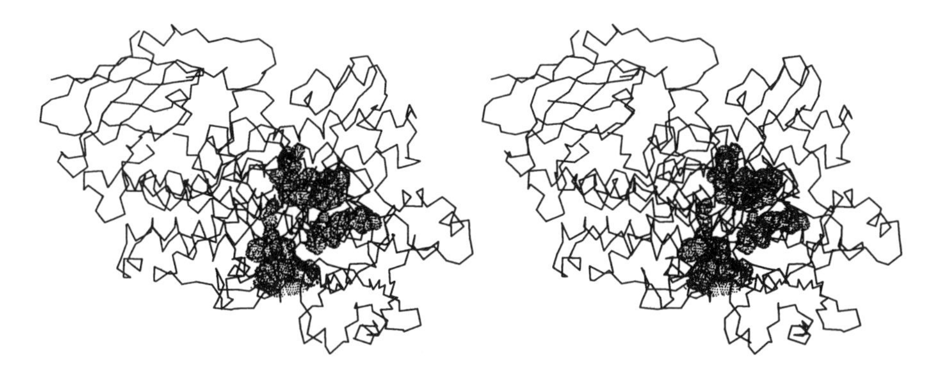

Figure 1.4. Location of cavities I and II in SBL-1. The surfaces of the cavities, represented by dots, are superimposed on the α–carbon trace of the enzyme. Cavity I is conical and opens toward the bottom of the enzyme. The iron atom is represented by a small sphere at the tip of cavity I and cavity II is the remaining surface. *Source:* Connolly (59).

four residues: E349, D490, H494, and R707. At the other bend, approximately in the middle, is where the cavity becomes narrow—primarily due to a buried salt bridge between D490 and R707.

The largest cavity is observed in the region between domains I and II. It is an oblate ellipsoid of 11 × 18 × 19 Å and is connected to the surface by a short, narrow elliptical channel with a cross section of approximately 5 × 7 Å. This cavity, much as the other two, is lined mostly by hydrophobic or neutral residues. It is neither close to nor connected with the other cavities or the iron site; it may be simply an interdomain space.

Structure of the Human Lipoxygenases

It is likely that plant lipoxygenases are structurally similar to SBL-1, due to the high pairwise sequence identity between plant lipoxygenases (43–86%). The more radical differences between plant and mammalian sequences (21–27% pairwise sequence identity) raise the possibility that the structures of the mammalian lipoxygenases differ significantly from the SBL-1 structure. Homology models of three mammalian lipoxygenases; human leukocyte 5-lipoxygenase (hl5lo), human platelet 12-lipoxygenase (hp12lo), and human reticulocyte 15-lipoxygenase (hr15lo); provide a means of rationalizing structural differences between plant and mammalian lipoxygenases and identifying structural similarities common to all lipoxygenases. In addition, models of three major classes of lipoxygenases can be used to probe the structural determinants of lipoxygenase positional specificity.

Lipoxygenase sequence numbers refer to the residue number in soybean lipoxygenase as aligned in Table 1.2, unless noted otherwise. Thus, H544 in hr15lo would be referred to as H694, or as H544 hr15lo.

Structure-Based Sequence Alignment

Several alignments of plant and animal lipoxygenase sequences have been published (21–23). The structure-based alignment agrees at the majority of residues, but there are four areas that differ significantly. These areas all occur in solvent-exposed regions of the structure where they can be accommodated without major changes in the overall structure. This suggests that the structure of domain II of SBL-1 can provide a good initial model for the interpretation of mutational and other data in the mammalian enzymes.

Gap 1. There is a large gap in mammalian lipoxygenases in the 32-residue loop region between Strand 12 and Helix 7 in SBL-1 (residues 363–394). Much of this region (about 40 residues including Helix 7) is on the surface of the protein where neighboring regions of the protein influence the conformation of the residues less. The main chain folds back on itself twice in this region, providing good sites for gaps to be placed (the distance between positions 385 and 395 is 6.6 Å; and the distance between positions 371 and 377 is 6.5 Å). Both sites were used to place the 13-residue gap in hp12lo and hr15lo, whereas only the former site was used in hl5lo due to the smaller gap size (9 residues).

Gap 2. The region between Helix 8 and Helix 9 (residues 424–473) contains three β-strands connected by small loops. The loop between Strand 15 and Strand 16 is the largest and allows up to six residues to be cut without affecting the β-strands (the distance between positions 454 and 461 is 5.4 Å). A second gap of three residues (472–474) shortens Helix 9, but optimizes the alignment of conserved residues—especially P469. It is possible that the residues comprising Strand 16 (between the two gaps) are not part of the β-sheet in the mammalian lipoxygenases, since prolines—especially adjacent prolines—tend to disrupt the hydrogen-bonding pattern of β-sheet structures.

Gap 3. The region between Helix 18 and Helix 21 (residues 702–740) contains a seven-residue gap (in hr15lo the gap is eight residues). There is a long loop after Helix 18 followed by Helix 19 and Helix 20—both short, eight-residue α-helices on the surface of the protein. It is difficult to cut residues from the loop for three reasons: most of the loop is buried in the protein core, the loop has a linear conformation (residues separated by seven positions are > 15 Å apart), and the sequence homology to SBL-1 is high. Removal of the most peripheral helix, Helix 19, allows Helix 20 to remain in place, whereas the removal of Helix 20 requires the drastic relocation of Helix 19. Since the position of Helix 19 is not conserved in either case, the gap was finally placed in the models of the human lipoxygenases at Helix 19.

Gap 4. The region from Helix 22 to the carboxy-terminus (residues 801–839) is 18 residues shorter in the human lipoxygenases. The main chain folds back on itself twice in this region, providing two sites for gaps to be placed (the distance between positions 800 and 816 is 7.6 Å; and the distance between positions 823 and 832 is 4.4 Å). If residues 801–815 are removed from the first site, conserved residues N800, P816, and Y817 shift position to reconnect the main chain. In the models, only 10 residues were removed from this site and eight from the second site (between positions 823 and 832), allowing the conserved residues to remain unperturbed.

Amino-Terminal Residues. The amino-terminal residues of the human lipoxygenases (138 residues in hl5lo and 140 residues in hp12lo and hr15lo) are 11–15.5% identical to the first 254 residues of SBL-1. The poor homology and large size difference suggest that major structural differences exist between the amino-termini of plant and mammalian lipoxygenases. One way to account for this large size difference is to assume that the β-barrel domain of SBL-1 (145 residues in size) is not present in the mammalian enzymes. Without the β-barrel, lipoxygenases would begin with a short loop (146–155) and Helix 1 (156–170) followed by a long 84-residue loop. The residues that appear to be important in organizing the 84-residue loop are rarely conserved between plant and mammalian lipoxygenases. This leads one to guess that these residues either form a small amino-terminal domain while foreshortening the catalytic domain, or that these residues fold onto the catalytic domain in some novel way.

TABLE 1.2 Structure-Based Sequence Alignment of SBL-1, h15lo, hp12lo, and hr15lo[a]

```
                      .         .         .         .         .         .
                  .........α2.........           ....α3....-β9- ...α4...   .α5
SB1         255 LEIGTKSLSQIVQPAFESAFDLKSTPIEFHSFQDVHDLYEGGIKLPRDVISTIIPLPVIK 314
h15lo       142 YRWMEWNPGFPLSIDAKCHKDLP*RDIQFDSEKGVDFVLNYSKAMENLFINRFMHMFQSS 200
hp12lo      139 YCWATWKEGLPLTIAADRKDDLP*PNMRFHEEKRLDFEWTLKAGALEMALKRVYTLL*SS 196
hr15lo      139 YRWGNWKDGLILNMAGAKLYDLP*VDERFLEDKRVDFEVSLAKGLADLAIKDSLNVL*TC 196
 Identity       3232232 3223 232    443  2224222323332  2 222  22322   22 22

                      .         .         .         .         .         .
                ..-β10  -β11-           ....α6.....      -β12
SB1         315 ELYRTDGQHILKFPQPHVVQVSQSAWMTDEEFAREMIAGVNPCVIRGLEEFPPKSNLDPA 374
h15lo       201 WNDFADFEKIFVKISNTISERVMNHWQEDLMFGYQFLNGCNPVLIRRCTELPEKLPVTTE 260
hp12lo      197 WNCLEDFDQIFWGQKSALAEKVRQCWQDDELFSYQFLNGANPMLLRRSTSLPSRLVL*** 253
hr15lo      197 WKDLDDFNRIFWCGQSKLAERVRDSWKEDALFGYQFLNGANPVVLRRSAHLPARLVF*** 253
 Identity       3222 43   432   22 223232   4224224233333424422243222234 2322

                      .         .         .         .         .         .
                                   ..α7...  --β13- .......α8......       --β14
SB1         375 IYGDQSSKITADSLDLDGYTMDEALGSRRLFMLDYHDIFMPYVRQINQLNSAKTYATRTI 434
h15lo       261 MVECSLERQLS*********LEQEVQQGNIFIVDFELLDGIDANKTDPCTLQFLAAPICL 311
hp12lo      254 **PSGMEELQA********QLEKELQNGSLFEADFILLDGIPANVIRGEK*QYLAAPLVM 302
hr15lo      254 **PPGMEELQA********QLEKELEGGTLFEADFSLLDGIKANVILCSQ*QHLAAPLVM 302
 Identity         2 2232223          2332332 3 342243 33333 3323     3 3343222

                      .         .         .         .         .         .
                ---    -----β15----       ---β16--     ....................
SB1         435 LFLREDGTLKPVAIELSLPHSAGDLSAAVSQVVLPAKEGVESTIWLLAKAYVIVNDSCYH 494
h15lo       312 LYKNLANKIVPIAIQLNQIP******GDENPIFLPSD***AKYDWLLAKIWVRSSDFHVH 362
hp12lo      303 LKMEPNGKLQPMVIQIQPPSP****SSPTPTLFLPSD***PPLAWLLAKSWVRNSDFQLH 355
hr15lo      303 LKLQPDGKLLPMVIQLQLPRT****GSPPPPLFLPTD***PPMAWLLAKCWVRSSDFQLH 355
 Identity       422 22333 422433223        222 22234423    22 244444 3432343224

                      .         .         .         .         .         .
                α9.....................     TT ..α10..TT TT.....α11....     α12
SB1         495 QLMSHWLNTHAAMEPFVIATHRHLSVLHPIYKLLTPHYRNNMNINALARQSLINANGIIE 554
h15lo       363 QTITHLLRTHLVSEVFGIAMYRQLPAVHPIFKLLVAHVRFTIAINTKAREQLICECGLFD 422
hp12lo      356 EIQYHLLNTHLVAEVIAVATMRCLPGLHPIFKFLIPHIRYTMEINTRARTQLISDGGIFD 415
hr15lo      356 ELQSHLLRGHLMAEVIVVATMRCLPSIHPIFKLIIPHLRYTLEINVRARTGLVSDMGIFD 415
 Identity       22224342343224322243242 43 24443433234 42322442244224322 4333
```

```
                   ..    ......α13.....          ...α14..                              ....
SB1         555  TTFLP*SKYSVEMSSAVYKNWVFTDQALPADLIKRGVAIKDPSTPHGVRLLIEDYPYAAD 613
hl5lo       423  KANATGGGGHVQMVQRAMKDLTYASLCFPEAIKARGMES**********KEDIPYYFYRDD 473
hp12lo      416  KAVSTGGGGHVQLLRRAAAQLTYCSLCPPDDLADRGL************LGLPGALYAHD 463
hr15lo      416  QIMSTGGGGHVQLLKQAGAFLTYSSFCPPDDLADRGL************LGVKSSFYAQD 463
 Identity        22 23333334322 23 2 333 3232423322442             3222 2243 4

                 .......α15..........  ..α16.  ......α17.....                  ..
SB1         614  GLEIWAAIKTWVQEYVPLYYARDDDVKNDSELQHWWKEAVEKGHGDLKDKPWWPKLQTLE 673
hl5lo       474  GLLVWEAIRTFTAEVVDIYYEGDQVVEEDPELQDFVNDVYVYGMRGRKSSGFPKSVKSRE 533
hp12lo      464  ALRLWEIIARYVEGIVHLFYQRDDIVKGDPELQAWCREITEVGLCQAQDRGFPVSFQSQS 523
hr15lo      464  ALRLWEIIYRYVEGIVSLHYKTDVAVKDDPELQTWCREITEIGLQGAQDRGFPVSLQARD 523
 Identity        24224324 2232224 324 242 43 43444 3223223 42 2223233323232222

                 ............α18.............                  ...α19.. .α20
SB1         674  DLVEVCLIIIWIASALHAAVNFGQYPYGGLIMNRPTASRRLLPEKGTPEYEEMINNHEKA 733
hl5lo       534  QLSEYLTVVIFTASAQHAAVNFGQYDWCSWIPNAPPTMRAPPPtak*******gvvTIEQ 586
hp12lo      524  QLCHFLTMCVFTCTAQHAAINQGQLDWYAWVPNAPCTMRMPPPttk*******edvTMAT 576
hr15lo      524  QVCHFVTMCIFTCTGQHASVHLGQLDWYSWVPNAPCTMRLPPPttk*******daTLET 575
 Identity        33222232233322334433324423322323434233 4 334323         223 22

                 ....  ......α21......                        ...........α22....
SB1         734  YLRTITSKLPTLISLSVIEILSTHASDEVYLGQRDNPHWTSDSKALQAFQKFGNKLKEIE 793
hl5lo       587  IVDTLPDRGRSCWHLGAVWALSQFQENELFLGMYPEEHF*IEKPVKEAMARFRKNLEAIV 645
hp12lo      577  VMGSLPDVRQACLQMAISWHLSRRQPDMVPLGHHKEKYF*SGPKPKAVLNQFRTDLEKLE 635
hr15lo      576  VMATLPNFHQASLQMSITWQLGRRQPVMVAVGQHEEEYF*SGPEPKAVLKKFREELAALD 634
 Identity        22 3332  22222222 3 432232223 3422 3223 322223222 243  42222

                 .......   .α23.
SB1         794  EKLVRRNNDPSLQGNRLGPVQLPYTLLYPSSEEGLTFRGIPNSISI              839
hl5lo       646  SVIAERNKKK**********QLPYYYLSPD********RIPNSVAI               673
hp12lo      636  KEITARNEQL**********DWPYEYLKPS********CIENSVTI               663
hr15lo      635  KEIEIRNAKL**********DMPYEYLRPS********VVENSVAI               662
 Identity        223  44 22           2244234 43          3244324
```

[a]Lower case refers to undetermined coordinates in the models (717–729 inclusive). Residues in bold type are conserved in all sequenced lipoxygenases (P213, R216, and G221 may also be conserved, but are not shown in this table). The three turns discussed below (V520–L521, P530–**H531**, and R533–N543) are denoted with "T."

The unmodeled amino-terminal region could be of importance to understanding substrate-binding in the human lipoxygenases. The unmodeled region forms part of the putative substrate-binding cavity (Cα–Fe distance is 11.6 Å for A254 in SBL-1); in the absence of these residues one end of cavity II is connected to the protein surface in the models. Even the residues at the very amino-terminus of the mammalian lipoxygenases could be critical in some way, since the truncation of only nine residues from hp12lo produces an inactive enzyme (24). Addition of a polyhistidyl tail, however, does not significantly affect activity (24).

The structure-based sequence alignment of SBL-1, hl5lo, hp12lo, and hr15lo from the beginning of Helix 2 (position 255) to the carboxy-terminus (position 839) is shown in Table 1.2. In this region, the sequence identity between SBL-1 and the three human lipoxygenases is 26.0, 26.1, and 25.8% for hl5lo, hp12lo, and hr15lo, respectively. There are 87 residues common to all four enzymes in this modeled region (positions 255–839), representing 14.9% of the SBL-1 sequence in this region and 16.4–16.6% of the human lipoxygenase sequences in this region. Hl5lo is 42.5% identical to hp12lo and 40.8% identical to hr15lo, whereas hp12lo and hr15lo are 66.6% identical.

Four Regions of Conserved Residues

Since all residues in the human enzymes—except for the three histidine iron ligands—were modeled in an unbiased manner, analysis of the conformations of conserved residues in the human lipoxygenase models was used to evaluate the reliability of the models. Among the 22 lipoxygenase sequences published to date (21,22,25–47), there are 56 conserved residues (including the four iron ligands). Figure 1.5 shows that 26 of these conserved residues are grouped in four regions that play structural roles in lipoxygenases. (Residues conserved in all lipoxygenase sequences will be in **bold** type in the following discussion.) Three of these regions are involved in anchoring the two longest helices: the 43-residue Helix 9, containing two iron ligands (**H499** and **H504**); and the 30-residue Helix 18, containing one iron ligand (**H690**). The fourth region is involved in stabilizing the conformation of the loop containing the fourth iron ligand–the carboxylate of carboxy-terminal **I839**. Hydrogen bonds and salt bridges found in these regions are included in Table 1.3.

Region I: Beginning of Helix 9. Near the amino-terminal end of Helix 9 there is a sequence of five highly conserved residues (**W479**, L480, L481, **A482**, and **K483**) that appears to anchor the end of Helix 9 to Helix 6 and to the β-sheet region involving Strands 15 and 16. The side chain of **K483** forms a salt bridge with **D343** in Helix 6 and forms a hydrogen bond with the carbonyl oxygen of residue **M341** (also in Helix 6); the side chain of **D343** forms a hydrogen bond with the indole ring nitrogen of **W479**. The side chain of **F346** (Helix 6) is also in this region in a small pocket of hydrophobic residues lined by **I448** from the β-sheet region, **V486** from Helix 9, and the aliphatic portion of the **K483** side chain. At the terminus of Strand 16, another conserved residue, **P469**, lies parallel to the indole ring of **W479**, strengthening inter-

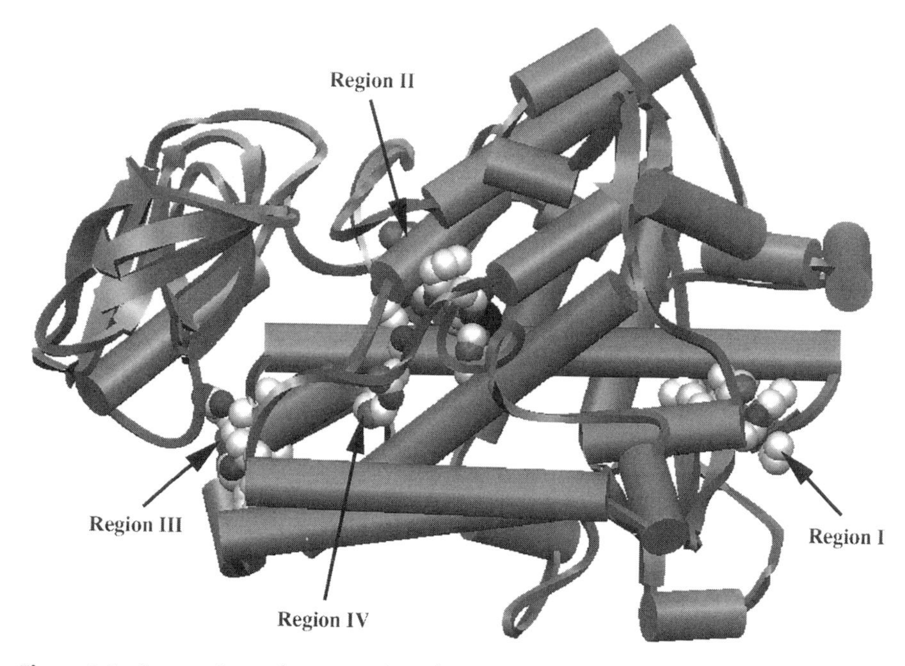

Figure 1.5. Four regions of conserved residues are shown superimposed on the schematic structure of SBL-1. The van der Waals surfaces of the side-chain atoms from the residues in Region I (beginning of Helix 9), Region II (middle of Helix 9), Region III (end of Helix 9), and Region IV (carboxy-terminal loop) are shown, as well as the iron atom. Helix 9 is horizontal in this view and traverses the catalytic domain.

actions between the β-sheet region and Helix 9. In addition to **I448** and **P469**, there are several highly conserved residues of the β-sheet that also interact in this region (L450 and P445 are conserved in all but two sequences).

Modeling this region in the human lipoxygenases was complicated by a three-residue gap at the beginning of Helix 9, between the conserved residues **P469** and **W479**. In the models this gap was placed in positions 472–4; regularization of adjacent residues in positions 470, 471, and 475 shortened Helix 9 by two residues, but did not shift the main-chain positions of any conserved residues. The side chains of all conserved residues in this region, except **W479**, have similar orientations in SBL-1 and the models, indicating that sequence conservation is critical for preserving the structural relationship between Helices 9 and 6 in human lipoxygenases.

The side-chain placement program tends to minimize clashes by orienting surface-accessible side chains, such as **W479**, away from the protein core. No serious clashes occurred when **W479** was modified manually to match the conformer found in the SBL-1 structure.

Region II: Middle of Helix 9. In SBL-1, interactions between Helix 9 and Helix 11 involve conserved residues found on the same side of the helices at intervals of *i* and

TABLE 1.3 Salt Bridges and Hydrogen Bonds in SBL-1 and the Human Lipoxygenase Models

Salt Bridges	First Residue Location	Second Residue Location
D343–K483	α6	α9
D508–R543	α9	α11
H522–D642	α9–α10 loop	α16–α17 loop
R162–**E644**	α1	α17
D490–R707 or R487 (models only)	α9	α18–α19 loop or α9
Hydrogen Bonds	**First Residue Location**	**Second Residue Location**
M341–**K483**	β11–α6 loop	α9
D343–W479	α6	α9
E508–N539	α9	α11
G218–**E508**	α1–α2 loop	α9
H504–N539 (2x)	α9	α11
S519–**H522** (type-I turn)	α9–α10 loop	a9–α10 loop
Y633–D642	α15	α16–α17 loop
D642–L645	α16–α17 loop	α17
Q646–L669	α17	α17–α18 loop
H522–E644 (weak)	α9–α10 loop	α17
T529–Y532 (type-I turn)	α10	α10–α11 loop
Y532–N535 (type-II turn)	α10–α11 loop	α11
P530–R533	α10–α11 loop	α10–α11 loop
H531–N835	α10–α11 loop	C-terminal loop
R533–**S836**	α10–α11 loop	C-terminal loop
R767–**N835**	α21–α22 loop	C-terminal loop
H690–Wat841	α18	Water
S836–Wat841	C-terminal loop	Water
H531–Y624 (models only)	α10–α11 loop	α15
N355–D490	α6–β12 loop	α9
H494–V693	α9	α18
N694–**Q697**	α18	α18
N694–**L754**	α18	α21

i + 4 (**H504** and **E508** from Helix 9; **I538**, **N539**, A542, and **R543** from Helix 11). There is a salt bridge formed between the helices by **E508** and **R543**. The orientation of the **E508** carboxylate is fixed by two hydrogen bonds: one from the Oε1 to the amide nitrogen of **N539**, and another from the Oε2 to the main chain of G218 (G218 is not included in the models). The **N539** side chain is oriented so that it makes two more hydrogen bonds with **H504** (B N–**H504** O and **N539** Oδ–**H504** Nδ). This network of polar interactions not only anchors Helix 11 to Helix 9, but also fixes the side chains of **E508**, **N539**, and **H504** in specific conformations. The orientation of **H504** is especially critical because its ε-nitrogen is an iron ligand.

In the three human lipoxygenase models, side-chain conformations of conserved residues are well-preserved (**H504** is frozen during modeling) with the exceptions of **R543** and the carboxylate of **E508**. (These two residues are free to move in the space generated by the absence of the unmodeled amino-terminal region that includes G218.) The side chain of **R543** was manually corrected in two of the models in order to maintain interaction with **E508**, but the carboxylate of **E508** was left as placed by the automatic modeling, since there is no way to predict what interaction will replace the loss of the **E508**–G218 hydrogen bond. The negative charge of **E508** may be necessary for lipoxygenase function since substitution with glutamine in hl5lo (E376Q hl5lo) yields an inactive enzyme (48,49), while substitution with aspartate (E376Q hl5lo) yields an enzyme with partial activity (50).

Region III: End of Helix 9. In the SBL-1 structure, a group of residues stabilizes the interactions between Helix 16, Helix 17, and the loop between Helices 9 and 10 (H517–**H522**). A hydrogen bond between the carbonyl oxygen of residue 519 and the amide nitrogen of **H522** (Type I turn: V520 $\phi,\psi \approx -60,-30$; L521 $\phi,\psi \approx -90,0$) organizes the loop. The residue after the turn, **P523**, caps the amino-terminal end of Helix 10. **H522** forms a salt bridge with **D642**, separating Helix 16 and Helix 17 with a one-residue bend. **D642** also forms hydrogen bonds with the side chain of **Y633** and the main chain of **L645**. The six residues of Helix 16 are anchored to this region by the interaction of **Y633** and **D642** that straddle the helix. Three conserved residues of Helix 17 (**E644**, **L645**, and **Q646**) anchor this helix to the region as well: **L645** and **Q646** form hydrogen bonds with **D642** and L669, respectively. The side chain of **E644** forms a salt bridge with R162 from Helix 1 and a weak hydrogen bond with Nδ of **H522** (3.2 Å).

The models of the human lipoxygenases show that **L645** and **L518** are part of a group of six or seven interacting hydrophobic residues that come from Helix 15, Helix 17, and Helix 18. The other residues in this hydrophobic pocket are not strictly conserved, but they are invariably hydrophobic, and they preserve the orientations of **L645** and **L518** by conserving a certain degree of side-chain bulk between them. In the models, orientations of the conserved residues are well-preserved, with the exception of **E644** and **Q646** that swing into the space generated by the absence of the unmodeled amino-terminal region. The hydrogen bond between **Q646** and L669 is probably preserved, since no other side chain shifts to compensate for the loss of this interaction. The salt bridge between **E644** and R162 is not preserved, due to the absence of Helix 1 in the models. The side chain of **Q646** was reoriented to match the SBL-1 structure, and **E644** was left as it is in the models.

In the models of the three human lipoxygenases, the orientation of **D642** varies only slightly, fixed in place by several hydrogen bonds. The general orientation of **H522** is well-preserved, but the angle of the ring varies in the models, suggesting that the role of **H522** is to compensate for the negative charge of **D642**. Results of mutagenesis experiments that substitute **H522** with uncharged amino acids support the conclusion that **H522** has a structural role, forming a conserved salt bridge with **D642** (low activities or low yields were found [24,26,48,51–53]). Mutagenesis of **D642** further supports this conclusion; substitution of **D642** with negatively charged glutamate (D502E hl5lo) preserves activity (49), while substitution with uncharged asparagine (D502N hl5lo) greatly reduces the activity (50).

Region IV: Carboxy-Terminal Loop. The carboxy-terminal 31 residues of SBL-1 form a long loop that ends in the center of the enzyme with conserved iron ligand **I839**. There are several conserved interactions between carboxy-terminal residues **N835** and **S836** and the five-residue loop between Helix 10 and Helix 11 (P530–N534) that play a role in organizing the two loops into the specific geometry necessary to orient iron ligand **I839**. The five-residue loop forms two hydrogen bonds with Helix 10 and one more with Helix 11, rigidly connecting the helices. The loop is composed of a type I turn (P530 $\phi,\psi \approx$ -60,-30; **H531** $\phi,\psi \approx$ -90,0) followed by a type II turn (R533 $\phi,\psi \approx$ -60,120; N534 $\phi,\psi \approx$ 80,0) with an additional hydrogen bond between the highly conserved residues R533 and P530, linking the turns. The two loops are locked together by two hydrogen bonds: one between the carbonyl oxygen of **N835** and the Nδ of **H531** and the other between the side chain of **S836** and the carbonyl oxygen of R533. The amide nitrogen of **N835** also forms a hydrogen bond with the main chain of R767. In addition, a buried water forms hydrogen bonds with the carbonyl oxygen of **S836** and the Nδ of iron ligand **H690**.

N835, **S836**, and iron ligand **I839** are not conserved in the sequence of rat leukocyte 5-lipoxygenase (rl5lo). Comparison of the gene from rl5lo (41) and the gene from hl5lo (hl5lo and rl5lo share 93% amino acid sequence identity, [39,40]) indicates a possible frame shift error in the six carboxy-terminal residues of rl5lo.

hl5lo gene	cca	gac	cgg	att	ccg	aac	agt	gtg	gcc	atc	tga
Sequence	P	D	R	I	P	**N**[a]	**S**[a]	V	A	**I**[a]	Stop
rl5lo gene	cca	gac	ag?	att	cca	aac	agt	gta	gcc	atc	taa
Sequence	P	D	R	F	Q	T	V	Stop			
Possible sequence	P	D	R?	I	P	**N**[a]	**S**[a]	V	A	**I**[a]	Stop

[a]Due to the uncertainty in the carboxy-terminal six residues of rl5lo, the three boldfaced residues—**N835**, **S836**, and **I839**—are considered to be conserved in all lipoxygenases.

The mammalian lipoxygenases are missing 18 residues in the carboxy-terminal region between positions 800 and 839. During model building, it was possible to remove two poorly conserved segments in the SBL-1 structure (10 and eight residues in length) without altering main-chain geometries in regions containing highly conserved residues. The side chains of **N835**, **S836**, and **I839** fell into similar positions in the models of the human enzymes. The side chain of R533 is solvent exposed, but retains the same general orientation in the models.

The three human lipoxygenase models show that the orientation of **H531** is almost identical to that in the SBL-1 structure, preserving the hydrogen bond between the Nδ of **H531** and the carbonyl oxygen of **N835**. Mutagenesis experiments on **H531** support these findings. Substitutions of **H531** with asparagine and with glutamine (H399Q hl5lo [51], H399N hl5lo [48], and H392Q hp12lo [24]) have the highest activities, perhaps due to the ability of side-chain amides to duplicate the hydrogen bond formed by **H531**. The models of the human lipoxygenases raise the possibility of another polar interaction involving **H531**: in the models of hp12lo and hr15lo, a tyrosine occupies position 624 and forms a hydrogen bond with the unprotonated Nε of **H531** (position 624 is a tyrosine only in 12- and 15-lipoxygenases).

Conservation of the D490–R707 Salt Bridge

In the SBL-1 structure, a buried salt bridge between R707 and **D490** forms a division in the large internal cavity lined by highly conserved residues. It has been noted that if these two residues shifted, the resulting 40 Å cavity would connect the iron site to within a few Ångstroms of the surface of the protein, making it possible for the cavity to be the substrate entry route and binding site (13). **D490** is conserved in all lipoxygenases, but in mammalian lipoxygenases the position of R707 is occupied by an alanine.

The absence of R707 in mammalian lipoxygenases would appear to leave the negative charge of **D490** uncompensated for and deeply buried in the protein, but the models show that the R707–**D490** salt bridge is effectively conserved by concerted changes of R707A and I487R. In the human lipoxygenase models, the arginine at position 487 is oriented so that the guanidinium moiety occupies the same region as the guanidinium of R707 from the SBL-1 structure (Fig. 1.6). This alternative salt bridge allows the carboxylate of **D490** to remain in roughly the same position and still maintain a hydrogen bond with the amide of conserved residue **N355**. Thus, the subdivision of the large internal cavity by an ion pair in the SBL-1 structure seems to be present in the mammalian lipoxygenases also. However, the salt bridge is not necessary for lipoxygenase activity, since substitution of **D490** with asparagine (D358N hl5lo) in hl5lo does not abolish activity (48).

The Role of Residue H494

The side chain of histidine 494 lies on the wall of the proposed substrate-binding cavity only 8 Å (Fe-Cε) from the iron atom. Its orientation allows the Nε of the imidazole ring to form a hydrogen bond with the carbonyl oxygen of residue V693, and exposes the Nδ side of the ring to the cavity. The proximity of H494 to the iron atom raises the possibility that it could be a positive counter ion involved in stabilizing the negatively charged carboxylate of fatty acid substrates.

Analysis of H494 in the human lipoxygenase models and the results of mutagenesis studies suggest that H494 has a structural role. The models give H494 an orientation almost identical to that in the SBL-1 structure, preserving the hydrogen bond between the Nε and the carbonyl oxygen of residue 693. This interaction helps to anchor Helix 9 (containing iron ligands **H499** and **H504**) to Helix 18 (containing iron ligand **H690**) and may have a role in initiating the π-helical section of Helix 9 that begins at H494 and ends at residue 506. Analysis of the models of hp12lo and hr15lo

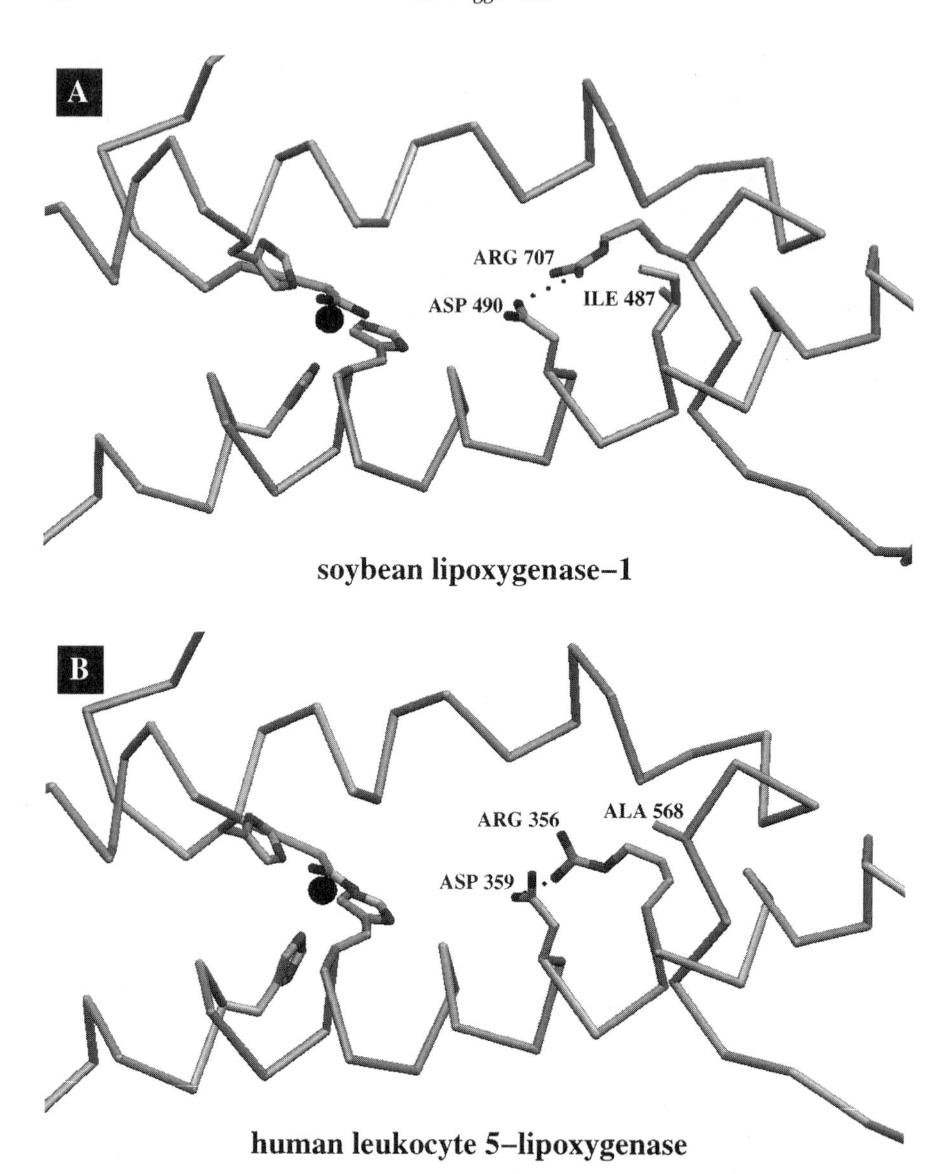

Figure 1.6. The backbones of Helix 9 and Helix 18 are shown with a conserved salt bridge and the active site iron ligands. *a*) The structure of SBL-1 in which residue 487 is an isoleucine and a salt bridge exists between **D490** and R707. *b*) A model of hl5lo in which residue 707 is an alanine (A568 hl5lo) and the salt bridge now exists between **D490** (D359 hl5lo) and R487 (R356 hl5lo).

raises the possibility that the carboxylate of an adjacent glutamate (position 495 is a glutamate in many mammalian lipoxygenases and in pea lipoxygenase-3) could be oriented back towards H494 to form a salt bridge. However, mutagenesis experiments in hp12lo and other lipoxygenases do not support the conclusion that H494 is a positively charged salt-bridge partner in these cases. Substitutions of H494 with glutamine (H494Q SBL-1 [26], H362Q hl5lo [51], and H355Q hp12lo [24]) show that an uncharged residue at this position does not abolish enzyme activity. In fact, the recently published sequence of murine platelet 12-lipoxygenase has a glutamine in this position (and a glutamate at position 495), making H494 no longer a conserved residue in lipoxygenase sequences (47). The loss of activity in the H494K (H363K hl5lo) variant in hl5lo and the activity of the H494L (H356L porcine leukocyte 12-lipoxygenase [pl12lo]) variant in pl12lo further demonstrate that a positive charge (lysine) at position 494 is deleterious to enzyme function (52,53), and an uncharged residue (leucine) is not—even in lipoxygenases with a glutamate at position 495. Thus, it appears that H494 is uncharged in lipoxygenases and does not interact ionicly with the carboxylates of fatty acid substrates.

The Role of Residue N694

Sequence alignments show that position 694 is occupied by an asparagine in plant lipoxygenases, mammalian 5-lipoxygenases, and hp12lo, but by a histidine in other mammalian 12-lipoxygenases and all mammalian 15-lipoxygenases. The orientation of this side chain is critical due to its proximity to the unoccupied iron-coordination position (opposite **H504**) at the terminus of the putative dioxygen entry channel. Small changes in this area could affect iron coordination and the size of the channel, only 2.5 Å wide in this region. In the SBL-1 structure, the amide oxygen of N694 is 2.8 Å from the amide of conserved **Q697**, while its amide nitrogen is 2.6 Å from the carbonyl oxygen of conserved **L754**. These hydrogen bonds orient the N694 side chain, making the distance between the amide oxygen and the iron atom 3.3 Å. Another structural study on SBL-1 reported N694 to be an iron ligand (54).

Two possibilities for the orientation of a histidine in position 694 arose during the modeling of hr15lo. The two histidine conformers differ in χ-1 by about 90°; one may be an iron ligand, while the other forms a salt bridge (Fig. 1.7). The first conformer is oriented in the same direction as the asparagine in the SBL-1 structure and may be capable of duplicating one or both of the hydrogen bonds formed by N694 in the SBL-1 structure. Small manual adjustments of this conformer show that the histidine Nδ could be an iron ligand, but the Nε could not be oriented so as to be an iron ligand. Ideal orientation of the histidine ring for iron-coordination geometry creates several severe clashes with other residues: one between the histidine Nε and the carbonyl oxygen of **L754**, and others between the histidine Cε and atoms of **I839**. Attempts to alleviate these clashes tilt the histidine ring with respect to the Fe–Nδ bond and force the histidine to adopt a nonideal iron-coordination geometry. The second histidine conformer is oriented away from the vacant iron-coordination position (opposite **H504**) and towards the glutamate that replaces P834 in all mammalian 12- and 15-lipoxygenases. Substitution of F695 with leucine seems to generate some of the space necessary to orient the histidine side chain in this direction. All lipoxygenases that have a histidine at position 694 also have a glutamate at position 834 and a leucine at position 695, making the salt bridge possible.

It is not clear from the models which conformer of H694 is most likely to be found in lipoxygenases. Mutagenesis of N694 indicates that residue 694 is important for lipoxygenase function, but not for iron binding. Mutant lipoxygenases in which N694 was substituted with aspartate, glutamine, histidine, alanine, or serine (N554D hl5lo and N554Q hl5lo [49]; N713H SBL-3, N713A SBL-3, and N713S SBL-3 [55]) all incorporated iron (the lowest iron content found was 51% of wild type), however, only the N694H variant had activity comparable to the wild-type enzyme (the N694D variant had barely detectable activity). The possibility has been raised that N694 in SBL-1 may shift position during dioxygen binding to interact with the distal oxygen atom of iron-bound dioxygen, helping to orient and stabilize the oxygen-addition step of the lipoxygenase reaction (13). The conserved residue **Q697** (discussed previously, the side chains of **Q697** and N694 form a hydrogen bond in the SBL-1 structure) may also be involved in some aspect of dioxygen binding—either alone or in conjunction with N/H694. Substitution of **Q697** with glutamate (Q557E hl5lo) in hl5lo results in very low activity (~3% [52]).

Lipoxygenase Positional Specificity

Positional Specificity in 12- and 15-Lipoxygenases

A series of mutagenesis studies of hp12lo (24), bovine epithelial 12-lipoxygenase (be12lo [56]), porcine leukocyte 12-lipoxygenase (pl12lo) (53), rat brain 12-lipoxygenase (rb12lo [57]), and hr15lo (58) attempted to shift the site of dioxygen addition from the 15th carbon to the 12th carbon of arachidonic acid and the reverse. Several residues were identified as possible determinants of 12/15 positional specificity from sequence comparisons of mammalian 12-lipoxygenases and mammalian 15-lipoxygenases; however, only substitutions at residues 556 and 557 (corresponding to positions 417 and 418 in hp12lo and hr15lo) affected the positional specificities of the enzymes (Table 1.4). Residues 556 and 557 seemed to shift positional specificity in mammalian 12- and 15-lipoxygenases by affecting where the methyl terminus of the substrate fatty acid binds. Bulky amino acids at positions 556 and 557 would not permit the substrate to bind as deeply, exposing C-15 of arachidonic acid to oxygenation instead of C-12 (58).

Of the sites chosen in the mutagenesis studies aimed at shifting positional specificity, only four (545, 547, 556, and 557) are close to the active site iron (Cβ-Fe distances < 15 Å) and have side chains facing towards the iron (Cα–Cβ–Fe angles >

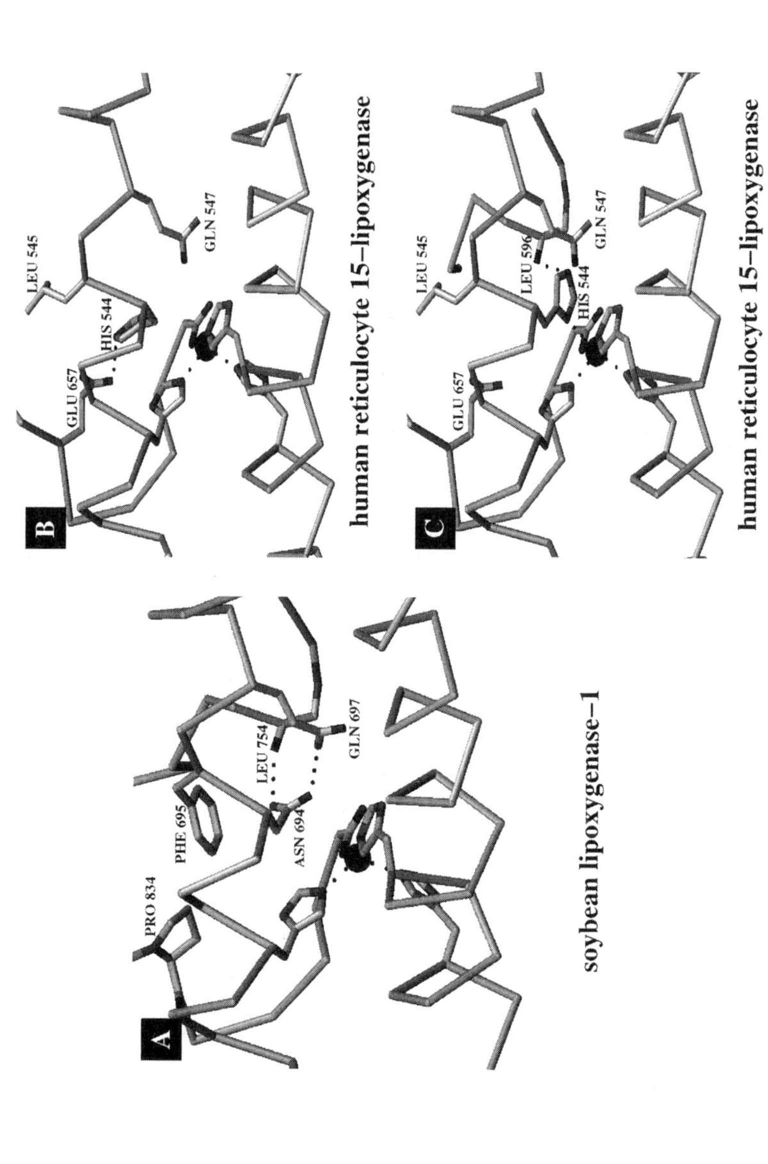

Figure 1.7. The backbones of Helix 9 and Helix 18 are shown with residues in the vicinity of N694 and active site iron ligands. *a*) The structure of SBL-1 in which N694 forms hydrogen bonds with **Q697** and the carbonyl oxygen of **L754**. Figures 1.7*b* and 1.7*c* show two alternate conformers of H694 (H544 hr15lo) in hr15lo. The first conformer of H694 (Figure 1.7*b*) forms a salt bridge with E834 (E657 hr15lo), while the second conformer of H694 (1.7*c*) has been manually oriented to be an iron ligand (poor geometry).

TABLE 1.4 Summary of Reported Effects of Mutagenesis on Residues 556 and 557 in Mammalian 12- and 15-lipoxygenases

Enzyme	Residue at Position 556	Residue at Position 557	Shift in Positional Specificity	Total Activity
hp12lo[a]	A417	V418		
	A	M	15-specificity 0 → 0%	94%
	I	V	15-specificity 0 → 10%	74%
	I	M	15-specificity 0 → 18%	67%
be12lo[b]	V418	V419		
	V	M	15-specificity 6 → 6%	n.a.
	I	M	15-specificity 6 → 25%	n.a.
pl12lo[c]	V418	V419		
	V	M	15-specificity 9 → 33%	n.a.
	I	V	15-specificity 9 → 68%	n.a.
	I	M	15-specificity 9 → 85%	n.a.
rb12lo[d]	A418	M419		
	A	V	12-specificity 86 → 71%	8%
	I	M	15-specificity 14 → 10%	150%
mp12lo	V417	V418		
ml12lo	V418	M419		
hr15lo[e]	I 417	M418		
	I	V	12-specificity 10 → 50%	112%
	A	V	12-specificity 10 → 94%	n.a.
rr15lo	I416	M417		

[a]A six amino acid substitution (T540I/Q545G/I547V/G550M/K555Q/V557M) including V557M increased 15-specificity to 33%; however the total activity of this variant was low (25%). Three substitutions, including A556I/V557M, were made; the 15-activities of K555Q/A556I/V557M, T540V/K555Q/A556I/V557M, and I547V/G550M/K555Q/A556I/V557M are 18, 21, and 20%, respectively; and the total activities are 67, 61, and 29%, respectively. *Source:* Chen and Funk (24).
[b]Sloane and Sigal (56).
[c]Reported reverse-phase HPLC data indicates that all three variants have total activities similar to the wild type. *Source:* Suzuki et al. (53).
[d]Watanabe and Haeggstrom (57).
[e]Sloane et al. (58).

90°). In the models of hp12lo and hr15lo, the β-carbon to iron distances for residues 545, 547, 556, and 557 are 14, 12, 15, and 15 Å respectively; the Cα–Cβ–Fe angles are 117, 118, 145, and 140°, respectively. Positions 545 and 547 are separated from positions 556 and 557 by more than 10 Å (Cβ–Cβ distance), making it impossible for both sites to interact with the methyl terminus of bound substrate.

The Connolly surface was calculated using a 1.2 Å probe radius (59,60). Analysis of the Connolly surface of the SBL-1 putative substrate-binding cavity superimposed on the models of hp12lo and hr15lo indicates that positions 545 and 547 and positions 556 and 557 have different roles in substrate binding. Positions 545 and 547 lie on the wall of the putative substrate-binding cavity opposite the iron atom, where they would be most likely be in the vicinity of the pentadienyl moiety of a bound substrate (position 545 especially), whereas positions 556 and 557 lie toward one end of the substrate-binding cavity, where they would be most likely to interact with the methyl terminus of a bound substrate.

The ratio of 15-activity to 12-activity increases with increased bulkiness of residues 556 and 557, and decreases with decreased bulkiness of these residues in all of the mutants except for the rb12lo variants (Fig. 1.8). In rb12lo, the M557V variant (M418V rb12lo) decreased 12-specificity from 86 to 71% and the A556I variant (A417I rb12lo) increased 12-specificity from 86 to 90%; the sequence numbers are those used by Watanabe and Haeggstrom (57). Analysis of residues in the putative substrate-binding pocket, especially residue 491, provides a rationale for why these mutations had different effects in rb12lo than they did in other lipoxygenases. In comparison to other 12-lipoxygenases, wild-type rb12lo contains a bulkier residue at position 557 (methionine instead of valine) and a less bulky residue at position 491 (leucine instead of phenylalanine [Table 1.5]). The methyl terminus of arachidonic acid may bind more in the vicinity of residue L491 in rb12lo due to the shift in bulk between residues 491 and 557. The M557V variant was designed to increase 12-specificity (as was done in hr15lo); however, in rb12lo the mutation created a loose binding pocket, with positions 491, 556, and 557 occupied by leucine, alanine, and valine, respectively. As a result, the positional specificity decreased along with the total activity (8% of wild type). The A556I variant in rb12lo was designed to increase 15-activity (as was done in hp12lo), however, with the reduced bulk of L491 taken into consideration, this mutation may have optimized substrate binding. As a result, the 12-specificity increased along with the total activity (150% of wild type). Since the recently sequenced murine leukocyte 12-lipoxygenase (ml12lo) has 93% sequence identity to rb12lo and contains identical residues at positions 491 (leucine) and 557 (methionine) (47), it could bind the methyl terminus of arachidonic acid in a manner similar to that of rb12lo.

Further analysis of the residues around positions 556 and 557 indicates that four residues (residues 552, 701, 750, and 751) are common to ml12lo and rb12lo, but differ in other mammalian 12-lipoxygenases (Table 1.5). These four residues, in conjunction with the residues discussed previously (491, 556, and 557), may help to differentiate the substrate-binding sites of 12-lipoxygenases. In the model of hp12lo, positions 552 and 750 face each other on one wall of the putative substrate-binding site. In ml12lo and rb12lo, increased bulk of phenylalanine at position 552 (usually isoleucine) is partially compensated for by reduced bulk of valine at position 750 (usually isoleucine). Similarly, changes in residue 701 seem to be compensated for by changes in residue 751. In lipoxygenases in which residue 701 is a tyrosine, residue 751 is a serine or threonine; when 701 is a phenylalanine, 751 is a valine. In the hp12lo and hr15lo models, Y701 is oriented into the cavity away from S/T751; however, it can be reoriented towards residue 751. If a hydrogen bond exists between the hydroxyls of Y701 and S/T751, it would anchor this wall of the putative substrate-binding cavity in a different way than a hydrophobic F701–V751 interaction would.

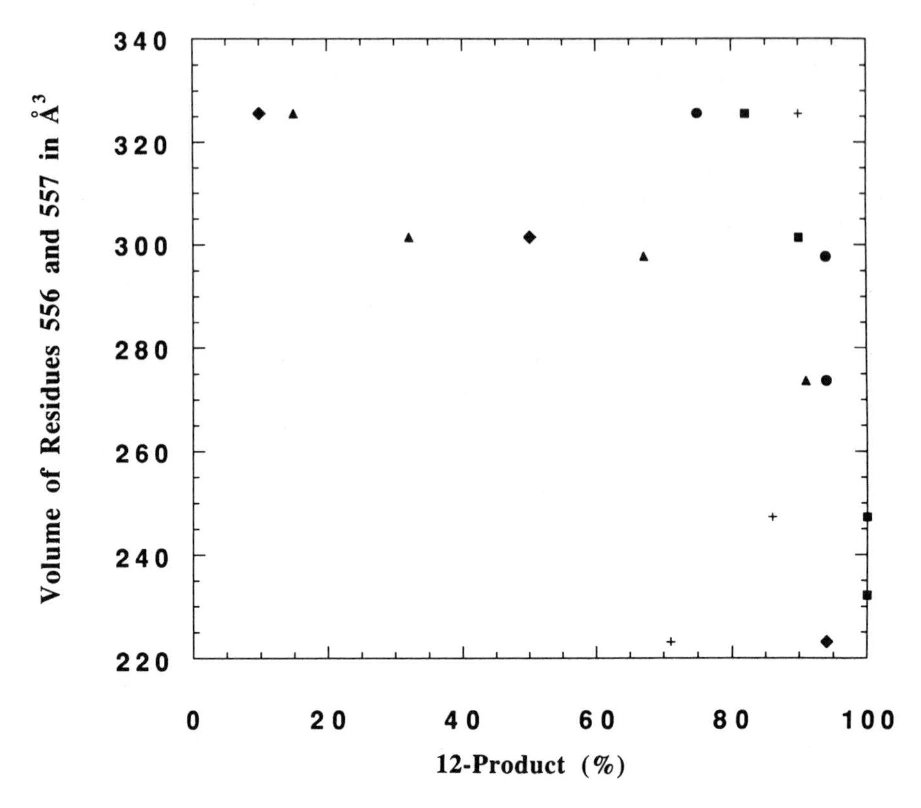

Figure 1.8. hp12lo (■), be12lo (●), pl12lo (▲), and hr15lo (◆) (58) produced a higher ratio of 15-product to 12-product with increased bulk at residues 556 and 557. rb12lo (+) was the only lipoxygenase that did not follow this trend. Values for amino acid bulk were taken from a table of amino acid partial specific volumes in solution. *Sources:* Chen and Funk (24), Matsumoto et al. (39), Dixon et al. (40), Suzuki et al. (53), Watanabe et al. (57), Sloane et al. (58), Creighton (87).

Thus, two pairs of interacting residues (552–750 and 701–751), common to wild type rb12lo and ml12lo, are different in other mammalian 12-lipoxygenases and may have contributed to the unexpected positional specificity of rb12lo mutants.

Model of Lipoxygenase Specificity

Mutagenesis studies suggest that binding the methyl end of a fatty acid substrate is responsible for positional specificity by aligning the substrate relative to the catalytic site. Other evidence supports this model of positional specificity in 12- and 15-lipoxygenases. Several experiments have shown that 12- and 15-lipoxygenases from plants and mammals are sensitive to the location of the reactive 1,4-diene relative to the methyl terminus, but not the carboxylate of fatty acid substrates (61–63). Fatty acids

TABLE 1.5 Residues Near Positions 556 and 557 Not Conserved in Wild-Type Mammalian Lipoxygenases[a]

Enzyme	491	552	556	557	701	747	750	751
SBL-1	S	I	T	F	G	S	V	I
hl5lo	F	L	A	<u>N</u>	C	<u>H</u>	A	V
rl5lo	F	L	A	<u>N</u>	C	<u>H</u>	A	V
hp12lo	F	I	A	V	Y	Q	I	S
be12lo	F	V	V	V	Y	Q	I	T
pl12lo	F	I	V	V	Y	Q	I	T
rb12lo	*L*	*F*	A	*M*	*F*	Q	*V*	*V*
mp12lo	F	I	V	V	Y	Q	I	T
ml12lo	*L*	*F*	V	*M*	*F*	Q	*V*	*V*
hr15lo	F	I	I	M	Y	Q	I	T

[a]References to these sequences can be found in the results and discussion section. Residues in italics type are common to ml12lo and rb12lo. Residues underlined are common to 5-lipoxygenases and may interact with the carboxylate of a bound substrate.

with modified carboxy termini (methylcarboxylate [64], sulfate [65], cholesterol esters and phospholipids [66]) are 15-lipoxygenase substrates, while fatty acids with altered methyl termini (19-methylarachidonic acid) are not substrates (62). In addition, hr15lo can catalyze oxygenation of LDL in addition to biological membranes (66), and rabbit reticulocyte 15-lipoxygenase (rr15lo) can catalyze the oxygenation of fatty acids in organelle membranes (67)—reactions thought to be important in the degradation of mitochondrial membranes in reticulocyte maturation and the processing of LDL in atherogenesis, although there is evidence against LDL oxidation (68). Together, these results are consistent with the model of positional specificity in 12- and 15-lipoxygenases based on the positioning of the substrate methyl terminus.

One final piece of evidence is the stereospecificity of dioxygen addition (S chirality) and proton abstraction (pro-S) in 12- and 15-lipoxygenase reactions. When a substrate containing a (E-Z)-1,4-diene moiety is used, the major product can be the (R) hydroperoxy fatty acid; however, this is the expected chirality for antarafacial addition to an olefin in the (E)-geometry (14). Partially purified extracts from several tissue sources catalyze oxygenation of arachidonic acid to produce (8R)-, (11R)-, or (12R)-hydro(pero)xyeicosatetraenoic acids (69–71); however, these enzymes have not been well characterized and some of them may be cytochromes or other oxygenases. In all well-characterized lipoxygenases, the chirality of 12- and 15-lipoxygenase reactions is identical, requiring only a 2.7 Å shift in the alignment of the arachidonic acid substrate to change positional specificity (Fig. 1.9).

The previously mentioned model of positional specificity in 12- and 15-lipoxygenase involves a translational shift of arachidonic acid in the substrate-binding cavity. Further translation of the fatty acid will not produce 5- or 8-lipoxygenase products with the correct stereospecificity of hydrogen abstraction and dioxygen addition. However, if arachidonic acid binds in the opposite direction (so that the carboxylate

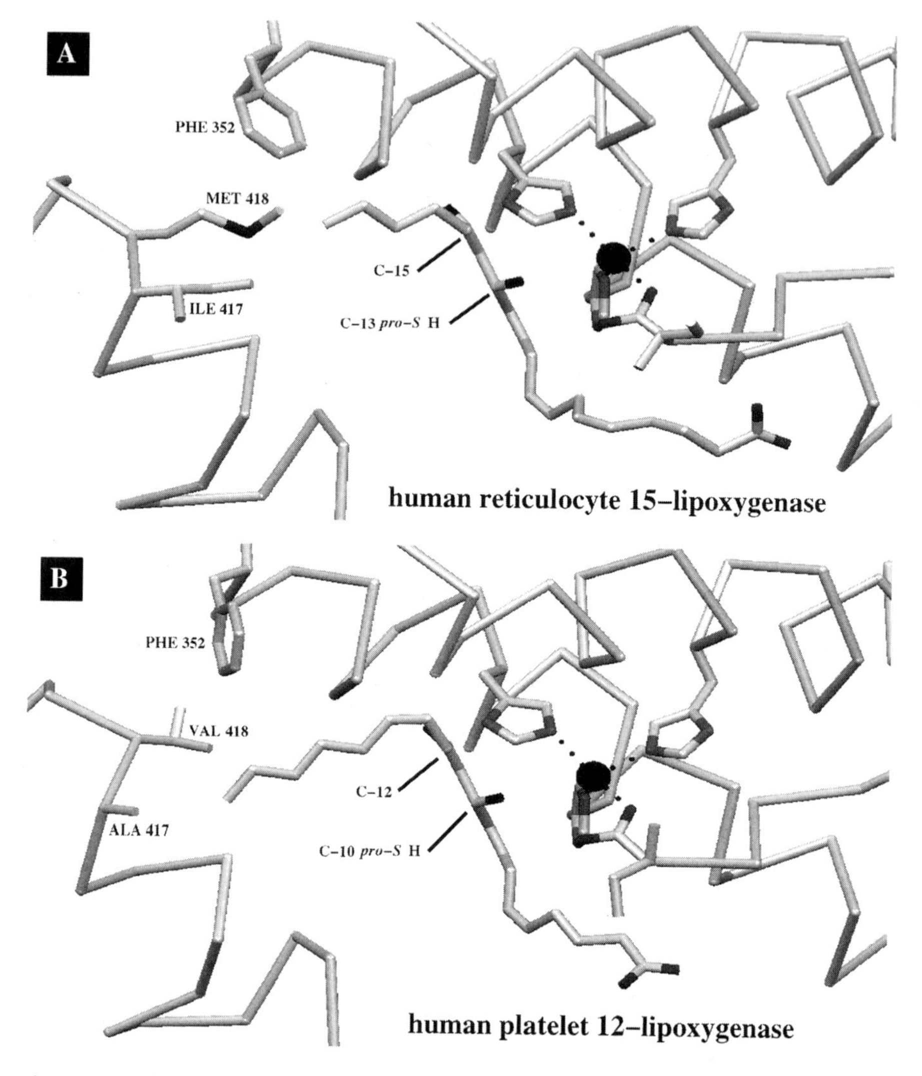

Figure 1.9. The side chains of active site iron ligands are shown with substrate arachidonic acid and three residues that interact with the methyl terminus of the arachidonic acid. *a*) A model of hr15lo in which F491 (F352 hr15lo), I556 (I417 hr15lo), and M557 (M418 hr15lo) interact with the methyl terminus of arachidonic acid. *b*) A model of hp12lo in which F491 (F352 hp12lo), A556 (A417 hp12lo), and V557 (V418 hp12lo) interact with the methyl terminus of arachidonic acid. The arachidonic acid was placed manually, based on hypothetical proton abstraction by an iron-bound hydroxyl. The two dark hydrogen atoms on the arachidonic acid mark the sites of oxygen addition and proton abstraction (C-15 and C-13, respectively, in *a*; and C-12 and C-10, respectively, in *b*).

and methyl positions are exchanged), the 5-lipoxygenase reaction becomes spatially identical to the 15-lipoxygenase reaction and the 8-lipoxygenase reaction becomes spatially identical to the 12-lipoxygenase reaction (Fig. 1.10). This model explains the observed stereospecificity of 5-, 12-, and 15-lipoxygenase reactions and predicts the stereospecificity of the 8-lipoxygenase reaction (abstraction of the C-10 pro-R proton and insertion of dioxygen, forming an (8S)-hydroperoxy product) by fusing two paradigms: the first pertaining to translation of substrate in the active site (56), and the second pertaining to the orientation of substrate in the active site (64,72).

The general model of lipoxygenase positional specificity and stereospecificity requires the substrate carboxylate to bind in 5- and 8-lipoxygenases in the same pocket as the substrate methyl terminus binds in 12- and 15-lipoxygenases. The carboxylate-binding pocket of 5- and 8-lipoxygenases is expected to position the substrate based on the binding of the substrate carboxylate instead of the substrate methyl terminus. This idea is supported experimentally by results that show that, unlike 12- and 15-lipoxygenases, 5- and 8-lipoxygenases are sensitive to the location of the reactive 1,4-diene relative to the carboxylate of fatty acid substrate (73-75). Analysis of the residues in the binding cavity shows that a histidine is located at position 747 (H600 hl5lo) and that asparagine occupies position 557 (N425 hl5lo—this position was thought to be a determinant of 15-specificity in the mutagenesis studies) in both of the sequenced 5-lipoxygenases (Table 1.5). In the model of hl5lo, the histidine and

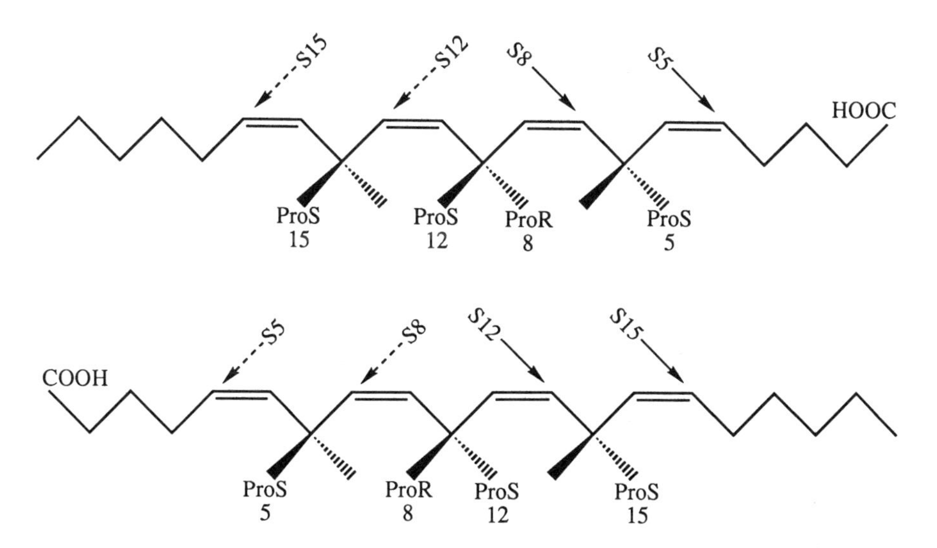

Figure 1.10. When arachidonic acid is arranged head to tail in opposite orientations, the 5- and 8-lipoxygenase reactions become spatially identical to the 15- and 12-lipoxygenase reactions. The stereospecificity of oxygen addition and stereo- selectivity of proton abstraction in 8-lipoxygenase is presumed from the model and is not currently known.

asparagine side chains are oriented toward the iron atom. When sequence information becomes available, it will be interesting to see if hydrophilic residues occupy nearby positions in 8-lipoxygenases.

Lipoxygenases often exhibit dual positional specificity—most commonly 12/15 or 5/8 dual positional specificity. Dual positional specificity may have arisen as a means of generating two products without expressing two independent enzymes. Lipoxygenases could produce the ratio of products required by the tissues in which they are expressed. For example, the four "leukocyte-type" 12-lipoxygenases produce 15-product, whereas the two "platelet-type" 12-lipoxygenases do not produce any detectable 15-product. Dual positional specificity in some lipoxygenases is also a means of synthesizing the epoxide moiety of leukotriene compounds from arachidonic acid. Mammalian 5-lipoxygenases and potato 5-lipoxygenase produce a minor 8-product and are capable of synthesizing 5,6-LTA4 (5S,6S-oxido-7E,9E,11Z,14Z-eicosatetraenoic acid) through successive 5- and 8-lipoxygenase reactions (76–78). Similarly, pl12lo exhibits dual 12/15 specificity and is capable of synthesizing 14,15-LTA4 (14S,15S-oxido-5Z,8Z,10E,12E-eicosatetraenoic acid) through successive 15- and 12-lipoxygenase reactions (79). Like pl12lo, hr15lo and rr15lo produce a 14R,15S-dihydroxy product that may be the result of 14,15-LTA4 synthase activity in the mammalian 15-lipoxygenases (66,80).

The mechanism used in leukotriene synthesis may also aid in the synthesis of the trioxygenated bioactive compounds lipoxin A4 (5S,6R,15S-trihydroxy-7E,9E,11Z,13E-eicosatetraenoic acid) and lipoxin B4 (5S,14R,15S-trihydroxy-6E,8Z,10E,12E-eicosatetraenoic acid). These compounds contain hydroxyls in the positions where the epoxide exists in 5,6-LTA4 and 14,15-LTA4 (lipoxin A4 contains hydroxyls at C-5 and C-6 and lipoxin B4 contains hydroxyls at C-14 and C-15). One pathway of lipoxin formation could be hydrolysis of the leukotriene epoxide followed by enzymatic oxygenation of C-15 (in the case of lipoxin A4) or C-5 (in the case of lipoxin B4) to produce the trioxygenated product.

Mechanism of Lipoxygenases

In the catalytic mechanism of lipoxygenases, the ferric form of the enzyme accepts one electron from the fatty acid substrate while a base abstracts a proton from the methylene carbon of the reactive 1,4-diene. The resulting intermediate could be described as having an Fe^{+3}—substrate organoiron bond or an Fe^{+2}—substrate free-radical complex. Dioxygen reacts with the intermediate, regenerating Fe+3 and producing a peroxidate anion that receives the proton from the base to give the hydroperoxide product.

Evidence in favor of an Fe^{+3}–carbon intermediate has been provided by kinetic studies with variations in substrate structure and by chemical analogy (81,82). Evidence for the Fe^{+2}–free radical pathway is based on anaerobic reduction of Fe^{+3} by fatty acid substrate (9), and on detection of substrate free radicals (83–85). The detection of substrate free-radicals does not rule out the organoiron intermediate because an Fe^{+3}–carbon bond could undergo anaerobic dissociation to form Fe^{+2} and a free radical.

The structure of SBL-1 suggests how some of these events might occur in the peroxidation of arachidonic acid, for example. The fatty acid enters cavity II and approaches the iron in such a way that the 11,14-diene system moves into the unoccupied octahedral position opposite to **H690**. Abstraction of the pro-S (L_S) proton from C-13 could result in either the interaction of Fe^{+3} with C-15 to form an Fe–C bond, or Fe^{+2} and a resonance-stabilized substrate free radical. Molecular oxygen enters the molecule through the tunnel (cavity I) and coordinates with iron in the unoccupied position opposite to the Nε of **H504**. (The side chain of N694 can reorient and form a hydrogen bond with the bound oxygen.) In this situation, molecular oxygen is in a location well-suited for reacting with C-15 of the arachidonic acid to produce peroxidate anion and Fe^{+3}. The peroxidate anion is protonated and released, and the coordination of the iron returns to what it was at the beginning of the catalytic cycle.

Arachidonic acid can be manually placed in the SBL-1 structure and in the human lipoxygenase models in such a way that the reactive pentadiene moiety fits into the putative substrate-binding cavity (cavity II) in the vicinity of the empty iron-coordination site (opposite **H690** [Fig. 1.9]). The substrate position is consistent with the organoiron intermediate, but it is consistent with the free-radical intermediate only if the proton-abstracting base is near the unoccupied iron-coordination site. There are three plausible candidates for the base in this area: the Nε of **H499**, the second oxygen of the terminal carboxylate (**I839**), or an iron-bound hydroxyl.

Analysis of unconserved residues in one pocket of the putative substrate-binding site (residues 491, 552, 556, 557, 701, 747, 750, and 751), combined with results of mutagenesis studies, provides the underpinnings of a model that accounts for lipoxygenase positional specificity and stereospecificity. The model predicts that the methyl terminus of arachidonic acid binds in this pocket in 12- and 15-lipoxygenases, and that the carboxy terminus of arachidonic acid binds in this pocket in 5- and 8-lipoxygenases. In 12- and 15-lipoxygenases, the bulkiness of hydrophobic residues in the pocket (especially residues 491, 556, and 557) affects how deeply substrate binds, thereby altering positional specificity. In 5- and 8-lipoxygenases, a similar steric mechanism would determine positional specificity—with the exception that some residues in the pocket would be hydrophilic to interact with the carboxy terminus of the substrate. In 5-lipoxygenases, H747 and N557 may fulfill this role (hydrophilic residues are expected to be found in this area in 8-lipoxygenases when sequences become available).

References

1. Vick, B.A., and Zimmerman, D.C. (1987) in *The Biochemistry of Plants,* Stumf, P.K., Conn, E.E., Academic, Orlando, pp. 53–90.
2. Hildebrand, D.F. (1989) *Physiol. Plant 76,* 249–253.
3. Siedow, J.N. (1991) *Ann. Rev. Plant Physiol. Plant Mol. Biol. 42,* 145–188.
4. Gardner, H.W. (1991) *Biochim. Biophys. Acta 1084,* 221–239.
5. Samuelsson, B. (1983) *Science 220,* 568–575.
6. Samuelsson, B., Dahlen, S.-E., Lindgren, J.Å., Rouzer, C.A., and Serhan, C.N. (1987) *Science 237,* 1171–1176.

7. Sigal, E. (1991) *Am. J. Physiol. 260,* L13-L28.
8. Yamamoto, S. (1992) *Biochim. Biophys. Acta 1128,* 117–131.
9. Cheesbrough, T.M., and Axelrod, B. (1983) *Biochemistry 22,* 3837–3840.
10. Dunham, W.R., Carroll, R.T., Thompson, J.E., Sands, R.H., and Funk, M.O., Jr. (1990) *Eur. J. Biochem. 190,* 611.
11. Navratnam, S., Feiters, M.C., Al-Hakim, M., Allen, J.C., Veldink, G.A., and Vliegthart, J.F. (1988) *Biochim. Biophys. Acta 956,* 70.
12. Gaffney, B.J., Mavrophilipos, D.V., and Doctor, K.S. (1993) *Biophys. J. 64,* 773.
13. Boyington, J.C., Gaffney, B.J., and Amzel, L.M. (1993) *Science 260,* 1482–1486.
14. Funk, M.O., Jr., Andre, J.C., and Otsuki, T. (1987) *Biochemistry 26,* 6880–6884.
15. Winkler, F.K., D'Arcy, A., and Hunziker, W. (1990) *Nature 343,* 771.
16. Kabsch, W., and Sander, C. (1983) *Biopolymers 22,* 2577–2637.
17. Finneh, D.C., Pinkerton, A.A., Dunham, W.R., Sands, R.H., and Funk, M.O., Jr. (1991) *Inorg. Chem. 30,* 3960.
18. Whittaker, J.W., and Solomon, E.I. (1988) *J. Am. Chem. Soc. 110,* 5329.
19. Van der Heijdt, L.M., Feiters, M.C., Navaratnam, S., Nolting, H.F., Hermes, C., Veldink, G.A., and Vliegenhart, J.F. (1992) *Eur. J. Biochem 207,* 793–802.
20. Stoddard, B.L., Howell, P.L., Ringe, D., and Pesko, G.A. (1990) *Biochemistry 29,* 8885–8893.
21. Sigal, E., Craik, C.S., Highland, E., Grunberger, D., Costello, L.L., Dixon, R.A.F., and Nadel, J.A. (1988) *Biochem. Biophys. Res. Commun. 157,* 457–464.
22. Fleming, J., Thiele, B.J., Chester, J., O'Prey, J., Janetzki, S., Aitken, A., Anton, I.A., Rapoport, S.M., and Harrison, P.R. (1989) *Gene 79,* 181–188.
23. Funk, C.D., Hoshiko, S., Matsumoto, T., Rådmark, O., and Samuelsson, B. (1989) *Proc. Natl. Acad. Sci. USA 86,* 2587–2591.
24. Chen, X.-S., and Funk, C.D. (1993) *Fed. Am. Soc. Exp. Biol. J. 7,* 694–701.
25. Shibata, D., Steczko, J., Dixon, J.E., Hermodson, M., Yazdanparast, R., and Axelrod, B. (1987) *J. Biol. Chem. 262,* 10080–10085.
26. Steczko, J., Donoho, G.P., Clemens, J.C., Dixon, J.E., and Axelrod, B. (1992) *Biochemistry 31,* 4053–4057.
27. Shibata, D., Steczko, J., Dixon, J.E., Andrews, P.C., Hermodson, M., and Axelrod, B. (1988) *J. Biol. Chem. 263,* 6816–6821.
28. Yenofsky, R.L., Fine, M., and Liu, C. (1988) *Mol. Gen. Genet. 211,* 215–222.
29. Shibata, D., Kato, T., and Tanaka, K. (1991) *Plant Mol. Biol. 16,* 353–359.
30. Bell, E., and Mullet, J.E. (1991) *Mol. Gen. Genet. 230,* 456–462.
31. Ealing, P.M., and Casey, R. (1989) *Biochem. J. 264,* 929–932.
32. Ealing, P.M., and Casey, R. (1988) *Biochem. J. 253,* 915–918.
33. Ohta, H., Shirano, Y., Tanaka, K., Morita, Y., and Shibata, D. (1992) *Eur. J. Biochem. 206,* 331–336.
34. Peng, Y.-L., Shirano, Y., Ohta, H., Hibino, T., Tanaka, K., and Shibata, D. (1994) *J. Biol. Chem. 269,* 3755–3761.
35. Melan, M.A., Nemhauser, J.L., and Peterman, T.K. (1994) *Biochim. Biophys. Acta 1210,* 377–380.
36. Hilbers, M.P., Rossi, A., Finazzi-Agrò, A., Veldink, G.A., and Vliegenthart, J.F.G. (1994) *Biochim. Biophys. Acta 1211,* 239–242.
37. Eiben, H.G., and Slusarenko, A.J. (1994) *Plant Journal 5,* 123–135.
38. Meier, B.M., Shaw, N., and Slusarenko, A.J. (1993) *Molecular Plant-Microbe Interactions 6,* 453–466.

39. Matsumoto, T., Funk, C.D., Rådmark, O., Hoog, J.-O., Jornvall, H., and Samuelsson, B. (1988) *Proc. Natl. Acad. Sci. USA 85,* 26–30.
40. Dixon, R.A.F., Jones, R.E., Diehl, R.E., Bennett, C.D., Kargman, S., and Rouzer, C.A. (1988) *Proc. Natl. Acad. Sci. USA 85,* 416–420.
41. Balcarek, J.M., Theisen, T.W., Cook, M.N., Varrichio, A., Hwang, S.-M., Strohsacker, M.W., and Crooke, S.T. (1988) *J. Biol. Chem. 263,* 13937–13941.
42. Funk, C.D., Furci, L., and FitzGerald, G.A. (1990) *Proc. Natl. Acad. Sci. USA 87,* 5638–5642.
43. De Marzo, N., Sloane, D.L., Dicharry, S., Highland, E., and Sigal, E. (1992) *Am. J. Physiol. 262,* L198–L207.
44. Yoshimoto, T., Suzuki, H., Yamamoto, S., Takai, T., Yokoyama, C., and Tanabe, T. (1990) *Proc. Natl. Acad. Sci. USA 87,* 2142–2146.
45. Watanabe, T., Medina, J.F., Haeggstrom, J.Z., Rådmark, O., and Samuelsson, B. (1993) *Eur. J. Biochem. 212,* 605–612.
46. Hada, T., Hagiya, H., Suzuki, H., Arakawa, T., Nakamura, M., Matsuda, S., Yoshimoto, T., Yamamoto, S., Azekawa, T., Morita, Y., Ishimura, K., and Kim, H.-Y. (1994) *Biochim. Biophys. Acta 1211,* 221–228.
47. Chen, X.-S., Kurre, U., Jenkins, N.A., Copeland, N.G., and Funk, C.D. (1994) *J. Biol. Chem. 269,* 13979–13987.
48. Ishii, S., Noguchi, M., Miyano, M., Matsumoto, T., and Noma, M. (1992) *Biochem. Biophys. Res. Comm. 182,* 1482–1490.
49. Rådmark, O., Zhang, Y.-Y., Hammarberg, T., Lind, B., Hamberg, M., Brungs, M., Steinhilber, D., and Samuelsson, B. (1995) in *Advances in Prostaglandin, Thromboxane, and Leukotriene Research,* Samuelsson, B., Ramwell, P.W., Paoletti, R., Folco, G., Granstrom, E., and Nicosia, S., Raven, New York, vol. 23, pp. 1–9.
50. Hammarberg, T., Zhang, Y.-Y., Rådmark, O., and Samuelsson, B. (1994) *Ninth International Conference on Prostaglandins and Related Compounds,* Florence, Italy, (Abstr.) June 6–10, 1994, 64.
51. Zhang, Y.-Y., Rådmark, O., and Samuelsson, B. (1992) *Proc. Natl. Acad. Sci. USA 89,* 485–489.
52. Nguyen, T., Falgueyret, J.-P., Abramovitz, M., and Riendeau, D. (1991) *J. Biol. Chem. 266,* 22057–22062.
53. Suzuki, H., Kishimoto, K., Yoshimoto, T., Yamamoto, S., Kanai, F., Ebina, Y., Miyatake, A., and Tanabe, T. (1994) *Biochim. Biophys. Acta 1210,* 308–316.
54. Minor, W., Steczko, J., Bolin, J.T., Otwinowski, Z., and Axelrod, B. (1993) *Biochemistry 32,* 6320–6323.
55. Kramer, J.A., Johnson, K.R., Dunham, W.R., Sands, R.H., and Funk, M.O., Jr. (1994) *Biochemistry 33,* 15017–15022.
56. Sloane, D.L., and Sigal, E. (1994) *Ann. N.Y. Acad. Sci. 744,* 99–106.
57. Watanabe, T., and Haeggstrom, J.Z. (1993) *Biochem. Biophys. Res. Commun. 192,* 1023–1029.
58. Sloane, D.L., Leung, R., Craik, C.S., and Sigal, E. (1991) *Nature 354,* 149–152.
59. Connolly, M.L. (1983) *J. Appl. Crystallogr. 16,* 548–558.
60. Rashin, A.A., Michael, I., and Honig, B. (1986) *Biochemistry 25,* 3619–3625.
61. Hamberg, M., and Samuelsson, B. (1967) *J. Biol. Chem. 242,* 5329–5335.
62. Holman, R.T., Egwim, P.O., and Christie, W.W. (1969) *J. Biol. Chem. 244,* 1149–1151.

63. Kühn, H., Sprecher, H., and Brash, A.R. (1990) *J. Biol. Chem. 265,* 16300–16305.
64. Gardner, H.W. (1989) *Biochim. Biophys. Acta 1001,* 274–281.
65. Allen, J.C. (1969) *J. Chem. Soc. Chem. Commun. 1969,* 906–907.
66. Kühn, H., Barnett, J., Grunberger, D., Baecker, P., Chow, J., Nguyen, B., Bursztyn-Pettegrew, H., Chan, H., and Sigal, E. (1993) *Biochim. Biophys. Acta 1169,* 80–89.
67. Kühn, H., Belkner, J., Wiesner, R., and Brash, A.R. (1990) *J. Biol. Chem. 265,* 18351–18361.
68. Sparrow, C.P., and Olszewski, J. (1992) *Proc. Natl. Acad. Sci. USA 89,* 128–131.
69. Corey, E.J., d'Alarcao, M., Matsuda, S.P.T., and Lansbury, P.T., Jr. (1987) *J. Am. Chem. Soc. 109,* 289–290.
70. Baer, A.N., Costello, P.B., and Green, F.A. (1991) *Biochim. Biophys. Acta 1085,* 45–52.
71. Baer, A.N., Costello, P.B., and Green, F.A. (1991) *J. Lipid Res. 32,* 341–347.
72. Kühn, H., Schewe, T., and Rapoport, S.M. (1986) *Adv. Enzymol. Relat. Areas Mol. Biol. 58,* 273–309.
73. Borgeat, P., Hamberg, M., and Samuelsson, B. (1976) J. Biol. Chem. 251, 7816–7820.
74. Ochi, K., Yoshimoto, T., Yamamoto, S., Taniguchi, K., and Miyamoto, T. (1983) *J. Biol. Chem. 258,* 5754–5758.
75. Kühn, H., Heydeck, D., Wiesner, R., and Schewe, T. (1985) *Biochim. Biophys. Acta 830,* 25–29.
76. Wiseman, J.S., Skoog, M.T., Nichols, J.S., and Harrison, B.L. (1987) *Biochemistry 26,* 5684–5689.
77. Riendeau, D., Falgueyret, J.-P., Meisner, D., Sherman, M.M., Laliberte, F., and Street, I.P. (1993) *J. Lipid Mediat. 6,* 23–30.
78. Shimizu, T., Rådmark, O., and Samuelsson, B. (1984) *Proc. Natl. Acad. Sci. USA 81,* 689–693.
79. Ueda, N., Yamamoto, S., Fitzsimmons, B.J., and Rokatch, J. (1987) *Biochem. Biophys. Res. Commun. 144,* 996–1002.
80. Kühn, H., Wiesner, R., Alder, L., Fitzsimmons, B.J., Rokach, J., and Brash, A.R. (1987) *Eur. J. Biochem. 169,* 593–601.
81. Corey, E.J., and Nagata, R. (1987) *J. Am. Chem. Soc. 109,* 8107–8108.
82. Corey, E.J., and Walker, J.C. (1987) *J. Am. Chem. Soc. 109,* 8108–8109.
83. de Groot, J.J.M.C., Garssen, G.J., Vliegenthart, J.F.G., and Boldingh, J. (1973) *Biochim. Biophys. Acta 326,* 279–284.
84. Chamulitrat, W., and Mason, R.P. (1989) *J. Biol. Chem. 264,* 20968–20973.
85. Nelson, M.J., Seitz, S.P., and Cowling, R.A. (1990) *Biochemistry 29,* 6897–6903.
86. Gaffney, B.G., Boyington, J.C., Amzel, L.M., Doctor, K.S., Prigge, S.T., and Yuan, S.M. in *Advances in Prostaglandin, Thromboxane, and Leukotriene Research,* Samuelsson, B., Ramwell, P.W., Paoletti, R., Folco, G., Granstrom, E., and Nicosia, S., Raven, New York, vol. 23, pp. 11–16.
87. Creighton, T.E. (1993) *Proteins,* W.H. Freeman, New York, p. 141.

Chapter 2

Genetics of Soybean Lipoxygenases

David Hildebrand

Agronomy and Plant Science, University of Kentucky, Lexington, KY 40546, USA.

Introduction

Lipoxygenase (linoleate: oxygen oxidoreductase, EC 1.13.11.12), also known as lipoxidase and carotene oxidase, refers to a class of dioxygenases that catalyzes the peroxidation of molecules containing *cis,cis*-1,4-pentadiene moieties and some related molecules (1). The substrate in higher plants is usually linoleic (C18:2, ω-6) and linolenic acid (C18:3, ω-3). Many plants contain multiple lipoxygenase isozymes encoded by different genes.

The genes for five lipoxygenases in soybeans have been cloned and sequenced (2–7). A sixth lipoxygenase in soybeans has been partially characterized, and there is evidence for at least one additional lipoxygenase gene in the soybean genome. The first lipoxygenase to be isolated and purified was soybean seed lipoxygenase-1 (SBL-1 [8]). Subsequently, two additional lipoxygenases from soybean seeds have been purified and characterized and have been designated lipoxygenase-2 and lipoxygenase-3 (SBL-2 and -3, respectively [9,10]).

In mature soybean seeds, SBL-3 is the most abundant of the isozymes on a protein basis, representing about 2% of the total seed protein. Soybean lipoxygenase-1 is also abundant being about 1% of the seed protein, while SBL-2 is the least abundant at about 0.1–1%. Soybean lipoxygenase-2, however, has the highest specific activity. Therefore, on an activity basis, there are similar amounts of the isozymes in mature seeds of typical commercial soybean cultivars (11–13, Hildebrand, et al.).

Lipoxygenase Characteristics

The soybean seed lipoxygenase isozymes were originally separated by ion-exchange chromatography, and ionic charge remains the principal characteristic by which the soybean seed lipoxygenase isozymes can be differentiated (14). The three seed lipoxygenases are monomeric proteins similar in size, ranging from 94 to 97 kD, but they have distinct isoelectric points ranging from about 5.7 to 6.3 (2–4,14). Roza and Francke reported observing a high molecular mass aggregate from soybean seeds that forms cyclic peroxides from methyl linolenate (15), but this has not been studied further.

Soybean lipoxygenase 1 has a pH optimum of 9-10, whereas SBL-2 and -3 have pH optima at pH 6–7 (11). In addition to their similar pH optima, SBL-2 and -3 have similar isoelectric points (pI) that results in a difficult chromatographic separation.

This difficulty in distinguishing SBL-2 from SBL-3 experimentally has led many researchers to group SBL-2 and -3 together as Type II lipoxygenase. Soybean lipoxygenase-1 is readily separated and therefore is referred to as Type I lipoxygenase. So-called Type I or II lipoxygenase activities in tissues other than soybean embryo tissue are due to lipoxygenase isozymes other than SBL-1, -2, or -3.

Crystallization and preliminary X-ray characterization were reported for SBL-1 (16) and -2 (17) and a three-dimensional structure was determined for SBL-1 (18). The active-site iron ligands have also been mapped by crystallography (19). Sequences of more than a dozen lipoxygenases from various plant and mammalian species contain a motif of 38 amino acids that includes five conserved histidines and a sixth histidine about 160 amino acids toward the carboxy terminus. The residues that occur at positions 499, 504, and 690 have been shown to be required for iron binding and catalytic activity (20,21). An arginine at position 713 in SBL-3 was found to be critical for catalytic activity but not for iron binding (22).

Seedling and leaf tissues in soybeans display lipoxygenases different from SBL-1, -2, and -3 found in the mature seeds (23–25). Seed coats and pod walls also appear to have different lipoxygenase isozymes than immature embryos (26). These new lipoxygenases have similar molecular weights but different pI from the seed isozymes. The seedling lipoxygenase isozymes (at least those in seedling cotyledons) have been designated soybean lipoxygenase-4 (SBL-4), -5 (SBL-5), and -6 (SBL-6 [5,6]).

At least some of the soybean leaf lipoxygenase isozymes appear to be different from those of seedling tissues (25). Saravitz and Siedow observed about a dozen lipoxygenases with different pI from soybean leaves; some of the lipoxygenases are wound-inducible (27). Some of these lipoxygenases with different isoelectric points may be different isoforms of the same isozyme, so there may be less than 12 different isozymes of lipoxygenase in soybean leaves. Multiple isoforms can be seen for the three embryo-specific lipoxygenases. For example, SBL-3 can display three different forms with slightly different native pI if the sample is not very fresh (25,28). These multiple forms or isoforms of individual isozymes may be due to different folding of the same gene product.

The relationship of the seed coat and pod wall lipoxygenase isozymes to those of seedlings and leaves is unknown. Preliminary evidence indicates that a minor isozyme is constitutively expressed in most or all tissues. Although SBL-1 through SBL-3 are the major isozymes expressed in cotyledons and embryonic axes from the maturation phase of seed development until maturity, they are not expressed in seed coats (26,29). The seed coat lipoxygenase isozymes are likewise not normally expressed in immature embryos (at least not at any appreciable level). Therefore, it is more precise to refer to SBL-1 through SBL-3 as embryo (cotyledon + embryonic axis) rather than seed lipoxygenases.

Soybean germplasm collections have been screened for lipoxygenase mutants and independent single null accessions have been found missing SBL-1, SBL-2, and SBL-3 (13,30). The SBL-1, -2, and -3 null mutants have been found to be inherited as simple recessive alleles (31,32,33). Soybean lipoxygenase 1 and SBL-2 have been found to be tightly linked, but the presence or absence of SBL-3 was found to be inherited independently from SBL-1 and -2 (33).

The linkage of SBL-1 and -2 has never been broken, suggesting that either the linkage is very tight or SBL-1, -2 nulls are lethal. Davies and Nielsen have backcrossed the lipoxygenase null alleles into the soybean cultivar “Century,” producing three single null backcross lines and two double null backcross lines (SBL-1 and -3 and SBL-2 and -3 nulls [34]). Neither the third double null (SBL-1 and -2 null) nor the triple null lines had been produced until recently because of the tight linkage of SBL-1 and -2. Hajika et al. reported obtaining a triple lipoxygenase null in soybeans by further mutagenesis of the double lipoxygenase nulls (35); this indicated that SBL-1, -2, or triple nulls are not lethal.

The lipoxygenase activity of the triple lipoxygenase null mutants during seed maturation is greatly reduced (36). The C6-aldehyde production of these “triple null” lines was greater than or equal to that of several of the double null lines (37). Tavares et al. reported that the SBL-3 nulls still have a certain amount of SBL-3 activity (38). Further studies by Hajika et al. showed no recombinants out of over 1000 F_2 progeny from crosses between SBL-1 and -2 nulls (39), indicating that SBL-1 and -2 were inherited as if they were on the same locus. Rapid techniques for screening soybean seeds for lipoxygenase nulls have recently been reported (40,41).

The complete coding regions of all three soybean embryo lipoxygenase cDNA have been sequenced (2–4,42). All three soybean seed lipoxygenases show a high degree of homology (at least 70% at the amino acid sequence level [3]) with SBL-1 and -2 being particularly homologous (81% at the amino acid sequence level [3,14]). The three new lipoxygenases expressed in soybean seedlings have been purified and partially characterized by Kato et al. (5,6). These were shown to be different from seed SBL-1 through SBL-3 by ionic charge, product specificity, and amino acid sequencing and are designated SBL-4, SBL-5, and SBL-6 (Table 2.1). Genomic DNA and cDNA of seedling lipoxygenases have also been cloned (43).

The seedling lipoxygenases are homologous to the embryo SBL-1 through SBL-3, but are clearly encoded by distinct genes. Exposure of soybean seedlings to a water deficit causes a rapid increase in seedling lipoxygenase mRNA levels suggesting a role of these enzymes in plant response to these stresses (44). Various designations for the cloned new lipoxygenase DNA of seedlings have been reported. Inspection of the sequences shows that *Loxa* and *Loxb*2 (44) = SC514 and SC501 (43). SC514 is now designated SBL-5 and SC501 is designated SBL-4.

The soybean leaf and stem vegetative storage protein, vsp94 is different from the previously characterized soybean embryo lipoxygenases, but appears to be very similar if not identical to seedling SBL-4 (45–47). Not only are the five sequenced lipoxygenases of soybeans very homologous, but most other plant lipoxygenases have high amino acid and nucleotide homology with these soybean lipoxygenases. All plant lipoxygenases with high homology to this group, including SBL-1, are grouped into one class in the genetic databank with a plantwide mnemonic designation of *Lox1* (48).

TABLE 2.1 Some Characteristics of Soybean Lipoxygenase Isozymes

	Soybean Lipoxygenase Isozyme					
Characteristic	L1	L2	L3	L4	L5	L6
pI[a]	5.85	6.25	6.4	4.65	4.5	7
9:13 18:2 Hydroperoxide product ratio[b]	19:81	42:58	?	54:46	15:85	15:85
Substrate Specificity (18:3/18:2) activity[c]	0.88	3.0	4.5	2.9	2.6	3.1

[a]Axelrod et al. (11); Wang unpublished results.
[b]Christopher et al. (10); Kato et al. (5).
[c]Kato et al. (5).

Novel lipoxygenases with low homology to the *Lox1* group that have been characterized from rice and *Arabidopsis* have been grouped into a second class with the designation *Lox2* (48). These two *Lox2* lipoxygenases appear to have a transit sequence for targeting to plastids. It is not yet known whether soybeans have any lipoxygenases of the *Lox2* class.

Original studies by Start et al. indicated that the null mutations of soybean seed lipoxygenases resulted in the loss of lipoxygenase transcript accumulation during seed development (42). At the time of that study neither the triple lipoxygenase null nor the SBL-1, SBL-2 null were available. This fact plus the high homology between SBL-1 through SBL-3 left the question ambiguous. Wang et al. addressed this question more thoroughly and found no SBL-1 or -3 transcript in developing seeds of the corresponding lipoxygenase mutant lines, but the SBL-2 transcript accumulated normally (36). They found that the SBL-2 null mutation is due to a missense mutation that results in the substitution of glutamine for His-532. This mutation apparently renders the SBL-2 translation product inactive and unstable, causing this protein to disappear during the soybean seed maturation process.

References

1. Kuhn, H., Eggert, L., Zabolotsky, O.A., Myagkova, G.I., and Schewe, T. (1991) *Biochemistry 30,* 10269–10273.
2. Shibata, D., Steczko, J., Dixon, J.E., Hermodson, M., Yazdanparast, R., and Axelrod, B. (1987) *J. Biol. Chem. 262,* 10080–10085.
3. Shibata, D., Steczko, J., Dixon, J.E., Andrews, P.C., Hermodson, M., and Axelrod, B. (1988) *J. Biol. Chem. 263,* 6816–6821.
4. Yenofsky, R.L., Fine, M., and Liu, C. (1988) *Molec. Gen. Genetics 211,* 215–222.
5. Kato, T., Ohta, H., Tanaka, K., and Shibata, D. (1992) *Plant Physiol. 98,* 324–330.
6. Kato, T., Terao, J., and Shibata, D. (1992) *Biosci. Biotechnol. Biochem. 56,* 1344.

7. Park, T.K., Holland, M.A., Laskey, J.G., and Polacco, J.C. (1994) *Plant Sci. 96,* 109–117.
8. Theorell, H., Holman, R.T., and Akeson, A. (1947) *Acta Chem. Scand. 1,* 571–576.
9. Christopher, J.P., Pistorius, E.K., and Axelrod, B. (1970) *Biochim. Biophys. Acta 198,* 12–19.
10. Christopher, J.P., Pistorius, E.K., and Axelrod, B. (1972) *Biochim. Biophys. Acta 284,* 54–62.
11. Axelrod, B., Cheesbrough, T.M., and Laakso, S. (1981) *Methods Enzymol. 71,* 441–451.
12. Yabuuchi, S., Lister, R.M., Axelrod, B., Wilcox, J.R., and Nielsen, N.C. (1982) *Crop Sci. 22,* 333–337.
13. Kitamura, K. (1984) *J. Agric. Biol. Chem. 48,* 2339–2343.
14. Hildebrand, D.F., Hamilton-Kemp, T.R., Legg, C.S., and Bookjans, G. (1988) *Curr. Top. Plant Biochem. Physiol. 7,* 201–219.
15. Roza, M., and Francke, A. (1978) *Biochim. Biophys. Acta 528,* 119–126.
16. Steczko, J., Muchmore, C.R., Smith, J.L., and Axelrod, B. (1990) *J. Biol. Chem. 265,* 11352–11354.
17. Stallings, W.C., Kroa, B.A., Carroll, R.T., Metzger, A.L., and Funk, M.O. (1990) *J. Mol. Biol. 211,* 685–687.
18. Boyington, J.C., Gaffney, B.J., and Amzel, L.M. (1993) *Science 260,* 1482–1486.
19. Minor, W., Steczko, J., Bolin, J.T., Otwinowski, Z., and Axelrod, B. (1993) *Biochemistry 32,* 6320–6323.
20. Steczko, J., and Axelrod, B. (1992) *Biochem. Biophys. Res. Comm. 186,* 686–689.
21. Steczko, J., Donoho, G.P., Clemens, J.C., Dixon, J.E., and Axelrod, B. (1992) *Biochemistry 31,* 4053–4057.
22. Kramer, J.A., Johnson, K.R., Dunham, W.R., Sands, R.H., and Funk, M.O., Jr. (1994) *Biochemistry 33,* 15017–15022.
23. Hildebrand, D.F., Snyder, K.M., Hamilton-Kemp, T.R., Bookjans, G., Legg, C.S., and Andersen, R.A. (1989) in *Biological Role of Plant Lipids,* Biacs, P.A., Gruiz, K., Kremmer, T., Akademiai Kiado, Budapest and Plenum Publishing Co., New York, pp. 51–56.
24. Park, T.K., and Polacco, J.C. (1989) *Plant Physiol. 90,* 285–290.
25. Grayburn, W.S., Schneider, G.R., Hamilton-Kemp, T.R., Bookjans, G., Ali, K., and Hildebrand, D.F. (1991) *Plant Physiol. 95,* 1214–1218.
26. Pfeiffer, T.W., Hildebrand, D.F., and Tekrony, D.M. (1992) *Crop Sci. 32,* 357–362.
27. Saravitz, D.M., and Siedow, J.N. (1995) *Plant Physiol. 107,* 535–543.
28. Hildebrand, D.F., Grayburn, W.S., Versluys, R.T., Hamilton-Kemp, T.R., Andersen, R.A., and Collins, G.B. (1990) in *Proceedings of the International Symposium on New Aspects of Dietary Lipids,* International Union of Food Sci. and Technology, Goteburg, p. 223–242.
29. Hildebrand, R.T., Versluys, R.T., and Collins, G.B. (1991) *Plant Sci. 75,* 1–8.
30. Hildebrand, D.F., and Hymowitz, T. (1981) *J. Am. Oil Chem. Soc. 58,* 583–586.
31. Hildebrand, D.F., and Hymowitz, T. (1982) *Crop Sci. 22,* 851–853.
32. Kitamura, K., Davies, C.S., Kaizuma, N., and Nielsen, N.C. (1983) *Crop Sci. 23,* 924–927.
33. Davies, C.S., and Nielsen, N.C. (1986) *Crop Sci. 27,* 460–463.
34. Davies, C.S., and Nielsen, N.C. (1987) *Crop Sci. 27,* 370–371.
35. Hajika, M., Igita, K., and Kitamura, K. (1991) *Jpn. J. Breed. 41,* 507–509.
36. Wang, W.H., Takano, T., Shibata, D., Kitamura, K., and Takeda, G. (1994) *Proc. Natl. Acad. Sci. USA 91,* 5828–5832.

37. Takamura, H., Kitamura, K., and Kito, M. (1991) *FEBS Lett. 292,* 42–44.
38. Tavares, J.T.Q., Tavares, D.Q., and Miranda, M.A.C. (1993) *J. Sci. Food Agric. 62,* 207–208.
39. Hajika, M., Kitamura, K., Igita, K., and Nakazawa, Y. (1992) *Jpn. J. Breed. 42,* 787–792.
40. Hammond, E.G., Duvick, D.N., Fehr, W.R., Hildebrand, D.F., Lacefield, E.C., and Pfeiffer, T.W. (1992) *Crop Sci. 32,* 820–821.
41. Evans, D.E., Nyquist, W.E., Santini, J.B., Bretting, P., and Nielsen, N.C. (1994) *Crop Sci. 34,* 1529–1537.
42. Start, W.G., Ma, Y., Polacco, J.C., Hildebrand, D.F., Freyer, G.A., and Altschuler, M. (1986) *Plant Molec. Biol. 7,* 11–23.
43. Shibata, D., Kato, T., and Tanaka, K. (1991) *Plant Mol. Biol. 16,* 353–359.
44. Bell, E., and Mullet, J.E. (1991) *Mol. Gen. Genet. 230,* 456–462.
45. Tranbarger, T.J., Francheschi, V.R., Hildebrand, D.F., and Grimes, H.D. (1991) *Plant Cell 3,* 973–987.
46. Grimes, H.D., Koetje, D.S., and Francheschi, V.R. (1992) *Plant Physiol. 100,* 433–443.
47. Kato, T., Shirano, Y., Iwamoto, H., and Shibata, D. (1993) *Plant Cell Physiol. 34,* 1063–1072.
48. Shibata, D., Slusarenko, A., Casey, R., Hildebrand, D., and Bell, E. (1994) *Plant Molec. Biol. Rep. 12,* S41–S42.

Chapter 3

Plant Lipoxygenase Genes

Daisuke Shibata

Mitsui Plant Biotechnology Research Institute, Sengen 2-1-6, TCI-D21, Tsukuba, Ibaraki 305, Japan.

Introduction

Plant lipoxygenases have been investigated for the past 60 years. Isozymes found in soybeans have had their enzymological properties studied intensively. Soybean lipoxygenases are monomeric proteins and contain a single atom of tightly bound nonheme iron per molecule (1). These features are thought to be common to other lipoxygenases. This plant lipoxygenase research has contributed to an understanding of the properties of mammalian lipoxygenases involved in arachidonic acid metabolism for the production of potent physiological effectors (eicosanoids), and to an understanding of the diverse metabolic functions of these effectors (2).

With the aid of molecular cloning, a number of amino acid sequences of plant and mammalian lipoxygenases have been determined, and comparison of these sequences revealed that significant conservation of certain amino acid residues occurs. This conservation suggests residue involvement in the lipoxygenase reaction mechanism common to both plant and mammalian enzymes. Indeed, site-directed mutagenesis of these residues in plant and mammalian enzymes has confirmed their crucial roles in lipoxygenase activity. Recently the X-ray crystallographic structure of soybean lipoxygenase-1 (SBL-1) has been reported by two groups (3,4). Site-directed mutagenesis and the tertiary structure analysis of SBL-1 are reviewed briefly in this chapter. A more detailed review for this subject is available (5).

Although the physiological functions of mammalian arachidonate lipoxygenases have been well documented, the individual roles of the various plant lipoxygenases remain to be determined. A number of compounds thought to be synthesized via lipoxygenation of linoleic (LA) or linolenic acid (LNA) have been found in various plants. (Since arachidonate is rarely found in higher plants, these two fatty acids constitute the natural substrates for plant lipoxygenases.) These lipoxygenase-pathway products are thought to be involved in plant defense, wound response, senescence, and development (6). In particular, jasmonic acid and its derivatives synthesized from LNA via lipoxygenation have been studied extensively with regard to their diverse physiological roles (7,8). However, the individual lipoxygenases responsible for biosynthesis of each compound have not been identified definitively, even for biosynthesis of jasmonic acid. Molecular cloning of lipoxygenase genes has helped illuminate the physiological functions of plant lipoxygenases. To date, more than 17 lipoxygenase genes have been cloned from at least eight plant species.

This chapter reviews research on lipoxygenase genes from a dicot plant, soybean, and a monocot plant, rice. This chapter also suggests a possible function in plant defense for a novel rice lipoxygenase gene, the product of which is active in the photosynthetic organelle, the chloroplast. Readers are also referred to Chapter 10 for involvement of dicot lipoxygenases in plant defense. Good reviews are available concerning the enzymological and physiological features of plant lipoxygenases (9,10).

Soybean Lipoxygenase Genes

Three lipoxygenase isozymes, SBL-1, SBL-2, and SBL-3, are found in mature soybean seeds, and have been studied extensively (1). They accumulate to high levels only in the mature embryo, are expressed only during seed maturation, and are different from lipoxygenase isozymes expressed in the embryo tissues at earlier development stages or in the seed coats (11). Vegetative tissues also have lipoxygenase isozymes distinct from those in the seed.

Molecular Cloning of Lipoxygenases Expressed in Seed

Start et al. isolated three partial cDNA clones for soybean lipoxygenases from a maturing seed library (12). Soon after, the first complete amino acid sequence of SBL-1 was reported by Shibata et al. (13). They isolated a full-length cDNA and unequivocally identified it as being for SBL-1 by Edman sequencing of selected peptides obtained by CNBr cleavage, by carboxyl-terminal amino acid sequencing of the intact protein, and by partial peptide analysis employing fast atom bombardment–mass spectroscopy. Based on the deduced amino acid sequence, SBL-1 contains 838 amino acid residues, and has an Mr of 94,038.

Molecular cloning of SBL-2 and SBL-3 was also reported (14,15). Soybean lipoxygenase-2 contains 865 residues and has an Mr of 97,035; the corresponding values for SBL-3 are 859 and 96,541, respectively. When compared to SBL-1, the conservation of amino acid sequences for SBL-2 and SBL-3 is 84 and 72%, respectively. The significant conservation in the nucleotide sequences of SBL-1 and SBL-2 cDNA, even in the noncoding sequences, indicates that these genes diverged relatively recently from a progenitor gene. In spite of high homology between these two genes, the gene products exhibit diversity in their enzymatic behavior: SBL-1 is active at pH 9, while SBL-2 is active at pH 7. Soybean lipoxygenase-1 prefers C-13 for the site of LA hydroperoxidation, but SBL-2 utilizes either C-13 or C-9. These features may provide a good model system for using site-directed mutagenesis to investigate the amino acid residues responsible for determining the pH optimum and regiospecificity.

Conserved Histidine Residues and the Tertiary Structure of Soybean Lipoxygenase-1. Shibata et al. suggested that a cluster of six histidines conserved in all three soybean lipoxygenases is a possible iron-binding region (14). Lipoxygenase sequences obtained from plants and mammals have shown that five of the six histidines and an additional histidine found about 160 amino acid residues downstream are strongly conserved (Fig. 3.1). Steczko et al. used site-directed mutagenesis to show that the conserved histidine residues at 499, 504, and 690 were essential for activity, while substituting glutamine or serine for any of the other three histidines resulted in greatly reduced activity (16). These results are consistent with those obtained for the corresponding histidine residues of mammalian lipoxygenases (17,18). Steczko and Axelrod expressed the mutated proteins in the presence of $^{59}FeCl_3$ in *E. coli* and showed that the inactive mutant forms, those carrying the substitutions at the three essential histidines, fail to bind iron (19).

Two groups reported the X-ray crystallographic structure of SBL-1 to a resolution of 2.6 Å (3,4). These models show that the three histidine residues 499, 504, and 690 are involved in iron binding. In addition, Asn-694 and the terminal carboxyl residue, Ile-893, are also required. These two residues are also conserved in the majority of the lipoxygenases sequenced. Recently, crystal structure determination has been improved to a resolution of 1.4 Å (20). The improved model produced data allowing the recognition of water as a sixth ligand.

Molecular Basis of Seed Lipoxygenase Deficiencies

Soybean mutants lacking lipoxygenases in their seeds have been isolated to improve soybean quality for food processing (21). Lipoxygenases are responsible for production of the unpleasant grassy-beany flavors that have limited the development of soybean-protein products for human consumption. Homogenization of soybeans results in hydroperoxidation of free polyunsaturated fatty acids by the action of lipoxygenase. Subsequently, hydroperoxide lyase cleaves the resulting hydroperoxides to form medium-chain-length aldehydes that are responsible for the undesirable flavors. Null

```
                         Accession         499  504                                690 694    839
          Soybean L-1    J02795  494  HQLMSHWLNTHAAMEPFVIATHRHLSVLHPIYKLLTPH...HAAVN...SISI
          Soybean L-2    J03211  522  ------------VI---I---N----A-----------...-----...----
          Soybean L-3    X06928  513  ---V--------VV---I---N-----V-------H--...-----...----
          Soybean L-4    D13999  520  ---V--------V----A---N-------------Y--...-----...----
          Soybean SC514  X56139  520  ---V--------V----A---N-------------Y--...-----...----
          Pea pPE1036    X07807  517  ---V--------VV-------N----C--------Y--...-----...----
          Pea pPE923     X17061  519  -------------I-------N-Q---V---N---A--...-----...----
          Bean L-1       X63525  517  ------------V--------N-------------L--...-----...----
          Lentil         X71344  520  ---V--------VV---I---N----V---HK---L--...-----...----
          Tomato 2       U09025  516  ---I--------VI-------N-Q-------H---Y--...-----...-V--
          Potato         X79107  517  ---I--------VI-------N-Q-------H---Y--...-----...----
A. thaliana LOX1         L04637  514  ---I---MQ---SI-------N-Q------VF---E--...-----...-V--
          Rice L-2       X64396  516  ---I--------V--------N-Q--A---VH---L--...-----...----
A. thaliana LOX2         L23968  549  ---I----R---CT--YI--AN-Q--AM----R--H--...-----...----
          Rice leaf-LOX  D14000  576  -E-IT---RY-C-V--YI--AN-Q--EM----Q--R--...-----...-T--
          Human 5-LOX    J03600  363  --TIT-L-R--LVS-V-G--MY-Q-PAV---F---VA-...-----...-VA-
          Human 12-LOX   M58704  355  -EIQY-L----LVA-VIAV--M-C-PG----F-F-I--...---I-...-VT-
          Human 15-LOX   M23892  355  -E-Q--L-RG-LMA-VI-V--M-C-PSI---F--II--...--S-H...-VA-
```

Figure 3.1. Amino acid residues involved in iron binding in SBL-1. His-499, -504, -690, Asn-694 and the carboxyl terminal residue, ILe-839, are involved in binding the iron in SBL-1. Amino acid sequences of other plant lipoxygenase and some mammalian lipoxygenases are also shown in this region. Conserved amino acid residues in all lipoxygenases are shown in bold. The accession numbers in the EMBL/GenBank/DDBJ Databases for these sequences are shown.

mutants for SBL-1 (22), SBL-2 (23,24), and SBL-3 (25) were isolated. The undesirable flavors are absent or less pronounced in food products made from seeds deficient in these isozymes, particularly those lacking SBL-2 (26). These mutants have been useful for understanding the molecular bases of null mutations and gene regulation (27,28).

Genetic studies of these mutant lines demonstrated that the absence of these isozymes is due to single recessive alleles (24,29). The symbols *lx1*, *lx2*, and *lx3* have been used to designate these alleles. Crossing of these mutants has produced two double mutant lines; one lacking both SBL-1 and SBL-3, the other, both SBL-2 and SBL-3. However, no double mutant lacking both SBL-1 and SBL-2 was obtained by these crossing efforts because of the tight linkage of *lx1* with *lx2*. Recently, Hajika et al. isolated a mutant lacking all three isozymes by γ-ray irradiation of seeds (30). It should be noted that these mutant alleles are not necessarily the structural genes for the corresponding lipoxygenase isozymes.

A Novel Mechanism for Soybean Lipoxygenase-2 Deficiency. Recently, Wang et al. indicated that *lx2* is a mutant allele of the SBL-2 structural gene (27). In an effort to clarify the molecular basis of these mutations, they carried out northern analyses using gene-specific probes. They found no detectable gene transcripts for SBL-1 and SBL-3 in the *lx1* and *lx3* mutants, respectively. However, the SBL-2 transcript was present in *lx2* mutants at a level and size unchanged from wild type (Fig. 3.2). Therefore, they isolated a nearly full-length cDNA from the *lx2* mutant line for nucleotide sequencing. The deduced amino acid sequence revealed two differences from the SBL-2 sequence reported by Shibata et al. (14); the protein from the mutant line has an inserted glutamine residue at 852, and a substitution of glutamine for His-532, an iron-binding ligand that corresponds to His-504 in SBL-1. The inserted Glu appears to constitute the norm, as SBL-3 contains a corresponding glutamic acid residue and SBL-1 contains a lysine residue at the equivalent position. It has been shown for SBL-1 that the corresponding His residue is required for iron binding. Western blot experiments showed that the enzymatically inactive gene product was readily detectable in midmaturation-stage seeds, but was absent from mature seeds (Fig. 3.2), indicating apparent instability of the inactive protein *in vivo* due to the inability of the protein to bind iron. This is a novel case of a spontaneous missense mutation that results in the rapid degradation of the protein product.

Impaired Promoter Function of the Soybean Lipoxygenase-3 Gene in the Soybean Lipoxygenase-3 Null Mutant. Recently, Wang et al. found three nucleotide substitutions in the 5′ upstream region of the SBL-3 gene in the *lx3* mutant by comparison with the wild-type sequence (28). These mutations were termed mut 1, mut 2, and mut 3. It is interesting that mut 1 and mut 2 each occur in separate AAATAC repeats that lie close to each other. The 5′ upstream region of the mutant and wild-type gene was fused separately with a reporter gene encoding β-glucuronidase (GUS), and the constructs were expressed transiently in tobacco suspension-cultured cells.

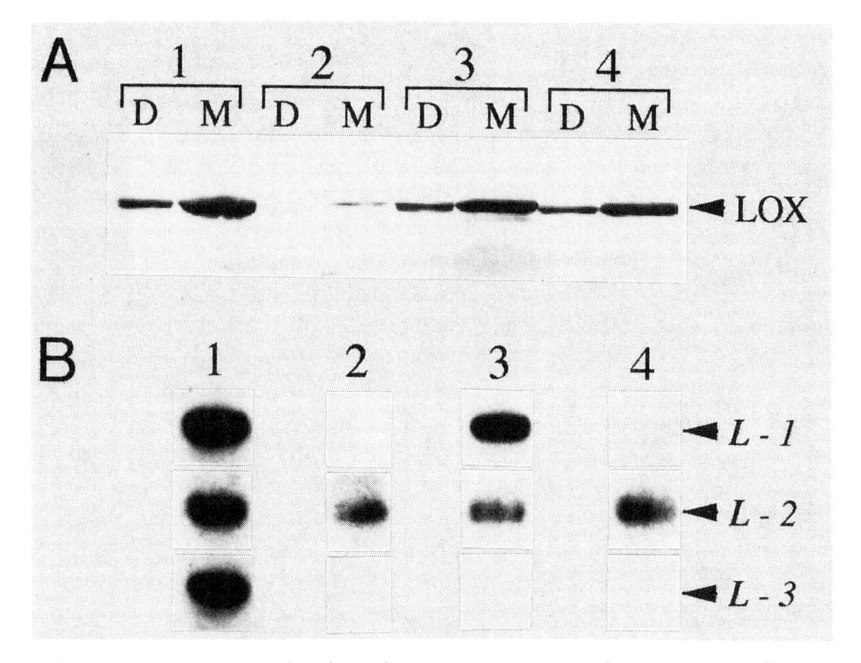

Figure 3.2. Expression of soybean lipoxygenase genes in lipoxygenase null mutants. *a*) Western blots of lipoxygenases extracted from midmaturation-stage seed (M) and mature dry seed (D). *b*) Northern hybridization of lipoxygenase mRNA extracted from midmaturing-stage seed. Gene specific probes for the SBL-1, SBL-2, and SBL-3 genes were used, respectively. Soybean lines; lane 1, wild type; lane 2, a line lacking SBL-1, SBL-2, and SBL-3; lane 3, a line lacking SBL-1 and SBL-3; lane 4, a line lacking SBL-2 and SBL-3.

The GUS activity expressed by the mutant promoter in the cells was 48% of the activity of the wild type, indicating that these nucleotide substitutions impair the promoter function. For evaluation of individual mutations, these mutated nucleotides were reverted to the corresponding wild-type nucleotides by site-directed mutagenesis. Reversion of either mut 1 or mut 2 restored promoter function, while reversion of mut 3 had no effect on the GUS gene expression in the tobacco cells. These results indicate that mut 1 and mut 2 occur in redundant *cis*-acting elements, probably the paired AAATAC repeats. The mut 3 substitution may be irrelevant to the absence of the SBL-3 transcript.

The reduced efficacy of the mutant SBL-3 gene promoter in driving transient expression in tobacco cells does not satisfactorily explain the complete lack of the SBL-3 gene transcript in the *lx3* mutant seeds (Fig. 3.2). Recently, Forster et al. showed that expression of a pea lipoxygenase promoter-driven GUS gene is not seed-specific in transgenic tobacco plants (31). Further studies of *lx3* mutant promoter activity, especially in transgenic soybean or legume plants, will be required to understand the molecular basis of the *lx3* mutation.

Physiological Roles of Seed Lipoxygenases. Interestingly, the mutants lacking all three seed lipoxygenases, SBL-1, SBL-2, and SBL-3, are physiologically normal; even when the mutants were grown in the field, no obvious difference was found in growth between the parental lines throughout the whole life cycle (30). These results indicate no functional importance of these isozymes in development. Pfeiffer et al. found that pod and stem blight disease should not be a more serious problem in the production of soybean genotypes with reduced seed lipoxygenase levels than in the production of soybean with a full seed lipoxygenase complement (32). However, one could not exclude the possibility that these isozymes have a role for plant defense against insects, microorganisms, and other threats that may not have been encountered in these tests.

The seed lipoxygenase proteins are stable during seed maturation, as shown in Figure 3.2, in contrast to the instability of the inactive SBL-2 protein in the *lx2* mutant. It is interesting to associate the stability of the lipoxygenases to a role of lipoxygenases as storage proteins. Actually, lipoxygenases found in vegetative tissues (distinct from the seed isozymes) are thought to be a class of storage proteins. It is not unprecedented that enzymes may function as storage proteins. For example, patatin is a family of glycoproteins that accounts for 30–40% of the total soluble protein in potato tubers, and it also exhibits lipid acyl hydrolase activity (33).

Deng et al. overexpressed the SBL-2 protein in transgenic tobacco plants by introducing SBL-2 cDNA and showed that the introduced lipoxygenase affected fatty acid oxidative metabolism; C6-aldehyde production increased 50–529% (34). However, their transgenic tobacco plants have not shown any apparent differences from control plants in growth and development. Their interpretation is that endogenous tobacco lipoxygenases are already sufficiently abundant to promote maximal growth and development.

Expression of Lipoxygenase Genes in Vegetative Tissues

New Lipoxygenases in Vegetative Tissues. In contrast to the extensive research on soybean seed lipoxygenases, information on lipoxygenase isozymes expressed in other soybean tissues is limited. In some cases, researchers have associated the lipoxygenase activity found in seedlings at pH 6.8 with SBL-2 and SBL-3 without giving evidence to prove that these two enzymes were the ones responsible for this activity. Recently, multiple lipoxygenase isozymes distinct from the seed isozymes have been found in other soybean tissues (35–37).

Kato et al. have characterized lipoxygenase activity in cotyledons after germination (37). They took advantage of the absence of seed lipoxygenases in a mutant soybean line, and identified new lipoxygenases that are active at pH 6.8 in 5-day-old cotyledons. These lipoxygenases were designated SBL-4, SBL-5, and SBL-6. They all have pH optima at 6.5, and none of them are active at pH 9.0. Soybean lipoxygenase-5 and -6 preferentially produce 13-hydroperoxy-linoleic acid over 9-hydroperoxy-linoleic acid, whereas SBL-4 produces both 13- and 9-hydroperoxy-linoleic acid in nearly equal amounts. Partial amino acid sequencing of SBL-4 and SBL-6 indicates that these isozymes were synthesized from genes distinct from those encoding SBL-1, SBL-2, and SBL-3 (37,38).

Lipoxygenase Genes Expressed in Vegetative Tissues. Vegetatively expressed lipoxygenase genes have been cloned (39–42). Shibata et al. isolated three distinct cDNA clones from seedlings after germination, called SC501, SC514, and SC500 (39). Using the SC514 clone as probe, the corresponding genomic clone was isolated (39). Partial amino acid sequences of selected polypeptides of SBL-6 are highly homologous with the deduced amino acid sequence from the SC514 gene (38), but there are some mismatches. It is possible that there is a gene subfamily for SBL-6, or that polymorphisms may exist between the soybean cultivars used for gene and protein isolation. Bell and Mullet reported isolation of partial cDNA clones, designated *lox*A, *lox*B1, and *lox*B2, from seedlings (40). The cDNA sequence (197 bp) of *lox*A is identical to the corresponding region of the SC514 gene. Park et al. reported a full-length cDNA clone (pTK11) for the SC514 gene (42). According to the guidelines of the Commission on Plant Gene Nomenclature, this gene is classified into a lipoxygenase gene category, termed *Lox1* (43). To identify lipoxygenase genes in this category, member numbers are given chronologically. The SC514 gene is designated *Lox1: Gm:4*, with the member number 4 and *Gm* indicating the species origin (*Glycine max* L.).

Interestingly, the *Lox1:Gm:4* gene is located just downstream of another lipoxygenase gene; there are only 987 bp between the stop codon of the upstream gene and the TATA box sequence of the *Lox1:Gm:4* gene. The intergenic region was fused with the GUS gene for transient expression in tobacco suspension-cultured cells as shown by Kato et al. (44). The construct showed a high level of expression similar to a GUS gene construct driven by the cauliflower mosaic virus 35S promoter (Kato and Shibata, unpublished data), indicating that the intergenic region contains the promoter for the *Lox1:Gm:4* gene.

Kato et al. isolated the genomic sequence for SC501 and showed that the gene encodes SBL-4 (41). Independently, Park et al. reported a full-length cDNA clone (pTK18) for this gene (42). The partial cDNA for *lox*B2 reported by Bell and Mullet also corresponds to the SBL-4 gene (40). This gene is designated *Lox1:Gm:5*.

Genomic sequences for *Lox1:Gm:4* gene (5,663 bp [39]) and *Lox1:Gm:5* (7,400 bp [41]) in addition to the SBL-3 gene have been sequenced (15). The length of these genes is shorter than mammalian lipoxygenase genes because of the shorter introns characteristic of plants (45). Even the first and longest intron found in *Lox1:Gm:5* is only 2,542 bp. The location of the eight introns is conserved among the plant lipoxygenase genes, but not between mammalian lipoxygenase genes, even in the regions that share amino acid sequence homology.

The nucleotides surrounding the translation start codon (ATG) of these soybean lipoxygenase genes are conserved; the consensus sequence is GCAAAG*ATG*TTT. Kato et al. took advantage of this conserved sequence to obtain an efficient translation of a bacterial gene in plant cells (44). They fused the consensus sequence in a frame with the

GUS gene originally derived from *E. coli* to create a fusion protein carrying 10 additional amino acid residues at the amino terminal end of the original enzyme. The fusion gene exhibits an increase of approximately tenfold in the GUS activity in tobacco and rice suspension-cultured cells when compared with the original construct. Forster et al. also demonstrated similar effects of the same conserved sequence from a pea lipoxygenase gene while using it to enhance transgene expression in tobacco plants (31).

Expression of these lipoxygenase genes in various tissues has been investigated. These genes are expressed in young seedlings at higher levels than in mature plants (40–42). Bell and Mullet showed that lipoxygenase gene expression is modulated by water deficit, wounding, and methyl jasmonate (40). Kato et al. also indicated that *Lox1:Gm:5* is induced to high levels in mature leaves by methyl jasmonate (41). *Lox1:Gm:5* is also induced by methyl jasmonate in suspension-cultured cells to a degree that it becomes a dominant lipoxygenase species (Igarashi, Kato, and Shibata, unpublished data). Recently, Liu et al. have investigated the expression of *Lox1:Gm:4* and *Lox1:Gm:5* (*Lox*B2) during the transition from maturation to the germination phase in soybean somatic embryos (46).

Lipoxygenases as Vegetative Storage Proteins. Tranbarger et al. showed that one of the vegetative storage proteins (VSP) in soybean leaves is a lipoxygenase (47). It has been shown that soybean leaves and immature organs accumulate three polypeptides of approximately 27, 29, and 94 kDa (termed vsp27, vsp29, and vsp94, respectively) as VSP, and several factors increase the level of accumulation of these VSP. Pod removal from the plants greatly increases VSP accumulation in leaves (Fig. 3.3). The cDNA for both vsp27 and vsp29 have been isolated and characterized (48). Tranbarger et al. determined partial amino acid sequences of vsp94 and showed that vsp94 is highly homologous to the lipoxygenase protein family (47). They also showed that the vsp94/lipoxygenase is primarily expressed in the paraveinal mesophyll cells and is subcellularly localized in the vacuole.

The partial amino acid sequences of vsp94 determined by Tranbarger et al. are highly homologous to the amino acid sequence of SBL-4 deduced from the nucleotide sequence of *Lox1:Gm:5* (41,47). To confirm that SBL-4 is vsp94 or a component of vsp94, Kato et al. purified lipoxygenases from leaves of depodded soybean plants (41). They showed that SBL-4 becomes the major lipoxygenase species in the leaves after depodding, and that at least three other lipoxygenase species also appear in the leaves. Northern blot experiments also demonstrate accumulation of the *Lox1:Gm:5* transcript in leaves of depodded plants (Fig. 3.3). The SBL-4 protein comprises 2.7% of the total soluble protein in leaves after pod removal. This fraction is consistent with the amount of vsp94 estimated by SDS-polyacrylamide gel electrophoresis.

Methyl jasmonate synthesized via the lipoxygenase pathway induces SBL-4 and vsp94 in leaves (41,49). A question related to this observation is whether or not this induction results in further biosynthesis of jasmonate-related messengers in plant cells (47). To answer this question, Kato et al. have characterized the enzymological properties of SBL-4 and demonstrated that the enzyme has an unusually high *Km*

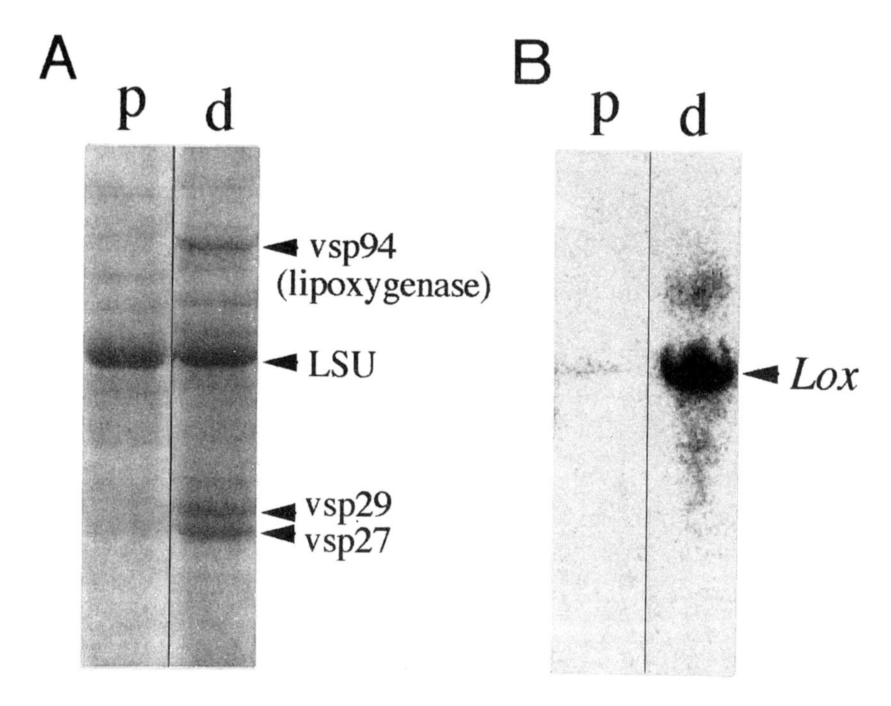

Figure 3.3. Accumulation of lipoxygenase proteins after removal of seed pods. *a*) SDS-PAGE of leaf proteins from normal plants (p) and depodded plants (d). After removal of pods, lipoxygenase protein (vsp94) and two other vegetative storage proteins, vsp27 and vsp29, appear in leaves. LSU, the large subunit of ribulose-1,5-bisphosphate carboxylase/oxygenase. *b*) Northern hybridization of lipoxygenase mRNA. The lipoxygenase SBL-4 gene transcript accumulates in leaves of depodded plants.

value for LNA (10.4 mM [41]). Since LNA is the biosynthetic precursor of jasmonate (7), it seems unlikely that the enzyme is involved in the biosynthesis of jasmonate and related compounds. However, as some other lipoxygenases that appear in the leaves of depodded plants show lower *Km* values for LNA (Iwamoto, Kato, and Shibata, unpublished data), we cannot exclude the possibility that these are involved in jasmonate biosynthesis.

An unusual feature of vsp94 is the localization of the protein in both vacuole and cytoplasm, but not in the ER or Golgi apparatus (47). In the case of vsp27 and vsp29, their precursor proteins have signal peptides for transportation into the ER; they actually accumulate in the vacuole and not in the cytoplasm (48), probably in a manner similar to the accumulation of seed storage proteins in protein bodies. However, as the amino acid sequence deduced from the SBL-4 gene sequence has no signal peptide sequence, it is unlikely that the SBL-4 protein is transported into the vacuole via the ER, suggesting an unknown mechanism for transportation of the cytoplasmic protein to the vacuole.

Rice Lipoxygenase Genes

Lipoxygenase Genes Expressed in Seeds and Seedlings

Rice lipoxygenases found in seeds and seedlings have been characterized (50–52). There are three isozymes in seeds, called L-1, L-2, and L-3 (53,54). Rice L-3 has been purified to homogeneity (53). Rice L-3 is the major isozyme in the rice embryo, and L-2 becomes predominant in seedlings 5 days after germination (54).

Molecular Cloning of Lipoxygenase from Monocot Plants. The first cloning of a lipoxygenase gene from a monocot plant, rice, was reported by Ohta et al. (55). Using dicot lipoxygenase genes as probes, no positive clone was obtained from rice cDNA libraries even under low stringency hybridization conditions, indicating low nucleotide sequence homology between the dicot and monocot genes. Thus, they prepared antibodies against rice lipoxygenase L-3 that was purified from seeds, screened an expression cDNA library using the antibodies, and found several candidate clones. Since the antibody used reacted with several polypeptides other than the L-3 protein, these clones were subjected to the epitope-selection method. In this manner a partial-length cDNA clone was identified and used to isolate a full-length cDNA clone.

A striking feature of the cDNA is a strong bias of the nucleotide sequence toward G and C, especially in the third codon position (93% G/C), while only about 50% of third position nucleotides in the soybean genes are comprised of G and C. Such a strong nucleotide bias is often seen in monocot plant genes. This bias may explain the previous failure to isolate the monocot lipoxygenase gene using dicot lipoxygenase gene probes. Another rice lipoxygenase cDNA isolated from leaves also shows a strong bias toward G/C-rich codons (56).

Characterization of Monocot Lipoxygenases Expressed as Active Enzymes in E. coli. The cDNA isolated from the seedlings was unequivocally identified as that for rice lipoxygenase L-2. Since the polyclonal antibody used for the screening was not specific to the L-3 isozyme, the cDNA was identified by characterizing an enzyme expressed from the cDNA in *E. coli* and sequencing portions of the purified L-2 protein (55).

An active enzyme from the cDNA was produced in *E. coli* by cultivation of the bacteria at low temperature (57). The full-length cDNA was inserted into the *Bam*HI site of the pET3a vector, resulting in a fusion protein with 18 extra amino acid residues at the amino-terminal end of the original enzyme. The fusion gene was expressed under the T7 RNA polymerase promoter in *E. coli.* When *E. coli* cells carrying the construct were incubated at 37°C, the fusion protein was produced as 27% of the total protein in inclusion bodies, but no lipoxygenase activity was detected. When the cells were grown at 15°C for 16 h, however, the active enzyme was found in the supernatant of disrupted cells. From the specific activity of the purified enzyme, the active enzyme was estimated to be 3% of the soluble protein in the supernatant of the disrupted cells (20 mg/L of culture). The enzyme showed the same pH optimum and regiospecificity toward LNA as rice lipoxygenase L-2. These properties differed from those of L-3.

Finally, the identity of the cDNA was confirmed by amino acid sequencing of rice lipoxygenase L-2. The L-2 protein was purified from SDS-PAGE gels, and digested with lysylendopeptidase. The proteolytic peptides were isolated by HPLC and sequenced. The amino acid sequences obtained from nine peptides (35 residues) corresponded perfectly with the cDNA. Rice lipoxygenase L-2 shares 52% homology with SBL-1. Hydropathy profiles of rice L-2 and SBL-1 reveal a conservation of secondary structures in these proteins.

The low-temperature cultivation method that was used to produce the active lipoxygenase L-2 in *E. coli* also has been applied to the pathogen-inducible rice lipoxygenase (56), SBL-1 (58), and a potato lipoxygenase (59). In addition to cultivation at low temperature, Steczko et al. also showed that addition of 3% ethanol to the *E. coli* culture increases the recovery of active enzyme by about 40% (58). Production of an active human 5-lipoxygenase in *E. coli* by cultivation at 20°C was reported by Noguchi et al. (60), but Zhang et al. reported no such temperature effect for production of the 5-lipoxygenase in bacteria (18).

A Novel Lipoxygenase Involved in Plant Defense

An important physiological role of the plant lipoxygenase pathway is for resistance to insects and pathogens. In Chapter 10, involvement of the lipoxygenase-pathway product jasmonic acid in insect protection is described. Jasmonic acid is also involved in the induction of the biosynthesis of antimicrobial compounds (phytoalexins [61,62]), and enhances the elicitation of activated oxygen species in plant defense (63). Some products from the lipoxygenase pathway have antimicrobial activities (64–68). Induction of lipoxygenases and lipoxygenase pathway enzymes has been observed in various plant species after inoculation with incompatible pathogens or after treatment with elicitors (69–76). Involvement of bean and tomato lipoxygenases in plant defense is reviewed in Chapter 10. Here, this chapter reviews a rice lipoxygenase pathway that may be involved in plant defense against the rice blast fungus, *Magnaporthe grisea.*

Oxygenated Unsaturated Fatty Acids as Antifungal Compounds. Kato and coworkers isolated a series of oxygenated unsaturated fatty acids from rice leaves that act as antifungal compounds against rice blast fungus (Fig. 3.4 [64,77–81]). These compounds are derivatives of either LA or LNA; monohydroxides, trihydroxides, epoxides, and epoxy monohydroxides. Li et al. indicated that hydroperoxides and monohydroxides derived from LA or LNA rapidly increase in rice leaves after inoculation with the rice blast fungus (82). In contrast, a terpenoid phytoalexin, momilactone A, accumulates much later than the fatty acid derivatives. Their results suggest that these oxygenated fatty acids are involved in an early response of the plant against the pathogen. Trihydroxides of LA and LNA were also isolated from taro tubers that had been inoculated with black root fungus (83).

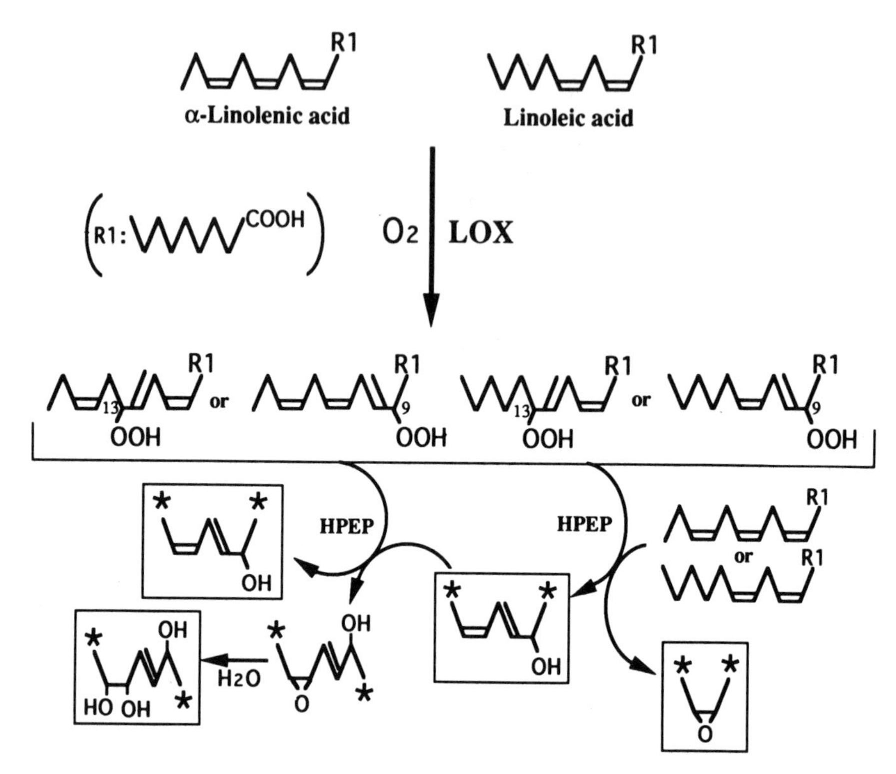

Figure 3.4. Proposed biosynthesis pathway of oxygenated unsaturated fatty acids isolated from rice leaves as antifungal compounds. Compounds that have antifungal activity against the rice blast fungus are boxed. The absolute configuration of the antifungal compounds is available in the papers of Kato et al. For simplicity, side structures of HPEP products are represented by *. *Abbreviations:* LOX, lipoxygenase; HPEP, hydroperoxide-dependent epoxygenase. *Sources:* Kato et al. (77–81).

Activation of a Lipoxygenase Pathway by Pathogen Attack in Rice Leaves. Ohta et al. found increased lipoxygenase activity in rice leaves after infection with an incompatible race of the rice blast fungus, but not with a compatible race (73). They also found that an activity that converts 9-hydroperoxy-linoleic acid to 9-hydroxy-linoleic acid and 9,12,13-trihydroxy-linoleic acid in vitro increases in the leaves after infection with the incompatible pathogen (84). This activity has been suggested to be a hydroperoxide-dependent epoxygenase that transforms hydroperoxide fatty acids into the corresponding hydroxy fatty acids and then co-oxidizes unsaturated fatty acids into the corresponding epoxy fatty acids (85). Epoxygenase (also called peroxygenase) activities found in the broad bean (*Vicia faba* L.) and in microsomes of soybean seedlings have been characterized well (Chapter 8, and 86). It is likely that most of the oxygenated unsaturated fatty acids isolated from rice leaves by Kato and co-workers are synthesized by the sequential action of lipoxygenase and hydroperoxide-dependent epoxygenase as shown in Figure 3.4. However one cannot exclude the possibility that a cytochrome P-450–dependent pathway synthesizes some of these compounds.

Involvement of a Novel Lipoxygenase in the Interaction Between Host and Pathogen. Peng et al. cloned a rice lipoxygenase gene that is expressed in leaves after inoculation with an incompatible race of the rice blast fungus, *M. grisea* (56) (Fig. 3.5). They took advantage of the amino acid sequence conserved among all plant lipoxygenases and some mammalian lipoxygenases, HAAVNFGQY, to prepare an oligonucleotide probe. Using this probe, they screened a cDNA library that had been prepared from rice leaves 30 h after inoculation with the incompatible race 131 of *M. grisea* and isolated a partial cDNA clone. Subsequently a full-length cDNA was isolated from this library. This cDNA shows a strong nucleotide bias toward G and C similar to that in the L-2 rice lipoxygenase cDNA. The gene product shares the least amino acid sequence homology of plant lipoxygenases. Even with L-2 rice lipoxygenase, there is only 42% amino acid sequence homology within the lipoxygenase homologous region (LHR), although the rice L-2 protein shares approximately 53% homology with soybean seed lipoxygenases.

A striking feature of the gene product is a putative transit sequence (approximately 100 amino acid residues) at the amino terminal end for targeting the protein into the

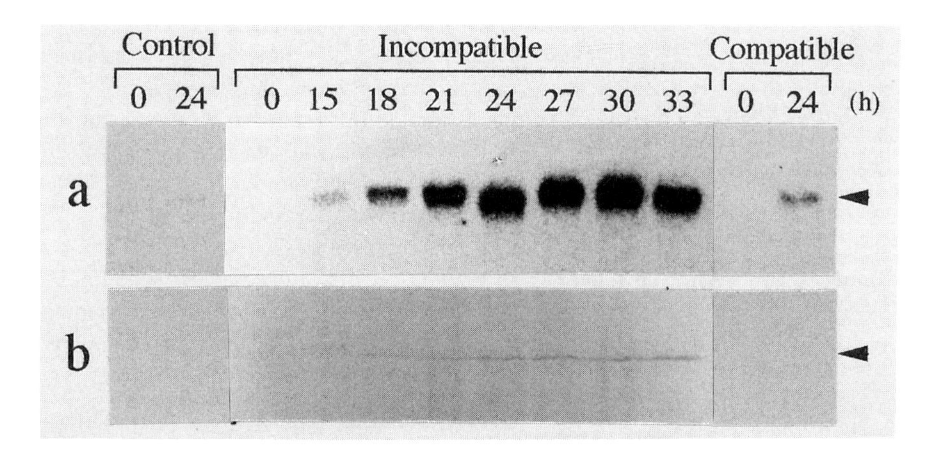

Figure 3.5. Expression of a novel lipoxygenase in rice leaves after inoculation with the rice pathogen, *M. grisea.* Poly(A)RNA and proteins were isolated from leaves that had been inoculated with the incompatible race 131, the compatible race 007, or left uninoculated. *a*) Northern blotting was carried out using *Lox2:Os:1* as probe. *b*) Western blotting was carried out using an antibody raised against the gene product expressed in *E. coli.*

chloroplast. Comparison of the deduced amino acid sequence with other plant lipoxygenases reveals a plant LHR that extends from Gly-111 to the carboxyl terminus Ile-923. The amino-terminal end outside the LHR (Met-1 to Ile-110) consists of the hydroxy amino acids Ser and Thr (33%), small hydrophobic residues, Ala and Val (24%), positively charged amino acids, Arg, Lys, and His (13%), negatively charged amino acids, Glu and Asp (5%), and other amino acids (25%). The amino acid composition of this sequence is consistent with those found in transit peptides that function to translocate nuclear-encoded proteins into chloroplasts from the cytosol. Another cDNA encoding this type of lipoxygenase has also been isolated from *Arabidopsis thaliana* (87).

To characterize the catalytic properties of the gene product, an active enzyme consisting of the LHR region was expressed in *E. coli* by the low-temperature cultivation method (56). The catalytic properties of this enzyme fuel speculation on its function within the chloroplast.

A novel rice lipoxygenase may be involved in biosynthesis of jasmonic acid. The enzyme expressed in *E. coli* has a threefold higher specific activity toward LA, the major unsaturated fatty acid in chloroplasts, than toward LNA, a minor component in this organelle. The enzyme introduces molecular oxygen exclusively into the C-13 position of LA and LNA. It is noteworthy that formation of 13-hydroperoxy-linolenic acid is a prerequisite for biosynthesis of jasmonic acid. Allene oxide synthase is the key enzyme in metabolizing 13-hydroperoxy-linolenic acid to jasmonic acid. Song et al. have cloned the allene oxide synthase gene from flaxseed, and found the protein to possess a mitochondria or chloroplast transit peptide (88). Although allene oxide synthase has not yet been shown to be localized in rice chloroplasts, it is interesting to speculate on the involvement of the rice chloroplast lipoxygenase in the biosynthesis of jasmonic acid, as well as oxygenated unsaturated fatty acids active as antifungal compounds.

Induction of rice lipoxygenase gene expression by abiotic stress, such as treatment with elicitor molecules or UV light, differs from that of other plant-defense-related genes (Shirano and Shibata, manuscript in preparation). Since most plant-defense-related genes are induced in either differentiated or undifferentiated tissues, undifferentiated suspension-cultured cells have been used to understand regulation of these genes. However, the rice lipoxygenase gene is expressed only in differentiated tissues that have been exposed to white light. No induction of gene expression is seen in etiolated tissues or undifferentiated suspension-cultured cells. However, after exposure of differentiated tissues to white light, the gene becomes inducible by abiotic stresses.

Other Plant Lipoxygenase Genes

Aside from soybean and rice, lipoxygenase genes have been isolated from pea (89,90), bean (91,92), lentil (93), tomato (94), potato (59), and *Arabidopsis thaliana* (87,95,96). Plant lipoxygenase genes are classified into two categories based on whether the gene products have a transit peptide at their amino terminal end, *Lox2*, or not, *Lox1*, (43). It should be noted that this criterion for classification is not based on enzymatic characteristics, although some researchers have used the terms "type 1" and "type 2" to distinguish lipoxygenases active at pH 9 (type 1) from those active at neutral pH (type 2). Most plant lipoxygenase genes isolated to date fall into the *Lox1* category. The two genes falling into the *Lox2* category have been isolated from *A. thaliana* (87) and rice (56).

References

1. Axelrod, B., Cheesbrough, T.M., and Laakso, S. (1981) *Methods Enzymol. 71,* 441–451.
2. Samuelsson, B., Dahlén, S.-E., Lindgren, J.Å., Rouzer, C.A., and Serhan, C.N. (1987) *Science 237,* 1171–1176.
3. Boyington, J.C., Gaffney, B.J., and Amzel, L.M. (1993) *Science 260,* 1482–1486.
4. Minor, W., Steczko, J., Bolin, J.T., Otwinowski, Z., and Axelrod, B. (1993) *Biochemistry 32,* 6320–6323.
5. Shibata, D., and Axelrod, B. (1995) *J. Lipid Mediators 12,* 213–228.
6. Hildebrand, D.F. (1989) *Physiol. Plant. 76,* 249–253.
7. Vick, B.A., and Zimmerman, D.C. (1984) *Plant Physiol. 75,* 458–461.
8. Hamberg, M., and Gardner, H.W. (1992) *Biochim. Biophys. Acta 1165,* 1–18.
9. Gardner, H.W. (1991) *Biochim. Biophys. Acta 1084,* 221–239.
10. Siedow, J.N. (1991) *Annu. Rev. Plant Physiol. Plant Mol. Biol. 42,* 145–188.
11. Hildebrand, D.F., Versluys, R.T., and Collins, G.B. (1991) *Plant Sci. 75,* 1–8.
12. Start, W.G., Ma, Y., Polacco, J.C., Hildebrand, D.F., Freyer, G.A., and Altschuler, M. (1986) *Plant Mol. Biol. 7,* 11–23.
13. Shibata, D., Steczko, J., Dixon, J.E., Hermodson, M., Yazdanparast, R., and Axelrod, B. (1987) *J. Biol. Chem. 262,* 10080–10085.
14. Shibata, D., Steczko, J., Dixon, J.E., Andrews, P.C., Hermodson, M., and Axelrod, B. (1988) *J. Biol. Chem. 263,* 6816–6821.
15. Yenofsky, R.L., Fine, M., and Liu, C. (1988) *Mol. Gen. Genet. 211,* 215–222.
16. Steczko, J., Donoho, G.P., Clemens, J.C., Dixon, J.E., and Axelrod, B. (1992) *Biochemistry 31,* 4053–4057.
17. Nguyen, T., Falgueyret, J.P., Abramovitz, M., and Riendeau, D. (1991) *J. Biol. Chem. 266,* 22057–22062.
18. Zhang, Y.Y., Rådmark, O., and Samuelsson, B. (1992) *Proc. Natl. Acad. Sci. USA 89,* 485–489.
19. Steczko, J., and Axelrod, B. (1992) *Biochem. Biophys. Res. Commun. 186,* 686–689.
20. Minor, W., Stec, B., Steczko, J., Axelrod, B., Bolin, J.T., Otwinowski, Z., and Walter, R. (1995) (Abstr) *Am. Crystallogr. Assoc.,* National Meeting, Atlanta, GA, p. 50.
21. Kitamura, K. (1993) *Trends Food Sci. Tech. 4,* 64–67.
22. Hildebrand, D.F., and Hymowitz, T. (1981) *J. Am. Oil Chem. Soc. 58,* 583–586.
23. Kitamura, K., Kumagai, T., and Kikuchi, A. (1985) *Jpn J. Breed. 35,* 413–420.
24. Davies, C.S., and Nielsen, N.C. (1986) *Crop Sci. 26*, 460–463.
25. Kitamura, K., Davies, C.S., Kaizuma, N., and Nielsen, N.C. (1983) *Crop Sci. 23,* 924–927.
26. Matoba, T., Hidaka, H., Narita, H. Kitamura, K., Kaizuma, N., and Kito, M. (1985) *J. Agric. Food Chem. 33,* 852–855.

27. Wang, W.H., Takano, T., Shibata, D., Kitamura, K., and Takeda, G. (1994) *Proc. Natl. Acad. Sci. USA 91,* 5828–5832.
28. Wang, W.H., Kato, T., Takano, T., Shibata, D., Kitamura, K., and Takeda, G. (1995) *Plant Sci. 109,* 67–73.
29. Hajika, M., Kitamura, K., Igita, K., and Nakazawa, Y. (1992) *Jpn J. Breed. 42,* 787–792.
30. Hajika, M., Igita, K., and Kitamura, K. (1991) *Jpn.J. Breed. 41,* 507–509.
31. Forster, C., Arthur, E., Crespi, S., Hobbs, S.L.A., Mullineaux, P., and Casey, R. (1994) *Plant Mol. Biol. 26,* 235–248.
32. Pfeiffer, T.W., Hildebrand, D.F., and TeKrony, D.M. (1992) *Crop Sci. 32,* 357–362.
33. Andrews, D.L., Beames, B., Summers, M.D., and Park, W.D. (1988) *Biochem. J. 252,* 199–206.
34. Deng, W., Grayburn, W.S., Hamilton-Kemp, T.R., Collins, G.B., and Hildebrand, D.F. (1992) *Planta 187,* 203–208.
35. Park, T.K., and Polacco, J.C. (1989) *Plant Physiol. 90,* 285–290.
36. Grayburn, W.S., Schneider, G.R., Hamilton-Kemp, T.R., Bookjans, G., Ali, K., and Hildebrand, D.F. (1991) *Plant Physiol. 95,* 1214–1218.
37. Kato, T., Ohta, H., Tanaka, K., and Shibata, D. (1992) *Plant Physiol. 98,* 324–330.
38. Kato, T., Terao, J., and Shibata, D. (1992) *Biosci. Biotech. Biochem. 56,* 1344.
39. Shibata, D., Kato, T., and Tanaka, K. (1991) *Plant Mol. Biol. 16,* 353–359.
40. Bell, E., and Mullet, J.E. (1991) *Mol. Gen. Genet. 230,* 456–462.
41. Kato, T., Shirano, Y., Iwamoto, H., and Shibata, D. (1993) *Plant Cell Physiol. 34,* 1063–1072.
42. Park, T.K., Holland, M.A., Laskey, J.G., and Polacco, J.C. (1994) *Plant Sci. 96,* 109–117.
43. Shibata, D., Slusarenko, A., Casey, R., Hildebrand, D., and Bell, E. (1994) *Plant Mol. Biol. Rep. 12,* S41–S42.
44. Kato, T., Shirano, Y., Kawazu, T., Tada, Y., Itoh, E., and Shibata, D. (1991) *Plant Mol. Biol. Rep. 9,* 333–339.
45. Hoshiko, S., Rådmark, O., and Samuelsson, B. (1990) *Proc. Natl. Acad. Sci. USA 87,* 9073–9077.
46. Liu, W., Hildebrand, D.F., Moore, P.J., and Collins, G.B. (1994) *Planta 194,* 69–76.
47. Tranbarger, T.J., Franceschi, V.R., Hildebrand, D.F., and Grimes, H.D. (1991) *Plant Cell 3,* 973–987.
48. Staswick, P.E. (1988) *Plant Physiol. 87,* 250–254.
49. Grimes, H.D., Koetje, D.S., and Franceschi, V.R. (1992) *Plant Physiol. 100,* 433–443.
50. Shastry, B.S., and Rao, M.R.R. (1975) *Cereal Chem. 52,* 597–603.
51. Yamamoto, A., Fujii, Y., Yasumoto, K., and Mitsuda, H. (1980) *Agric. Biol. Chem. 44,* 443–445.
52. Ida, S., Masaki, Y., and Morita, Y. (1983) *Agric. Biol. Chem. 47,* 637–641.
53. Ohta, H., Ida, S., Mikami, B., and Morita, Y. (1986) *Agric. Biol. Chem. 50,* 3165–3171.
54. Ohta, H., Ida, S., Mikami, B., and Morita, Y. (1986) *Plant Cell Physiol. 27,* 911–918.
55. Ohta, H., Shirano, Y., Tanaka, K., Morita, Y., and Shibata, D. (1992) *Eur. J. Biochem. 206,* 331–336.
56. Peng, Y.-L., Shirano, Y., Ohta, H., Hibino, T., Tanaka, K., and Shibata, D. (1994) *J. Biol. Chem. 269,* 3755–3761.
57. Shirano, Y., and Shibata, D. (1990) *FEBS Letters 271,* 128–130.

58. Steczko, J., Donoho, G.A., Dixon, J.E., Sugimoto, T., and Axelrod, B. (1991) *Protein Expre. Purif. 2,* 221–227.
59. Geerts, A., Feltkamp, D., and Rosahl, S. (1994) *Plant Physiol. 105,* 269–277.
60. Noguchi, M., Matsumoto, T., Nakamura, M., and Noma, M. (1989) *FEBS Letters 249,* 267–270.
61. Gundlach, H., Muller, M.J., Kutchan, T.M., and Zenk, M.H. (1992) *Proc. Natl. Acad. Sci. USA 89,* 2389–2393.
62. Choi, D., Bostock, R.M., Avdiushko, S., and Hildebrand, D.F. (1994) *Proc. Natl. Acad. Sci. USA 91,* 2329–2333.
63. Kauss, H., Jeblick, W., Ziegler, J., and Krabler, W. (1994) *Plant Physiol. 105,* 89–94.
64. Kato, T., Yamaguchi, Y., Uyehara, T., and Yokoyama, T. (1983) *Tetrahedron Letters 24,* 4715–4718.
65. Vaughn, S.F., and Gardner, H.W. (1993) *J. Chem. Ecology 19,* 2337–2345.
66. Doehlert, D.C., Wicklow, D.T., and Gardner, H.W. (1993) *Phytopathology 83,* 1473–1477.
67. Croft, K.P.C., Jüttner, F., and Slusarenko, A.J. (1993) *Plant Physiol. 101,* 13–24.
68. Deng, W., Hamilton-Kemp, T.R., Nielsen, M.T., Andersen, R.A., Collins, G.B., and Hildebrand, D.F. (1993) *J. Agric. Food Chem. 41,* 506–510.
69. Yamamoto, H., and Tani, T. (1986) *J. Phytopathol. 116,* 329–337.
70. Ocampo, C.A., Moerschbacher, B., and Grambow, H.J. (1986) *Z. Naturforsch. 41c,* 559–563.
71. Peever, T.L, and Higgins, V.J. (1989) *Plant Physiol. 90,* 867–875.
72. Croft, K.P.C., Voisey, C.R., and Slusarenko, A.J. (1990) *Physiol. Mol. Plant Pathol. 36,* 49–62.
73. Ohta, H., Shida, K., Peng, Y.-L., Furusawa, I., Shishiyama, J., Aibara, S., and Morita, Y. (1991) *Plant Physiol. 97,* 94–98.
74. Koch, E., Meier, B.M., Eiben, H.-G., and Slusarenko, A. (1992) *Plant Physiol. 99,* 571–576.
75. Kato, T., Maeda, Y., Hirukawa, T., Namai, T., and Yoshioka, N. (1992) *Biosci. Biotech. Biochem. 56,* 373–375.
76. Fournier, J., Pouénat, M.-L., Rickauer, M., Rabinovitch-Chable, H., Rigaud, M., and Esquerré-Tugaye, M.-T. (1993) *Plant J. 3,* 63–70.
77. Kato, T., Yamaguchi, Y., Uyehara, T., Yokoyama, T., Namai, T., and Yamanaka, S. (1983) *Naturwissenschaften 70,* 200–201.
78. Kato, T., Yamaguchi, Y., Hirano, T., Yokoyama, T., Uyehara, T., Namai, T., Yamanaka, S., and Harada, N. (1984) *Chem. Lett.,* 409–412.
79. Kato, T., Yamaguchi, Y., Abe, N., Uyehara, T., Namai, T., Kodama, M., and Shiobara, Y. (1985) *Tetrahedron Lett. 26,* 2357–2360.
80. Kato, T., Yamaguchi, Y., Ohnuma, S., Uyehara, T., Namai, T., Kodama, M., and Shiobara, Y. (1986) *Chem. Lett.,* 577–580.
81. Kato, T., Yamaguchi, Y., Ohnuma, S., Uyehara, T., Namai, T., Kodama, M., and Shiobara, Y. (1986) *J. Chem. Soc., Chem. Commun.,* 743–744.
82. Li, W.-X., Kodama, O., and Akatsuka, T. (1991) *Agric. Biol. Chem. 55,* 1041–1047.
83. Matsui, H., Kondo, T., and Kojima, M. (1989) *Phytochemistry 28,* 2613–2615.
84. Ohta, H., Shida, K., Peng, Y.-L., Furusawa, I., Shishiyama, J., Aibara, S., and Morita, Y. (1990) *Plant Cell Physiol. 31,* 1117–1122.
85. Hamberg, M., and Fahlstadius, P. (1992) *Plant Physiol. 99,* 987–995.

Chapter 4

Exploring the Structure and Function of Mammalian Lipoxygenases by Site-Directed Mutagenesis

David L. Sloane

Targeted Genetics Corporation, 1100 Olive Way, Seattle, WA 98101, USA.

Introduction

The role of lipoxygenases in mammalian organisms has been an area of active research since their discovery in the mid-1970s. Recently, molecular biological techniques have aided in the characterization of different isoforms of lipoxygenases. Particularly, molecular cloning, sequencing, and site-directed mutagenesis have contributed to our understanding of the structure–function relationship of lipoxygenases.

Mammalian lipoxygenases catalyze the formation of inflammatory mediators, such as leukotrienes and lipoxins, from arachidonic acid (1). It is currently thought that mammals contain three lipoxygenases, a 5-, a 12-, and a 12/15-lipoxygenase, named for their positional specificity on arachidonic acid. Inhibition of lipoxygenase activity may be of therapeutic value since a role for the various lipoxygenases has been proposed in the pathogenesis of a variety of conditions, such as arthritis, psoriasis, bronchial asthma, atherosclerosis, and tumor metastasis (2–6).

The 5-lipoxygenase is found in neutrophils. It catalyzes the oxygenation of carbon-5 of arachidonic acid, the first step in the biosynthesis of leukotrienes. The 12-lipoxygenase has been classified into two categories, based on its tissue localization. The "platelet-type" 12-lipoxygenase has a stringent substrate and product profile, forming only 12-hydroperoxyeicosatetraenoic acid (12-HPETE) from free arachidonic acid. The "leukocyte-type" 12-lipoxygenase has a broader substrate preference, acting on both free and esterified fatty acids, and will form both 12- and 15-HPETE from arachidonic acid (7).

The "platelet-type" 12-lipoxygenase was the first mammalian lipoxygenase discovered (8). It has since been found in the platelets of a variety of mammals (9). Recently, the appearance of both the "platelet-type" and the "leukocyte-type" 12-lipoxygenase in the same organism has been documented on a molecular level by cloning both forms from mice (10).

It may be reasonable to consider the "leukocyte-type" 12-lipoxygenase and the 15-lipoxygenases "12/15-lipoxygenases" for the following reasons. Molecular cloning and sequence characterization of these enzymes has demonstrated that the "leukocyte-type" 12-lipoxygenase is more closely related to the human and rabbit 15-lipoxygenase than to the "platelet-type" 12-lipoxygenase. The "leukocyte-type" 12-lipoxygenase sequences and the 15-lipoxygenase sequences are greater than 70% identical (Table 4.1). The tissue localization of the 15-lipoxygenase in some mammals matches that of the "leukocyte-type" 12-lipoxygenase in others; thus, these two enzymes may be functionally interchangeable from one organism to another. Furthermore, molecular cloning has not demonstrated the occurrence of both a "leukocyte-type" 12-lipoxygenase and a 15-lipoxygenase in the same organism, although there is recent evidence by immunoblotting and by polymerase chain reaction for a "leukocyte-type" 12-lipoxygenase in human adrenal glomerulosa cells (11). Therefore, the close relationship of the "leukocyte-type" 12-lipoxygenase and the 15-lipoxygenase, together with the tissue localization of each and the broad substrate and product specificities of these two enzymes make it reasonable to consider the "leukocyte-type" 12-lipoxygenase to be a 12/15-lipoxygenase.

The 12/15-lipoxygenase was first found in developing red blood cells (12), and porcine leukocytes (13). This enzyme has also been found in human eosinophils (14), airway epithelial cells (15–17), rat and canine brain (18,19), and mouse leukocytes (10). The 12/15-lipoxygenase has the unique property of being capable of oxidizing complex membrane lipids (20–22).

Through a combination of enzymology, mutagenesis, crystallography, and spectroscopy, the structural basis of the enzymatic reaction is just beginning to be understood. This chapter will review the current advances in our understanding of the structure–function relationship in mammalian lipoxygenases as they have been studied by

TABLE 4.1 Amino Acid Sequence Identity Comparison of Mammalian Lipoxygenases with Soybean Lipoxygenase-1[a]

	soy1	rat5	hum5	hum12	mpl12	mlu12[b]	rat12[b]	pig12[b]	cow12[b]	rab15[b]
hum15	26	40	41	66	64	74	75	86	86	81
rab15	25	38	39	62	60	71	72	79	78	
cow12	24	40	41	65	64	72	73	89		
pig12	26	40	42	66	64	71	71			
rat12	25	41	42	59	58	94				
mlu12	26	41	42	58	58					
mpl12	26	41	42	84						
hum12	26	41	43							
hum5	28	93								
rat5	26									

[a]All sequences were accessed from GenBank and compared using the BestFit program in the GCG Sequence Analysis Software Package, Version 7.3 on a VAX computer (79).
[b]Indicates the identities between the "leukocyte-type" 12-lipoxygenases and the 15-lipoxygenases.
Abbreviations: human reticulocyte 15-lipoxygenase (hum15 [60]), rabbit reticulocyte 15-lipoxygenase (rab15 [81]), bovine tracheal epithelial 12-lipoxygenase (cow12 [82]), porcine leukocyte 12-lipoxygenase (pig12 [83]), rat brain 12-lipoxygenase (rat12 [18]), murine leukocyte 12-lipoxygenase (mlu12 [10]), murine platelet 12-lipoxygenase (mpl12 [10]), human platelet 12-lipoxygenase (hum12 [84]), human leukocyte 5-lipoxygenase (hum5 [85]), rat leukocyte 5-lipoxygenase (rat5 [59]), soybean lipoxygenase isoform-1 (soy1 [86]), soybean lipoxygenase isoform-2 (soy2 [58]), soybean lipoxygenase isoform-3 (87), soybean cotyledon lipoxygenase (soyC [88]), barley lipoxygenase 1 (bar1 [89]), pea seed lipoxygenase, "soybean-2-like" (pea2 [90]), pea seed lipoxygenase, "soybean-3-like" (pea3 [91]), lentil seedling lipoxygenase (lent [92]), tomato LoxA (tomA), tomato LoxB (tomB [93]), rice lipoxygenase-1 (ric1 [94]), rice lipoxygenase-2 (ric2 [95]), *Solanum tuberosum* lipoxygenase (tubd [96]), *Arabidopsis thaliana* lipoxygenase-1 (ara1 [97]), *Arabidopsis thaliana* lipoxygenase-2 (ara2 [98]), and *Porphyra purpurea* lipoxygenase (alga [99]).

site-directed mutagenesis. Site-directed mutagenesis has been used to study the iron-binding ligands of the lipoxygenases and the structural requirements for the positional specificity of the various lipoxygenases. The most informative insights into the structure–function relationship of lipoxygenases has come from sequence alignments and a comparison of the conservations and differences between the different lipoxygenases. In the absence of a three-dimensional structure, the sequence conservations were the only clues to important amino acids in the enzyme. Amino acids that have been conserved in evolution are considered potentially important in the activity of the enzyme. Now that the three-dimensional structure of soybean lipoxygenase is known, more sophisticated predictions can be made concerning which amino acids may play a role in the enzyme's activity. However, until the three-dimensional structure of a mammalian lipoxygenase is known, we must rely on sequence alignments with the soybean enzyme to make testable hypotheses about the structure–function relationship in mammalian lipoxygenases.

Catalytic Mechanism

A possible catalytic scheme has emerged from many kinetic and spectroscopic studies (Fig. 4.1). This scheme has mainly come from work on the soybean enzyme, due to its abundance and ease of purification. Experimentally, the activities of the rabbit reticulocyte 15-lipoxygenase (23), the porcine leukocyte 12-lipoxygenase (24), and the human 5-lipoxygenase (3) have been consistent with this scheme. Thus, it is likely that the catalytic mechanism of the mammalian lipoxygenases and the soybean lipoxygenase is similar. Several key features of this scheme need to be considered. This chapter will first describe the overall scheme, then go through each step in more detail. Many aspects of the catalytic mechanism of lipoxygenases have been reviewed elsewhere (3,23,25–27). Therefore, this chapter is intended to provide a framework for the discussion of the mutagenesis that has been done on the iron-binding site and on the positional specificity of the enzyme.

The lipoxygenases contain a catalytically essential iron cofactor in its ferrous form in the resting enzyme (28). The ferrous iron must first be oxidized to its ferric form by the hydroperoxide product of the enzyme (25). This oxidized enzyme can now abstract a hydrogen atom from a bis-allylic methylene (29,30). Possibly a base in the active site accepts the proton and the iron accepts the electron of the hydrogen, becoming transiently reduced. The resulting fatty acid intermediate has been proposed to be either an alkyl radical, a vinylallyl radical (27), or an organoiron compound (31). Oxygen is now inserted two carbons away from the site of the initial hydrogen abstraction, resulting in a peroxyl radical that contains a conjugated diene. This intermediate can accept an electron from the iron (returning the enzyme to its ferric form) and the initial proton, leaving the enzyme as a hydroperoxide (Fig. 4.1). The product of the 15-lipoxygenase reaction is a fatty acid that contains a hydroperoxide at the ω-6 carbon. Furthermore, the *cis,cis*-pentadiene becomes a conjugated *cis,trans*-pentadienyl hydroperoxide.

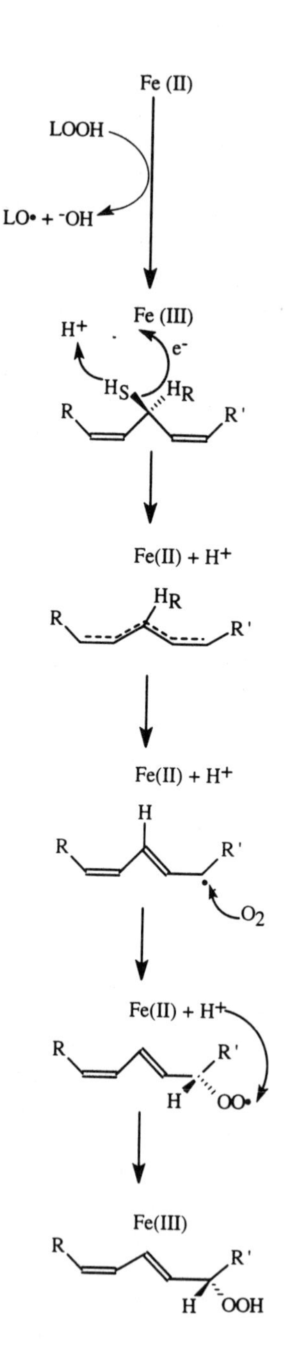

Figure 4.1. A simplified illustration of the steps in the dioxygenase reaction of 15-lipoxygenase. The initial step in the reaction is the oxidation of the iron cofactor by the lipid hydroperoxide product. This is followed by hydrogen abstraction from a bis-allylic methylene, concomitant with reduction of the iron. Insertion of molecular oxygen occurs stereospecifically antarafacial to the hydrogen abstraction and two carbons away, resulting in a conjugated diene. Transfer of an electron from the iron to the lipid–peroxy radical and the addition of a proton results in the lipid hydroperoxide product and returns the enzyme to its ferric state.

A lag phase in the kinetics of the dioxygenase reaction is characteristic of all lipoxygenases. It is likely that the lag phase represents oxidation of the ferrous iron to its ferric form in the enzyme for the following reasons. This lag phase can be shortened by the addition of the hydroperoxy-fatty acid product (32,33) and extended by the enzymatic removal of the peroxide (34). The "resting" ferrous enzyme can be completely converted to the ferric form by the addition of the hydroperoxy-fatty acid product (35). Furthermore, the fatty acid hydroperoxide is decomposed during the oxidation of the enzyme. Decomposition of the hydroperoxide is known as the "pseudoperoxidase" activity of the enzyme (36,37). The pseudoperoxidase activity can be stimulated by certain compounds that reduce the iron to its ferrous form. Such compounds are inhibitors of the dioxygenase activity of the enzyme. A kinetic analysis of the soybean enzyme using rapid kinetic techniques is consistent with a scheme in which the dioxygenation rate is dependent on the amount of enzyme in the ferric form and the lag phase is due to the oxidation of the iron from ferrous to ferric (38,39). Thus, the initial oxidation of the iron should be considered a crucial first step in the enzyme's kinetics.

Once the iron is oxidized, it is poised to act by a redox cycle to first accept an electron from the substrate, then donate an electron to the lipid-peroxy intermediate. Clearly, the redox potential of the iron is of crucial importance in the reaction mechanism. The coordination of the iron to the protein must "fine-tune" the redox potential for the iron's role in the reaction. Thus, the iron environment of the enzyme is an attractive target for mutagenesis, and many studies have focused on the iron center.

The abstraction of hydrogen from a bis-allylic methylene is the rate-limiting step of the reaction. Evidence in favor of this first came from early work on the soybean enzyme (29,30), in which a kinetic isotope effect was noticed with stereospecifically tritium-labeled 8,11,14-eicosatrienoic acid. The isotope effect and loss of tritium was seen only when the label was in the pro-S conformation. This demonstrated that the hydrogen abstraction is not only rate-limiting, but also stereospecific. Recently, it has been observed that the soybean enzyme exhibits an unusually large kinetic isotope effect with deuterated linoleic acid as a substrate (40,41). The mechanism underlying this isotope effect has yet to be fully explained. The occurrence of such an isotope effect in mammalian enzymes awaits experimental proof.

The structure of the reaction intermediate after hydrogen abstraction is one area of controversy. A pentadienyl radical is one possibility for this intermediate, due to its relative stability. However, a pentadienyl radical could lead to various racemic products if it were to leave the enzyme before oxygen insertion, and such a mixture of racemic products has not been seen. Therefore, if the intermediate is a pentadienyl radical, it would have to stay bound to the enzyme in such a way to accept oxygen in the correct stereoconfiguration.

Pentadienyl radical binding could be performed by using the structure of the binding pocket of the substrate to constrain the pentadienyl radical intermediate and force the localization of the radical favoring the insertion of oxygen at the correct carbon and in the correct stereoconfiguration. Since the substrate is all *cis* and the product is *cis,trans,* the enzyme might constrain the pro-*trans* bond to facilitate the double bond formation forcing the radical to the appropriate carbon for oxygen insertion. Another possibility is that the enzyme sterically directs the approach of oxygen (i.e., the enzyme may have an oxygen "tunnel," as suggested by the structure of the soybean isoform-1 [42]) in a way that favors the "correct" insertion. Full understanding of the structure of the binding pocket of the enzyme and alteration of this pocket through mutagenesis could help discriminate between these possibilities.

Another proposed reaction intermediate is an organoiron complex (31). The organoiron mechanism differs from the pentadienyl radical mechanism in several aspects. Rather than having complete electron migration from the bis-allylic methylene carbon to the iron, upon hydrogen abstraction the electrons shift along the pentadiene to the soon-to-be oxygenated carbon. A covalent bond is formed between the iron and the carbon to be oxygenated.

Several key issues distinguish the pentadienyl radical mechanism from the organoiron mechanism. In the radical mechanism, the iron interacts with the bis-allylic methylene carbon, whereas the iron interacts with the oxygen-reactive carbon in the organoiron mechanism. Another key issue is whether the electron fully migrates to the ferric iron, reducing it, or is the electron shared between the iron and the carbon in a covalent bond. Such fine-structure localization of iron and electrons will be difficult to determine.

A vinylallyl radical as the reaction intermediate at this step has recently been proposed based on electron paramagnetic resonance (EPR) spectroscopy of specifically deuterated substrates (27). In this mechanism, after hydrogen abstraction and iron reduction, the electron remains delocalized over only one-half of the pentadiene. The ferrous iron then directs the insertion of oxygen across the π-bonds of the other half of the pentadiene. This mechanism places the iron in a position relative to the substrate that is similar to that proposed by the organoiron mechanism. An understanding of the structural elements responsible for the binding of the substrate to the enzyme and the positioning of the substrate relative to the iron may help distinguish between these different mechanisms.

The penultimate step in the reaction mechanism results in the formation of the lipid–peroxy anion and returns the enzyme to its ferric form. This anion now accepts a proton to leave the enzyme as a lipid hydroperoxide. The shuttling of a proton is required, for both the hydrogen abstraction step and the product release step, implying the presence of an essential base in the active site. It has been proposed that one of the amino acids that is coordinated with the iron could act as a base (42). The mutagenesis of an essential iron ligand that also acts as a base would result in an inactive enzyme by virtue of destroying the iron-binding capability of the enzyme, in addition to removing an essential base. The replacement of the ligand with an amino acid that can still bind iron and has a different pKa (i.e., a His-to-Asn change) may change the catalytic activity. Kinetic studies of such a mutant, together with spectroscopy, would be required to make conclusions about the mechanism. Such work has recently been done on the soybean isoform-3 (43). It has been proposed that one of the iron ligands is water or a hydroxide (44,45), which would be an attractive candidate for the active site base. The hydroxide could be displaced by substrate binding. During the displacement, the hydroxide could accept a proton from the substrate and leave as water. In the final step of the reaction another water molecule could replace the product, donating a proton to the lipid–peroxy anion and coordinating again with the iron.

An interesting feature of the mammalian lipoxygenases is the self-inactivation of the enzyme (12). It has been calculated for rabbit reticulocyte 15-lipoxygenase that an irreversible inactivation occurs every 600 turnovers (23). Furthermore, this inactivation is dependent on the concentration of the hydroperoxy product of linoleic acid dioxygenation (13-HPODE [46]). Cyanogen bromide cleavage of the 13-HPODE-treated and native lipoxygenase, followed by two-dimensional thin-layer chromatography revealed a correlation between this product-based inactivation and 1 mole of methionine sulfoxide formed in the enzyme (47). It was technically difficult to determine which methionine (of the 16 methionines in the protein) was oxidized, thus the mechanistic basis for the inactivation could not be tested by mutagenesis. Such work awaits the identification of the oxidized methionine, followed by mutating this methionine to a similar amino acid that cannot be oxidized and analyzing the self-inactivation kinetics of the variant. The oxidized methionine has recently been identified by tryptic digestion, followed by HPLC–MS. Mutagenesis of this methionine demonstrated that it is not essential for activity and the oxidation is not required for inactivation (101).

Clearly, there are many issues remaining to be resolved concerning the reaction mechanism of lipoxygenases. Since the iron ligands have been studied by mutagenesis, additional site-directed mutagenesis can be applied to help resolve some of these issues. Now that the iron ligands have been identified by X-ray crystallography, more sophisticated experiments can be done to analyze how the redox potential of the iron affects the reaction. Another aspect of the catalysis that has been analyzed by site-directed mutagenesis is the positional specificity of the enzyme. These two areas of the structure–function relationship of the enzyme will be discussed.

Iron-Binding Ligands

All lipoxygenases analyzed to date contain a catalytically essential iron cofactor coordinated directly to the protein. This was first found in the soybean enzyme (48) and has since been found to be true for mammalian lipoxygenases as well (49,50). Spectroscopic studies, including extended X-ray absorption fine structure (EXAFS [51]), X-ray absorption near-edge spectroscopy (XANES [45]), Mössbauer (52), and circular and magnetic circular dichroism (53), have been aimed at defining the geometric structure of the iron site in the soybean enzyme. The studies are consistent with a six-coordinate distorted octahedral geometry. A recent study has indicated that the iron coordination is flexible, appearing as five-coordinate in the ferrous form and six coordinate with a hydroxide complexed to the ferric iron (45). No heme or iron–sulfur clusters have been found in the soybean enzyme, suggesting that the iron is bound directly to the protein (54). The EXAFS data suggested that the iron is bound by 4 ± 1 nitrogen ligands and 2 ± 1 oxygen ligands. Electron paramagnetic resonance, done on the soybean enzyme (27), the human 5-lipoxygenase (55), and the rabbit reticulocyte 15-lipoxygenase (56,57) have demonstrated that the iron cycles between the ferrous and ferric states.

A conservation of histidines in the sequences of lipoxygenases was first noted when the second soybean isoform was cloned and sequenced (58). Shortly after the publication of the soybean isoform-2 sequence, the conservation of histidines was also noticed in the sequences of the rat 5-lipoxygenase (59) and the human 15-lipoxygenase (60). Each of these researchers suggested that these histidines may be the iron ligands for the enzyme. Most of these histidines have since been found to be conserved in all lipoxygenases sequenced to date (Fig. 4.2). The first set of histidines are clustered in a highly conserved region in the enzyme in a motif His-$(X)_4$-His-$(X)_4$-His-$(X)_{17}$-His-$(X)_8$-His. The one exception to this motif is mouse platelet 12-lipoxygenase, that contains a glutamine in place of the first histidine (Fig. 4.2). The final conserved histidine in all lipoxygenases is in another highly conserved region of the sequences, nearer to the carboxy terminus.

The determination of the X-ray crystal coordinates has directly identified which amino acids are iron ligands. However, two structures have been published that differ in the iron coordination (42,61). In both structures, three of the conserved histidines (His-499, His-504, and His-690) are iron ligands. A fourth ligand in both structures is an oxygen of the carboxy terminus of the polypeptide chain. There is a high level of sequence conservation at the carboxy terminus of the lipoxygenases. With one exception (rat 5-lipoxygenase), all lipoxygenases end with an isoleucine. It has been pointed out that the rat 5-lipoxygenase could end with an isoleucine if a stop codon occurring near the end of the clone is removed, suggesting the possibility that this stop codon may have been a cloning artifact (61).

The major difference between the two reported structures is the presence of a fifth ligand. The structure from Minor et al. has an asparagine (Asn-694) as a fifth ligand (61). In the structure from Boyington et al. (42), this asparagine is too far away (3.3 Å) to be coordinated with the iron. It is interesting to note that in all lipoxygenase sequences this asparagine is either conserved or is a histidine (Fig. 4.2); thus, it is possible that this residue is an iron ligand in all lipoxygenases. Since both structures are of the ferrous form of the enzyme and both were solved in the absence of substrate, the possibility of rearrangements at the iron-coordination site during enzyme activity cannot be ruled out. Electron paramagnetic resonance studies, and recent EXAFS and XANES studies have been suggestive that a change in the iron coordination occurs during product binding, by indicating a change in the geometry of the iron coordination when the soybean enzyme is treated with 13-HPODE (35,45,54).

Prior to identification of the iron ligands in lipoxygenase by X-ray crystallography, the role of the conserved histidines in the structure–function relationship of the enzyme had been studied by site-directed mutagenesis. The first site-directed mutagenesis done on any lipoxygenase enzyme was done on two of these conserved his-

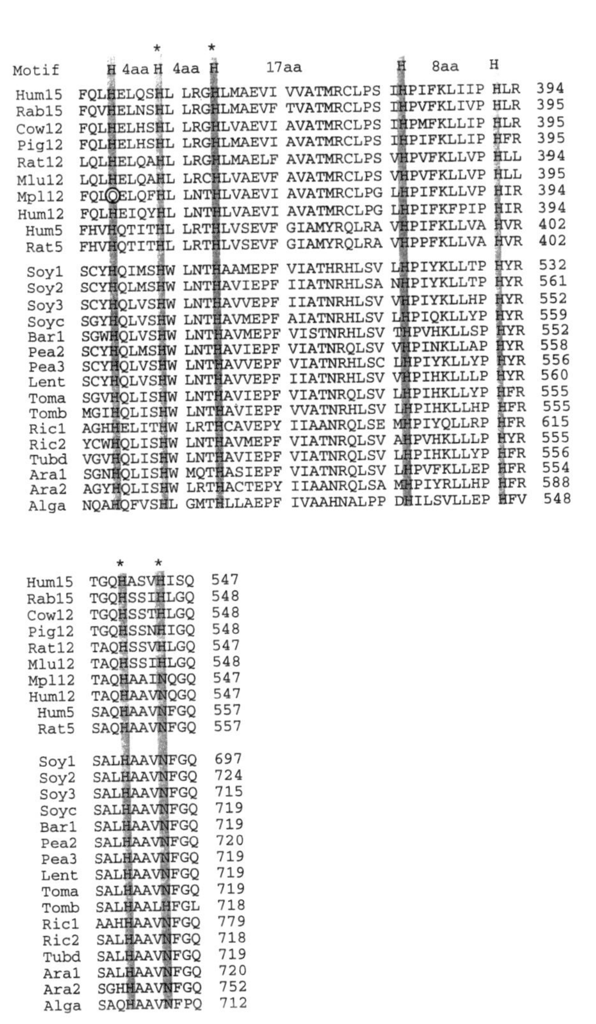

Figure 4.2. Sequence comparisons of the regions containing the conserved histidines in the lipoxygenase sequences. Only the lipoxygenases for which the full-length coding region of the cDNA have been cloned are presented. The sequence abbreviations are listed in Table 4.1. The numbers flanking each sequence correspond to the number of the final amino acid in that region, where the initial methionine of the protein is number 1. The single letter amino acid code is used. The conserved histidines and the site that corresponds to soybean N694 are indicated in grey. The amino acids that have been identified as iron ligands in the crystal structure of the soybean enzyme are indicated with asterisks. *a*) The highly conserved region that matches the sequence motif His-$(X)_4$-His-$(X)_4$-His-$(X)_{17}$-His-$(X)_8$-His. The only exception to this motif is the glutamine in the mouse platelet 12-lipoxygenase sequence. *b*) The later highly conserved region of the protein in which two other iron ligands are located. The amino acid that corresponds to N694 is either a histidine or an asparagine in each sequence. Figure is adapted from Funk (9).

tidines in the human 5-lipoxygenase (62). The conclusion of this study was not completely consistent with the assignment of ligands by X-ray crystallography (i.e., the mutation of histidine-373 to serine resulted in an active enzyme); however, it has been speculated that the activity was due to accidental contamination of the mutated sample with a wild-type sample (9). The conserved histidines have also been mutated in the soybean enzyme (63), the porcine leukocyte 12-lipoxygenase (64), and further mutagenesis of the human 5-lipoxygenase (65–67 [Table 4.2]). No mutagenesis studies on the histidines of human 15-lipoxygenase have been published. Our unpublished results have shown a loss of activity when the histidines at position 360 (H360), H365, and H383 are mutated to alanine; however, we cannot conclude any role for these histidines in iron binding, since the iron content of the variant enzymes was not determined. It is reasonable to assume that the mutagenesis work done on the 5- and 12-lipoxygenases also explains the iron binding of the human 15-lipoxygenase. The consensus of these studies agrees with the crystallographic determination of the iron-binding ligands, that is when three of the five conserved histidines are mutated, the result is an inactive enzyme.

Two groups have also examined the contribution of the carboxy terminus to the enzyme's activity. Before the crystal structure was reported, Zhang et al. showed that deletion of six amino acids at the carboxyl terminus of the human 5-lipoxygenase resulted in an inactive mutant (66). After publication of the crystal structure, a series of mutations was done on the carboxy terminal isoleucine of the two murine 12-lipoxygenases (10). It was found that when the isoleucine was deleted from either enzyme, the resulting mutant was inactive. Furthermore, when the isoleucine of the murine platelet 12-lipoxygenase was replaced with a charged residue (lysine or aspar-

TABLE 4.2 Iron-Binding Ligands Determined by Site-Directed Mutagenesis and Iron-Content Analysis[a]

Protein	Iron-Binding Ligands				
	1	2	3	4	5
soy-1	H499[b]	H504[b]	H690[b]	I839	N694
hum15	H360	H365	H540	I662	H544
pig12	H361[c]	H366[c]	H541[c]	I663	H545
hum5	H368[d]	H373[d]	H551[d]	I675	N556
mpl12	H360	H365	H540	I663[e]	N544

[a]Additional ligands are assigned by sequence alignment with the soybean enzyme. The ligands for the soybean enzyme, determined by X-ray crystallography are also indicated (Boyington et al. [42], and Minor et al. [61]). For reference, the amino acids of human 15-lipoxygenase that align with the soybean ligands are also shown. Mutagenesis and iron-content studies are referenced in other footnotes for this table.
[b]Steczko et al. (63).
[c]Suzuki et al. (64).
[d]Ishii et al.(67), Percival (68), and Zhang et al. (69).
[e]Chen et al. (10).
Sequence abbreviations are listed in Table 4.1.

tic acid), no activity could be detected. Replacing the isoleucine with a valine had no effect on enzyme activity. Other amino acids (i.e., leucine, arginine, asparagine, serine, and glycine) at this position resulted in severely reduced enzymatic activity.

A limitation of the mutagenesis studies is that the mutation of an essential histidine results in an inactive enzyme. In order to conclude a role for this histidine in iron binding, direct measurements of the iron content need to be correlated with activity. Studies that have made such a correlation are referenced in Table 4.2. The first such study analyzed the activity and iron content of five histidine-to-serine mutants (68). The five histidines that were mutated were at position 363, 368, 373, 391, and 400 of the primary sequence (if one counts the initiator methionine). This group did not look at the effects of mutating the histidine at position 551.

Previously, in a crude cell lysate the histidine-to-serine changes at positions 368 and 373 had been demonstrated to have no 5-lipoxygenase activity (65). This observation was extended with enzymes purified to greater than 98% homogeneity from a baculovirus/insect cell culture system. Furthermore, the ability of the variant enzymes to degrade 13-HPODE (the "pseudoperoxidase" activity), as well as the ability of the enzymes to convert 5-HPETE to LTA_4 (the LTA_4 synthase activity) were analyzed. The mutated enzymes were then analyzed for iron content by a colorimetric assay, using FerroZine as a chromophore (49). This group found that replacing the histidines at positions 363, 391, and 400 with serines had very little effect on any of the three enzymatic activities. The iron contents of each of these variants were similar to wild-type. However, changing the histidine at position 373 to serine (H373S) completely abolished enzymatic activity. In correlation with the activity, the H373S variant was found to contain undetectable levels of iron. When the H368S variant was analyzed, it was found to have no lipoxygenase or LTA_4 synthase activity, however it had approximately 20% of the wild-type pseudoperoxidase activity. Both preparations of this variant contained iron, albeit a decreased amount (0.15 and 0.43 mol Fe/mol enzyme). This result implied that H368 is not essential for iron binding, yet is essential for enzyme activity.

The conclusions of this study were supported by another careful examination of the iron contents of mutated 5-lipoxygenase from a different laboratory (69). This group expressed eight separate variants in *E. coli* and purified each variant multiple times. The lipoxygenase activity of each variant was assayed. The iron content of each was determined by atomic absorption spectrometry. When the histidines at positions 363, 391, and 400 were replaced with glutamine, the expressed proteins contained a wild-type level of iron. The H391Q and H400Q variants had a slight decrease in activity, whereas the activity of the H363Q variant was decreased tenfold even though it had wild-type levels of iron. The result of the activity and iron content of the H363Q variant illustrates the importance of measuring both the enzymatic activity and the iron content to conclude a role for the histidine in iron binding. The mutations done on the histidine at position 373 completely abolished both the activity and the iron content.

This study also included mutations of the histidine at 551. All mutations at H551 completely abolished both iron binding and enzymatic activity. Finally, mutating the histidine at position 368 to either glutamine, asparagine, or serine resulted in the expression of an inactive protein. In both the H368Q and the H368S variants, the iron content was approximately one-half that of the wild-type enzyme. The H368N variant had its iron content decreased to about 17% of wild-type levels. This result is similar to that found in the experiments described from the baculovirus/insect cell culture (68). The histidine at 368 does not appear to be essential for iron binding, yet it is essential for activity.

Assuming that the structural similarity of 5-lipoxygenase and soybean lipoxygenase-1 is strong enough that they have the same iron ligands, then H368, H373, and H551 are coordinating the iron. Therefore, considering the mutagenesis results in light of the soybean structure, it may be possible that the H368 variants are capable of binding iron by virtue of the remaining ligands, yet the redox potential of this bound iron is not properly poised for full enzymatic activity. A spectroscopic analysis of this mutated form of the enzyme would address this issue.

The mutagenesis of histidines and iron content of the 12-lipoxygenase from porcine leukocytes has also been done (64). The eight histidines that were found to be conserved in all mammalian lipoxygenases then known were mutated to leucine. The histidine at 361 (corresponding to H368 in 5-lipoxygenase and H499 in soybean, Table 4.2) was also mutated to a glutamine. The variant enzymes were expressed in *E. coli* and purified by immunoaffinity chromatography.

The specific enzymatic (lipoxygenase) activity and iron content of each variant were measured. The iron content was determined by atomic absorption spectrometry. The variants of 12-lipoxygenase that had no detectable lipoxygenase activity were H361L/Q, H366L, H393L, and H541L. Iron content was measured for the mutations at H361, H366, and H541. Unfortunately, the levels of expression of the H393L variant were too low to measure the iron content, but apparently not too low to accurately measure the enzymatic activity. The H361L variant had 20% wild-type levels of iron, thus, as in the case for 5-lipoxygenase, variants of H361 may be able to bind iron, yet be unable to catalyze the lipoxygenase reaction. However, unlike the situation with the 5-lipoxygenase, mutating H361 to glutamine reduced the levels of iron further, to 2% of wild-type. The H426L variant also had severely reduced enzymatic activity, however, it had close to wild-type levels of iron, that again illustrates the importance of measuring both the enzymatic activity and the iron content. The conclusions that were reached from this study are that H361, H366, and H541 are iron ligands. Due to the low levels of expression, no conclusion could be made regarding the role of H393 in iron binding.

The mutagenesis studies on the iron-binding ligands of lipoxygenase continues to be of interest. Rather than putting to rest the issue of the role particular amino acids have in the iron binding of the lipoxygenases, X-ray crystallography of the soybean enzyme has opened up new questions about the role of these amino acids. In light of the three-dimensional structure of the soybean enzyme, the focus of the mutagenesis studies has shifted from the assignment of certain histidines as iron ligands to the analysis of the way in which the specific ligands are coordinating the iron and what effect particular amino acids may have on the catalytic activity of the enzyme. Of particular interest are H368 variants of 5-lipoxygenase, that still contain iron, yet have no activity. A spectroscopic analysis of such variants may reveal the role of this particular ligand in coordinating the iron.

Furthermore, several questions result from the two structures of the soybean enzyme: What is the role of N694 in the enzyme's activity? Does this ligand come off and on the iron in catalysis? Is there a catalytic or spectroscopic difference between the lipoxygenases that contain a histidine at this position and those that contain asparagine? These questions can be addressed by mutagenesis. Recently, the contribution of the asparagine to the activity of the soybean isoform-3 was examined by site-directed mutagenesis (43). In this study, the asparagine was replaced with alanine, serine, and histidine. The alanine and histidine substitutions contained wild-type levels of iron; the serine variant contained one-half the normal levels, indicating that asparagine at this position is not required for iron binding. Also, the EPR spectra of the histidine-substituted enzyme were similar to the wild-type. While the serine- and alanine-substituted enzymes had no detectable catalytic activity, the histidine substitution resulted in an enzyme with kinetics very similar to the wild-type. This result casts some doubt on asparagine acting as an active-site base involved in the initial hydrogen abstraction and may imply a role for the amino acid in a hydrogen bond or some interaction with the substrate. The clear implication from this result is that the amino acid at this position is not required for iron binding, yet it has an important role in catalysis. Such a role for this amino acid in the mammalian lipoxygenases will be interesting to test by mutagenesis. The soybean structure has given us a model to test the structural similarity of the various lipoxygenases. An obvious starting point for the testing of the structural similarity of the lipoxygenases is the iron-binding site.

Positional Specificity

Several groups have used site-directed mutagenesis to analyze the structural features of the 12- and 15-lipoxygenases that account for their positional specificities. Much of this work has recently been reviewed (70). Early work on soybean lipoxygenase resulted in a model for the reaction mechanism that is currently believed to describe all lipoxygenase enzymes (29,30). An analysis of the substrate requirements of the soybean enzyme showed that the enzyme requires a bis-allylic methylene at the ω-8 position of the unsaturated fatty acid substrate (the eighth carbon from the methyl terminus). An important feature of the reaction that was noticed in this work was the requirement for the methylene to be at a proper distance from the methyl terminus of the substrate, implying that the enzyme "counts" from the methyl terminus. It is reasonable to suppose that the enzyme has a binding pocket that is specific for an aliphatic hydrocarbon. Furthermore, this binding pocket would position the substrate so that the bis-allylic ω-8 methylene is aligned with the reaction center.

The 12/15-lipoxygenases exhibit dual positional specificities, forming both 12- and 15-HPETE in varying ratios, characteristic of the particular isoform of the enzyme (Table 4.3). The activities of these 12/15-lipoxygenases imply that these enzymes can abstract hydrogen from both the ω-8 carbon and the ω-11 carbon in varying ratios. This feature of rabbit reticulocyte enzyme has been studied in detail by analyzing the activity of the enzyme on fatty acids of various lengths and of various degrees of saturation (71). Particular attention was paid to the position of the bis-allylic methylenes on these fatty acids, since it was known that the abstraction of hydrogen is the first, rate-limiting step. This study concluded that to achieve single positional specificity, the optimal placement of the bis-allylic methylene for the rabbit enzyme would be at the ω-9 position. Arachidonic acid has bis-allylic methylenes at the ω-8 and the ω-11 positions. Thus, the reaction center of the rabbit enzyme is shifted away from the ω-8 methylene slightly and abstracts hydrogen predominantly from the ω-8 carbon and to a lesser extent the ω-11 carbon, resulting in 15- and 12-HPETE at a ratio of approximately 9:1. Considered together, all of the data from this study are consistent with the model in which the enzyme binds the methyl terminus of the substrate.

Further evidence that implies that the key enzyme–substrate interaction of the 12/15-lipoxygenases is at the methyl terminus of the substrate comes from the observation that these enzymes are able to oxygenate membrane phospholipids. This has been observed for rabbit (20,72) and human 15-lipoxygenases (73), as well as the porcine leukocyte 12-lipoxygenase (22).

By using site-directed mutagenesis, it has been possible to alter the positional specificity of the human 15-lipoxygenase (74), the human platelet 12-lipoxygenase (75), and the porcine leukocyte 12-lipoxygenase (64). Such alterations have identified important determinants of the enzyme–substrate interaction. A comparison of the sequences of mammalian 12- and 15-lipoxygenases indicated four potential amino acid determinants of positional specificity. These four amino acids were conserved differences between the sequences of the 12- and 15-lipoxygenases. In other words, these amino acids were conserved in the sequences of the 15-lipoxygenases and differed from amino acids found at those positions of the 12-lipoxygenases; however, the difference was conserved among the 12-lipoxygenase sequences. We reasoned that these four amino acids, or a subset of the four, could define a part of the substrate-binding pocket responsible for aligning the substrate relative to the reaction center of the enzyme.

TABLE 4.3 Product Profiles of 12/15-Lipoxygenases Using Arachidonic Acid as Substrate[a]

Enzyme	417/418	15H(P)ETE:12H(P)ETE	Ref.
hum15	ile/met	9:1	79
rab15	ile/met	9:1	100
cow12	val/val	1:11	16
pig12	val/val	1:10	22
rat12	ala/met	1:6	18
mlu12	val/met	1:3	10

[a]The amino acids that align with positions 417 and 418 in the primary sequence of human 15-lipoxygenase are also indicated. Enzyme abbreviations are listed in Table 4.1.

Each of the four amino acids were specifically mutated in human 15-lipoxygenase to their 12-lipoxygenase counterparts. The mutated proteins were produced in *E. coli* and analyzed for their positional specificities on arachidonic acid. It was found that mutating the methionine at position 418 to valine (15LOX M418V) resulted in an enzyme that performed equal 12- and 15-lipoxygenation. This was found when this methionine was mutated alone, or in combination with the other three amino acids. We reasoned that the methionine is a primary determinant of the positional specificity of the enzyme and that it may interact with the substrate to place it in the active site.

The equal 12- and 15-lipoxygenation of the 15LOX M418V variant implied a shift in the binding of the substrate to the enzyme so that the site of hydrogen abstraction was now equidistant between the ω-8 and the ω-11 carbons of arachidonic acid. This prediction was further tested by analyzing the activity of the 15LOX M418V on an eicosatetraenoic acid substrate that had its double bonds shifted by one carbon toward the methyl terminus (6Z,9Z,12Z,15Z-eicosatetraenoic acid [20:4 ω-5]). It had been shown previously that the rabbit 15-lipoxygenase oxygenates this substrate at the ω-5 and the ω-8 carbons in a ratio of 1:1 (71), implying that the wild-type enzyme has its site of hydrogen abstraction aligned equidistant between the ω-7 and the ω-10 methylenes. If the mutation M418V caused a shift in the binding of the substrate, then the 6Z,9Z,12Z,15Z-eicosatetraenoic acid substrate should yield oxygenation predominantly at the ω-8 carbon. When this fatty acid was tested on 15LOX M418V, it was found that the enzyme did predominantly (in a ratio of 75:25) catalyze the oxygenation of the ω-8 carbon (74).

This result is consistent with the methionine at 418 defining a part of the methyl terminal–binding pocket of the enzyme. We next looked at the sequence alignments again, this time focusing on the region around 418. We found that this region is highly conserved in all mammalian lipoxygenases. The human platelet 12-lipoxygenase, known to have a stringent product profile, performing exclusively 12-lipoxygenation, differs from the 15-lipoxygenase in this region by only three amino acids. Furthermore, the other two 12-lipoxygenases that were known at the time of this study (porcine leukocyte 12-lipoxygenase and bovine tracheal epithelial 12-lipoxygenase) differed by only two amino acids. Further mutagenesis on the human 15-lipoxygenase to introduce the 12-lipoxygenase sequences showed that both the triple mutation (15LOX Q416K, I417A, M418V) and the double mutation (15LOX I417V, M418V) performed 12-lipoxygenation predominantly (at a ratio of 20:1 [76]). Thus, the 15-lipoxygenase can be completely converted to a 12-lipoxygenase by changing only two amino acids. By mutating the isoleucine at 417 to valine alone, we saw that this amino acid and the methionine at 418 contribute equally to the positional specificity of the enzyme (Fig. 4.3).

Based on this work, we proposed a model for the enzyme–substrate interaction, in which amino acids 417 and 418 define the binding region for the methyl terminus of the substrate. One prediction of this model is that the amino acid side-chain bulk and hydrophobicity are important features in defining the depth of the binding pocket, so replacing the methionine at 418 and the isoleucine at 417 with the smaller valines

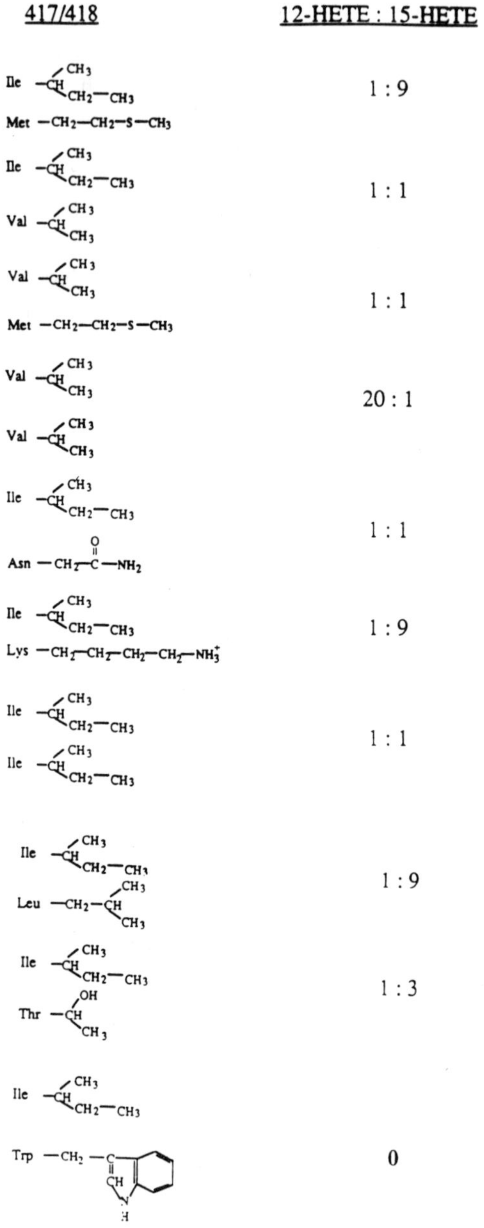

417/418	12-HETE : 15-HETE
Ile $-CH(CH_3)-CH_2-CH_3$ Met $-CH_2-CH_2-S-CH_3$	1 : 9
Ile $-CH(CH_3)-CH_2-CH_3$ Val $-CH(CH_3)-CH_3$	1 : 1
Val $-CH(CH_3)-CH_3$ Met $-CH_2-CH_2-S-CH_3$	1 : 1
Val $-CH(CH_3)-CH_3$ Val $-CH(CH_3)-CH_3$	20 : 1
Ile $-CH(CH_3)-CH_2-CH_3$ Asn $-CH_2-C(=O)-NH_2$	1 : 1
Ile $-CH(CH_3)-CH_2-CH_3$ Lys $-CH_2-CH_2-CH_2-CH_2-NH_3^+$	1 : 9
Ile $-CH(CH_3)-CH_2-CH_3$ Ile $-CH(CH_3)-CH_2-CH_3$	1 : 1
Ile $-CH(CH_3)-CH_2-CH_3$ Leu $-CH_2-CH(CH_3)-CH_3$	1 : 9
Ile $-CH(CH_3)-CH_2-CH_3$ Thr $-CH(OH)-CH_3$	1 : 3
Ile $-CH(CH_3)-CH_2-CH_3$ Trp $-CH_2-$ (indol-3-yl: C=CH–NH fused to benzene ring)	0

Figure 4.3. The product profile of variants of human 15-lipoxygenase that contain the indicated amino acids at positions 417 and 418. Ratios of products were calculated from an HPLC analysis of the products of incubation of arachidonic acid with *E. coli* expressing the variant lipoxygenases. Data are from Sloane et al. (76).

would define a deeper pocket, allowing the substrate to shift relative to the reaction center of the enzyme, and shifting the site of hydrogen abstraction from carbon 13 to carbon 10 of arachidonic acid. We tested this prediction by introducing seven different amino acids at position 418 (Fig 4.3). The result of this series of mutations is that the side-chain bulk is important in positioning the substrate in the active site.

Surprisingly, the chemical characteristics of the side chain are not important. When the hydrophobic methionine is replaced with the charged, isosteric lysine, no change in the enzyme's activity is seen. Furthermore, changing the methionine to an asparagine resulted in the same change in activity as seen for the M418V mutation (asparagine and valine have the same bulk [77], but asparagine is polar and valine is hydrophobic).

Another important feature of the side chains at 418 that was noticed is their geometry. When the methionine is replaced with an isoleucine, the resulting variant performs equal 12- and 15-lipoxygenation. However, when the methionine to leucine change is made, no effect on the activity of the 15-lipoxygenase is seen.

Another prediction of the model is that the mutations that alter the positional specificity of the enzyme from 15- to 12-lipoxygenase do not change the kinetic parameters of the enzymes. In order to test this prediction, we expressed the I417V, M418V mutated DNA in a baculovirus/insect cell culture system and purified this variant enzyme to greater than 95% homogeneity by ion-exchange chromatography (76). When the kinetics of dioxygenation of arachidonic acid were analyzed with the wild-type 15-lipoxygenase and the 15LOX I417V, M418V, the two enzymes were found to have identical K_M, (approximately 10 μM) and the Vmax of the reaction was decreased by a very small amount (threefold) in the variant enzyme. Thus, rather than changing an important energetic interaction between the substrate and the enzyme, the mutations I417V and M418V allow the enzyme to bind the substrate in a different catalytic register, a finding that is consistent with the model that we have proposed. The activity of the wild-type and variant enzymes on a variety of fatty acid substrates was also analyzed. Consistent with the model, it was found that the wild-type enzyme prefers substrates that contain a bis-allylic methylene at the ω-8 position and the 15LOX I417V, M418V variant prefers substrates that contain a bis-allylic methylene at the ω-11 position (76).

Several groups have analyzed the importance of this region in 12-lipoxygenases. Interestingly, the results from different groups have shown that the 12-lipoxygenases differ in their structural determinants of positional specificity. Chen and Funk did a series of mutations on the human platelet 12-lipoxygenase (75). They found that a single mutation of this enzyme (V418M, the converse of the 15LOX M418V mutation) had no effect on the enzyme's activity. The triple mutation (K416Q, A417I, V418M) increased the amount of 15-lipoxygenation of the enzyme, producing 12-H(P)ETE and 15-H(P)ETE in a ratio of 4:1. This result is significant since the wild-type human platelet 12-lipoxygenase produces no detectable 15-H(P)ETE. The enzyme produces predominantly 15-H(P)ETE (in a ratio of 12-HETE:15-HETE equal to 1:2) when all the amino acids between positions 398 and 429 are changed to those found in human

15-lipoxygenase (a total of seven amino acids are different in this span). This variant had considerably reduced total activity (25% of the wild-type enzyme). The conclusion from this study was that the valine at position 418 was not as important in the positional specificity of the human platelet 12-lipoxygenase as in the human 15-lipoxygenase. However, it was found that the adjacent alanine (417) is important in the positional specificity, since mutating it to an isoleucine resulted in 15-lipoxygenase products. Thus the region around amino acid 417 seems to be an important part of the active site in both the human 15-lipoxygenase and the human platelet 12-lipoxygenase.

Our unpublished results are consistent with the findings of Chen and Funk (75). We mutated the bovine tracheal epithelial 12-lipoxygenase. The single V419M (the converse of 15LOX M418V) mutation had no effect on enzyme activity, yet the double V418I, V419M variant had an increase in the amount of 15-lipoxygenase products formed (12-HETE:15-HETE equal to 3:1 [D.L. Sloane and R. Leung, unpublished results]). The conclusion from this, together with the results from Chen and Funk are that the 12-lipoxygenases contain other structural elements that are the key features for positional specificity. However, one possible problem with the attempt to convert a 12-lipoxygenase to a 15-lipoxygenase is that these mutations introduce large bulk, whereas in the case of mutating the 15-lipoxygenase bulk was removed. Structurally, the enzyme will be more tolerant of replacing larger amino acids with smaller amino acids.

Another group has demonstrated that the primary determinants of the porcine leukocyte 12-lipoxygenase are similar to the human 15-lipoxygenase (64). When the single V418I or V419M changes were made (the converse of 15LOX I417V and M418V), the resulting variants produced 12-H(P)ETE and 15-H(P)ETE in a ratio of 2:1 and 1:2, respectively. The double mutation (12LOX V418I, V419M) produced 12-H(P)ETE and 15-H(P)ETE in a ratio of 1:6, demonstrating that mutations around amino acid 419 can change the positional specificity of the porcine leukocyte 12-lipoxygenase to that of a 15-lipoxygenase.

Two 12-lipoxygenases that contain a methionine at the analogous position to the human 15-lipoxygenase M418 (from rat brain [18] and from mouse leukocytes [10]) have been cloned and sequenced. One group mutated the rat brain 12-lipoxygenase to see what effect changes of this methionine, as well as neighboring amino acids would have on the activity of this enzyme (78). Mutating the methionine to a valine reduced the overall activity of the enzyme tenfold, yet did not significantly change the positional specificity of the enzyme. The other mutations of the neighboring amino acids had no effect on either the overall activity or the positional specificity of the rat brain 12-lipoxygenase. Thus, this enzyme has other structural features that are the key determinants for its positional specificity. It should be pointed out that the two 12-lipoxygenases that contain methionine at this position each have a lower ratio of 12-lipoxygenation to 15-lipoxygenation, similar to the mutagenesis data on both the bovine tracheal epithelial 12-lipoxygenase and the human platelet 12-lipoxygenase (Table 4.3).

The two amino acids in human 15-lipoxygenase (I417 and M418) are analogous by sequence alignment, to two amino acids that are in a solvent-accessible cavity of soybean lipoxygenase-1 (42). These two amino acids are located approximately 12.5 Å away from the iron. Therefore, the mutagenesis data imply that this cavity is the fatty-acid-binding cavity of the enzyme. Since the soybean structure was solved in the absence of substrate, or substrate analogs, an interaction between these amino acids and the substrate cannot be deduced. Furthermore, the amount of sequence identity between the human and soybean enzymes is low (25%), making it difficult to confidently assign structural homology from sequence alignment. Further mutagenesis of the mammalian enzymes, based on the structure of the soybean enzyme, will test the structural similarities of these enzymes.

Conclusion

Site-directed mutagenesis has been applied to two important features of the structure–function relationship of lipoxygenases. The studies have primarily been directed by sequence alignments of the deduced amino acid sequences of the lipoxygenases. The first efforts were focused on the identification of the iron-binding ligands of the lipoxygenase. Work on the human 5-lipoxygenase and the porcine 12-lipoxygenase in which not only the enzyme activity but also the iron-binding capabilities of the variant enzymes were analyzed revealed two histidines that were required for activity as well as iron binding. The three-dimensional models of the soybean lipoxygenase-1 have identified four, and potentially five ligands (H499, H504, H690, I839-COOH, and potentially N694) that include the two homologs of the two histidines required for the 5- and 12-lipoxygenases. A third histidine that was studied by mutagenesis was required for activity, yet apparently not for iron binding. This histidine (in the human 5-lipoxygenase, H368, which corresponds to the soybean H499) may be required for maintaining the iron in the proper redox potential.

The positional specificity of the 12/15-lipoxygenase has been studied by several groups, using site-directed mutagenesis. The studies have shown that the amino acids at 417 and 418 of the primary sequence are important in determining the positional specificity of the enzyme. Replacement of these amino acids in the human 15-lipoxygenase results in a variant enzyme that performs 12-lipoxygenation predominantly. The results are consistent with a model in which amino acids 417 and 418 define a structural element that positions the substrate in the binding pocket of the enzyme in the correct catalytic register for either 15- or 12-lipoxygenation. The mutagenesis of the 12-lipoxygenases has shown that changes in this region affect the positional specificity, yet do not always completely convert the enzyme to a 15-lipoxygenase. This implies that the 12-lipoxygenases contain other structural elements that determine its substrate binding.

Many features of the reaction mechanism of lipoxygenases have yet to be resolved. A combination of various disciplines, including site-directed mutagenesis, can be applied to this problem. The three-dimensional structure of the soybean enzyme should allow us to make more sophisticated hypotheses about the structure–function relationship of all lipoxygenases. However, in the absence of a three-dimensional structure of a mammalian lipoxygenase these hypotheses will continue to rely on the sequence alignments between mammalian and soybean lipoxygenase. Until the structure of a mammalian lipoxygenase has been solved, the site-directed mutagenesis studies will test how predictive the soybean structure is of the mammalian structures.

References

1. Samuelsson, B., Dahlen, S.E., Lindgren, J.A., Rouzer, C.A., and Serhan, C.N. (1987) *Science 237,* 1171–1176.
2. Lewis, R.A., Austen, K.F., and Soberman, R.J. (1990) *N. Engl. J. Med. 323,* 645–655.
3. Ford-Hutchinson, A.W., Gresser, M., and Young, R.N. (1994) *Ann. Rev. Biochem. 63,* 383–417.
4. O'Byrne, P.M. (1994) *Ann. NY Acad. Sci. 744,* 251–261.
5. Sigal, E., Laughton, C.W., and Mulkins, M.A. (1994) *Ann. NY Acad. Sci. 714,* 211–224.
6. Honn, K.V., and Tang, D.G. (1992) *Can. Meta. Rev. 11,* 353–375.
7. Takahashi, Y., Ueda, N., and Yamamoto, S. (1988) *Arch. Biochem. Biophys. 266,* 613–621.
8. Hamberg, M., and Samuelsson, B. (1974) *Proc. Natl. Acad. Sci. USA 71,* 3400–3404.
9. Funk, C.D. (1993) *Prog. Nucl. Acids Res. Mol. Biol. 45,* 67–98.
10. Chen, X.-S., Kurre, U., Jenkins, N.A., Copeland, N.G., and Funk, C.D. (1994) *J. Biol. Chem. 269,* 13979–13987.
11. Gu, J.-L., Nataragan, R., Ben-Ezra, J., Valente, G., Scott, S., Yoshimoto, T., Yamamoto, S., Rossi, J.J., and Nadler, J.L. (1994) *Endocrinology 134,* 70–74.
12. Rapoport, S.M., Schewe, T., Wiesner, R., Halangk, W., Ludwig, P., Janicke-Hohne, M., Tannert, C., Hiebsch, C., and Klatt, D. (1979) *Eur. J. Biochem. 96,* 545–561.
13. Yokoyama, C., Shinjo, F., Yoshimoto, T., Yamamoto, S., Oates, J.A., and Brash, A.R. (1986) *J. Biol. Chem. 261,* 16714–16721.
14. Turk, J., Mass, R.L., Brash, A.R., Roberts II, L.J., and Oates, J.A. (1982) *J. Biol. Chem. 257,* 7068–7076.
15. Hunter, J.A., Finkbeiner, W.E., Nadel, J.A., Goetzl, E.J., and Holtzman, M.J. (1985) *Proc. Natl. Acad. Sci. USA 82,* 4633–4637.
16. De Marzo, N., Sloane, D.L., Dicharry, S., Highland, E., and Sigal, E. (1992) *Am. J. Physiol. 6,* L198–L207.
17. Sigal, E., Dicharry, S., Highland, E., and Finkbeiner, W.E. (1992) *Am. J. Physiol. 6,* L392–L398.
18. Watanabe, T., Medina, J.F., Haeggström, J.Z., Rådmark, O., and Samuelsson, B. (1993) *Eur. J. Biochem. 212,* 605–612.
19. Nishiyama, M., Okamoto, H., Watanabe, T., Hori, T., Hada, T., Ueda, N., Yamamoto, S., Tsukamoto, H., Watanabe, K., and Kirino, T. (1992) *J. Neurochem. 58,* 1395–1400.
20. Murray, J.J., and Brash, A.R. (1988) *Arch. Biochem. Biophys. 263,* 514–523.
21. Kühn, H., and Brash, A.R. (1990) *J. Biol. Chem. 265,* 1454–1458.
22. Takahashi, Y., Glasgow, W.C., Suzuki, H., Taketani, Y., Yamamoto, S., Anton, M., Kühn, H., and Brash, A.R. (1993) *Eur. J. Biochem. 218,* 165–171.

23. Schewe, T., Rapoport, S.M., and Kühn, H. (1986) *Adv. Enzymol. 58,* 191–271.
24. Maas, R.L., and Brash, A.R. (1983) *Proc. Natl. Acad. Sci. USA 80,* 2884–2888.
25. Petersson, L., Slappendel, S., Feiters, M.C., and Vliegenthart, J.F.G. (1987) *Biochim. Biophys. Acta 913,* 228–237.
26. Gardner, H.W. (1991) *Biochim. Biophys. Acta 1084,* 221–239.
27. Nelson, M.J., Cowling, R.A., and Seitz, S.P. (1994) *Biochemistry 33,* 4966–4973.
28. Pistorius, E.K., and Axelrod, B. (1974) *J. Biol. Chem. 249,* 3183–3186.
29. Hamberg, M., and Samuelsson, B. (1965) *Biochem. Biophys. Res. Commun. 21,* 531.
30. Hamberg, M., and Samuelsson, B. (1967) *J. Biol. Chem. 242,* 5329–5335.
31. Corey, E.J., and Nagata, R. (1987) *J. Am. Chem. Soc. 109,* 8107–8108.
32. Haining, J.L., and Axelrod, B. (1958) *J. Biol. Chem. 232,* 193–202.
33. Finazzi-Agrò, A., Avigliano, L., Veldink, G.A., Vliegenthart, J.F.G., and Boldingh, J. (1973) *Biochim. Biophys. Acta 326,* 462–470.
34. Smith, W.L., and Lands, W.E.M. (1972) *J. Biol. Chem. 247,* 1038–1047.
35. Slappendel, S., Veldink, G.A., Vliegenthart, J.F.G., Åasa, R., and Malmström, B.G. (1983) *Biochim. Biophys. Acta 747,* 32–36.
36. De Groot, J.J.M.C., Veldink, G.A., Vliegenthart, J.F.G., Boldingh, J., Wever, R., and VanGelder, B.F. (1975) *Biochim. Biophys. Acta 377,* 71–79.
37. Falgueyret, J.-P., Desmarais, S., Roy, P.J., and Riendeau, D. (1992) *Biochem. Cell Biol. 70,* 228–236.
38. Schilstra, M.J., Veldink, G.A., Verhagen, J., and Vliegenthart, J.F.G. (1992) *Biochemistry 31,* 7692–7699.
39. Schilstra, M.J., Veldink, G.A., and Vliegenthart, J.F.G. (1993) *Biochemistry 32,* 7686–7691.
40. Glickman, M.H., Wiseman, J.S., and Klinman, J.P. (1994) *J. Am. Chem. Soc. 116,* 793–794.
41. Hwang, C.-C., and Grissom, C.B. (1994) *J. Am. Chem. Soc. 116,* 795–796.
42. Boyington, J.C., Gaffney, B.J., and Amzel, L.M. (1993) *Science 260,* 1482–1486.
43. Kramer, J.A., Johnson, K.R., Dunham, W.R., Sands, R.H., and Funk, M.O., Jr. (1994) *Biochemistry 33,* 15017–15022.
44. Nelson, M.J. (1988) *J. Am. Chem. Soc. 110,* 2985–2986.
45. Scarrow, R.C., Trimitsis, M.G., Buck, C.P., Grove, G.N., Cowling, R.A., and Nelson, M.J. (1994) *Biochemistry 33,* 15023–15035.
46. Hartel, B., Ludwig, P., Schewe, T., and Rapoport, S.M. (1982) *Eur. J. Biochem. 126,* 353–357.
47. Rapoport, S., Hartel, B., and Hausdorf, G. (1984) *J. Biochem. 139,* 573–576.
48. Chan, H.W.-S. (1973) *Biochim. Biophys. Acta 327,* 32–35.
49. Percival, M.D. (1991) *J. Biol. Chem. 266,* 10058–10061.
50. Kroneck, P.M., Cucrou, C., Ullrich, V., Ueda, N., Suzuki, H., Yoshimoto, T., Matsuda, S., and Yamamoto, S. (1991) *FEBS Lett. 287,* 105–107.
51. Navaratnam, S., Feiters, M.C., Al-Hakim, M., Allen, J.C., Veldink, G.A., and Vliegenthart, J.F.G. (1988) *Biochim. Biophys. Acta 956,* 70–76.
52. Dunham, W.R., Carroll, R.T., Thompson, J.F., Sands, R.H. and Funk, M.O., Jr. (1990) *Eur. J. Biochem. 190,* 611–617.
53. Whittaker, J.W., and Solomon, E.I. (1988) *J. Am. Chem. Soc. 110,* 5329–5339.
54. Veldink, G.A., and Vliegenthart, J.F.G. (1984) *Adv. Inorg. Biochem. 6,* 139–162.

55. Chasteen, N.D., Grady, J.K., Skorey, K.I., Neden, K.J., Riendeau, D., and Percival, M.D. (1993) *Biochemistry 32,* 9763–9771.
56. Carroll, R.T., Muller, J., Grimm, J., Dunham, W.R., Sands, R.H., and Funk, M.O., Jr. (1993) *Lipids 28,* 241–244.
57. Boyington, J.C., Gaffney, B.J., Amzel, L.M., Doctor, K.S., Mavrophilipos, D.V., Mavrophilipos, Z.V., Colom, A., and Yuan, S.M. (1994) *Ann. NY Acad. Sci. 744,* 310–313.
58. Shibata, D., Steczko, J., Dixon, J.E., Andrews, P.C., Hermodson, M., and Axelrod, B. (1988) *J. Biol. Chem. 263,* 6816–6821.
59. Balcarek, J.M., Theisen, T.W., Cook, M.N., Varrichio, A., Hwang, S.-M., Strohsacker, M.W., and Crooke, S.T. (1988) *J. Biol. Chem. 263,* 13937–13941.
60. Sigal, E., Craik, C.S., Highland, E., Grunberger, D., Costello, L.L., Dixon, R.A.F., and Nadel, J.A. (1988) *Biochem. Biophys. Res. Commun. 157,* 457–464.
61. Minor, W., Steczko, J., Bolin, J.T., Otwinowski, Z., and Axelrod, B. (1993) *Biochemistry 32,* 6320–6323.
62. Funk, C.D., Gunne, H., Steiner, H., Izumi, T., and Samuelsson, B. (1989) *Proc. Natl. Acad. Sci. USA 86,* 2592–2596.
63. Steczko, J., Donoho, G.P., Clemens, J.C., Dixon, J.E., and Axelrod, B. (1992) *Biochemistry 31,* 4053–4057.
64. Suzuki, H., Kishimoto, K., Yoshimoto, T., Yamamoto, S., Kanai, F., Ebina, Y., Miyatake, A., and Tanabe, T. (1994) *Biochim. Biophys. Acta 1210,* 308–316.
65. Nguyen, T., Falgueyret, J.-P., Abramovitz, M., and Riendeau, D. (1991) *J. Biol. Chem. 266,* 22057–22062.
66. Zhang, Y.Y., Rådmark, O., and Samuelsson, B. (1992) *Proc. Natl. Acad. Sci. USA 89,* 485–489.
67. Ishii, S., Noguchi, M., Miyano, M., Matsumoto, T., and Noma, M. (1992) *Biochem. Biophys. Res. Commun. 182,* 1482–1490.
68. Percival, M.D., and Ouellet, M. (1992) *Biochem. Biophys. Res. Commun. 186,* 1265–1270.
69. Zhang, Y.-Y., Lind, B., Rådmark, O., and Samuelsson, B. (1993) *J. Biol. Chem. 268,* 2535–2541.
70. Sloane, D.L., and Sigal, E. (1994) *Ann. NY Acad. Sci. 744,* 99–106.
71. Kühn, H., Sprecher, H., and Brash, A.R. (1990) *J. Biol. Chem. 265,* 16300–16305.
72. Kühn, H., Belkner, J., Wiesner, R., and Brash, A.R. (1990) *J. Biol. Chem. 265,* 18351–18361.
73. Kühn, H., Barnett, J., Grunberger, D., Baecker, P., Chow, J., Nguyen, B., Burstyn-Pettegrew, E., Chan, H., and Sigal, E. (1993) *Biochim. Biophys. Acta 1169,* 80–89.
74. Sloane, D.L., Leung, R., Craik, C.S., and Sigal, E. (1991) *Nature 354,* 149–152.
75. Chen, X.-S., and Funk, C. (1993) *Fed. Am. Soc. Exp. Biol. J. 7,* 694–701.
76. Sloane, D.L., Leung, R., Barnett, J., Craik, C.S., and Sigal, E. (1995) *Protein Engineering 8,* 275–282.
77. Ponder, J.W., and Richards, F.M. (1987) *J. Mol. Biol. 193,* 775–791.
78. Watanabe, T., and Haeggström, J.Z. (1993) *Biochem. Biophys. Res. Commun. 192,* 1023–1029.
79. Devereux, J., Haeberli, P., and Smithies, O. (1984) *Nucleic Acids Res. 12,* 387–395.
80. Bryant, R.W., Bailey, J.M., Schewe, T., and Rapoport, S.M. (1982) *J. Biol. Chem. 257,* 6050–6055.

81. Fleming, J., Thiele, B.J., Chester, J., O'Prey, J., Janetzki, S., Aitken, A., Anton, I.A., Rapoport, S.M., and Harrison, P.R. (1989) *Gene 79,* 181–188.
82. De Marzo, N., Sloane, D.L., Dicharry, S., Highland, E., and Sigal, E. (1992) *Am. J. Physiol. 6,* L198–L207.
83. Yoshimoto, T., Suzuki, H., Yamamoto, S., Takai, T., Yokoyama, C., and Tanabe, T. (1990) *Proc. Natl. Acad. Sci. USA 87,* 2142–2146.
84. Funk, C.D., Furci, L., and FitzGerald, G.A. (1990) *Proc. Natl. Acad. Sci. USA 87,* 5638–5642.
85. Dixon, R.A.F., Jones, R.E., Diehl, R.E., Bennett, C.D., Kargman, S., and Rouzer, C.A. (1988) *Proc. Natl. Acad. Sci. USA 85,* 416–420.
86. Shibata, D., Steczko, J., Dixon, J.E., Hermodson, M., Yazdanparast, R., and Axelrod, B. (1987) *J. Biol. Chem. 262,* 10080–10085.
87. Yenofsky, R.L., Fine, M., and Liu, C. (1988) *Mol. Gen. Genet. 211,* 215–222.
88. Shibata, D., Kato, T., and Tanaka, K. (1991) *Plant Molecular Biology 16,* 353–359.
89. van Mechelen, J.R., Smits, M., Douma, A.C., Rouster, J., Cameron-Mills, V., Heidekamp, F., and Valk, B.E. (1994) GenBank #L35931.
90. Ealing, P.M., and Casey, R. (1989) *Biochem. J. 264,* 929–932.
91. Ealing, P.M., and Casey, R. (1988) *Biochem. J. 253,* 915–918.
92. Hilbers, M.P., Rossi, A., Finazzi-Agrò, A., Veldink, G.A., and Vliengenthart, J.F.G. (1994) *Biochim. Biophys. Acta 1211,* 239–242.
93. Ferrie, B.J., Beaudoin, N., Burkhart, W., Bowsher, C.G., and Rothstein, S.J. (1994) *Plant Physiol.* in press.
94. Peng, Y.L., Shirano, Y., Ohta, H., Tanaka, K., and Shibata, D. (1994) *J. Biol. Chem. 269,* 3755–3761.
95. Ohta, H., Shirano, Y., Tanaka, K., Morita, Y., and Shibata, D. (1992) *Eur. J. Biochem. 206,* 331–336.
96. Casey, R. (1994) GenBank #X79107.
97. Melan, M.A., Nemhauser, J.M., and Peterman, T.K. (1994) *Biochim. Biophys. Acta 1210,* 377–380.
98. Bell, E., and Mullet, J.E. (1993) *Plant Physiol. 103,* 1133–1137.
99. Liu, Q., and Reith, M.E. (1994) GenBank #U08842.
100. Sigal, E., Grunberger, D., Highland, E., Gross, C., Dixon, R.A., and Craik, C.S. (1990) *J. Biol. Chem. 265,* 5113–5120.
101. Gan, Q.-F., Witkop, G.L., Sloane, D.L., Straub, K.M., and Sigal, E. (1995) *Biochemistry 34*, 7069–7079.

Chapter 5

Fatty Acid Radicals and the Mechanism of Lipoxygenase

Mark J. Nelson

Central Research and Development, E.I. DuPont de Nemours & Co., Wilmington, DE 19880-0328, USA.

Introduction

Lipoxygenases catalyze the addition of oxygen to polyunsaturated fatty acids with 1,4(*Z,Z*)-dienes, linoleic acid (LA) and arachidonic acid (AA), to yield fatty acid hydroperoxides (1,2). The 11(*proS*) proton is removed from LA by soybean lipoxygenase isozyme 1 (SBL-1), and is followed by dioxygen addition to yield 13(*S*)-hydroperoxy-9,11(*Z,E*)-octadecadienoic acid (13(*S*)-HPOD) primarily. Despite the simplicity of the reaction and the extent of study of these enzymes, no clear consensus as to the mechanism of the dioxygenation reaction exists.

Lipoxygenases are unique among nonheme iron dioxygenases in that the O–O bond is not cleaved in the reaction. Two primary mechanisms have been proposed to account for this novel reaction (Fig. 5.1). The first proposes oxidation and deprotonation of the substrate diene to a pentadienyl radical (3). This alkyl radical adds dioxygen to form a fatty acid peroxyl radical that is reduced by the nonheme ferrous ion to the product. The second proposes deprotonation of the diene and coordination to the metal ion to form a σ-organometallic complex (4). Insertion of dioxygen into the Fe–C bond gives a coordination complex between the product and the iron ion. A substantial body of evidence supports the notion of radical intermediates; nevertheless it has proven impossible to rule out closed shell pathways.

Precedents for Dioxygen Addition to Alkenes

Autoxidation

The pathway of autoxidation of LA, for example, is analogous to the first mechanism proposed previously (Fig. 5.2*a* [5]). It begins with abstraction of a hydrogen atom from C-11 to form a pentadienyl radical that reacts with dioxygen. The resulting peroxyl radicals are quenched by hydrogen atom abstraction from another fatty acid molecule, continuing the radical chain process. The intermediate pentadienyl radicals have C_{2v} symmetry and the resulting hydroperoxides are formed without stereo- or regiospecificity. Further, the reversibility of the dioxygen addition results in a loss of stereochemical integrity of the double bonds (6). In contrast, SBL-1 predominantly produces 13(*S*)-HPOD, with a small amount of both 9(*R*)- and 9(*S*)-hydroperoxy-10,12(*E,Z*)-octadecadienoic acid (7,8).

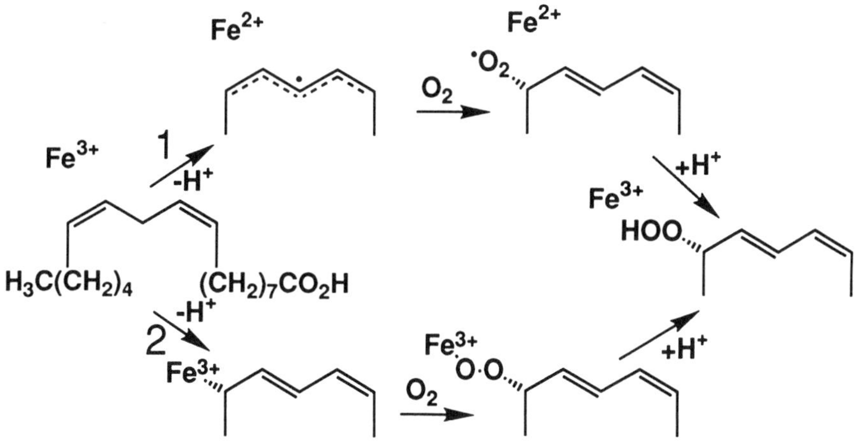

Figure 5.1. Alternate mechanisms proposed for the lipoxygenase reaction. 1. *Source:* de Groot et al. (3). 2. *Source:* Correy and Nagata (4).

Note that the autoxidation reaction requires no activation of the dioxygen. If lipoxygenase were to use an analogous mechanism, the role of the catalyst would be to activate the substrate, and to direct the reaction of dioxygen to one face and one end of the pentadienyl radical.

Carbanionic Routes

Dioxygen reacts with carbanions or carbanion equivalents (e.g., Grignard and organolithium reagents) to yield hydroperoxides (Fig. 5.2*b* [9]). This is analogous to the second mechanism for lipoxygenase proposed previously. A mechanism that includes insertion of dioxygen into a σ-organometallic bond immediately rationalizes the stereospecificity of dioxygen addition in the formation of 13-hydroperoxide by the soybean enzyme. There is significant precedent for this reaction: alkyl complexes of ferric porphyrins insert dioxygen to yield ferric peroxide complexes that have been characterized at low temperature (10).

Singlet Dioxygen

The reaction of singlet dioxygen with alkenes generates allylic hydroperoxides by a mechanism that is thought to go through either a perepoxide or dioxetane intermediate (Fig. 5.2*c* [11]). Indeed, singlet oxygen reacts with methyl linoleate to generate a mixture of isomeric hydroperoxy–diene fatty acids, including 13(*R,S*)-hydroperoxy-9,11-octadecadienoic acid (12). However, singlet dioxygen is an excited state, being some 23 kcal/mol higher in energy than triplet dioxygen, the ground state. If the enzyme is to utilize an analogous mechanism, it must be able to generate this highly excited species as well as to control its reaction with the diene to enforce the stereospecificity and regioselectivity of the reaction.

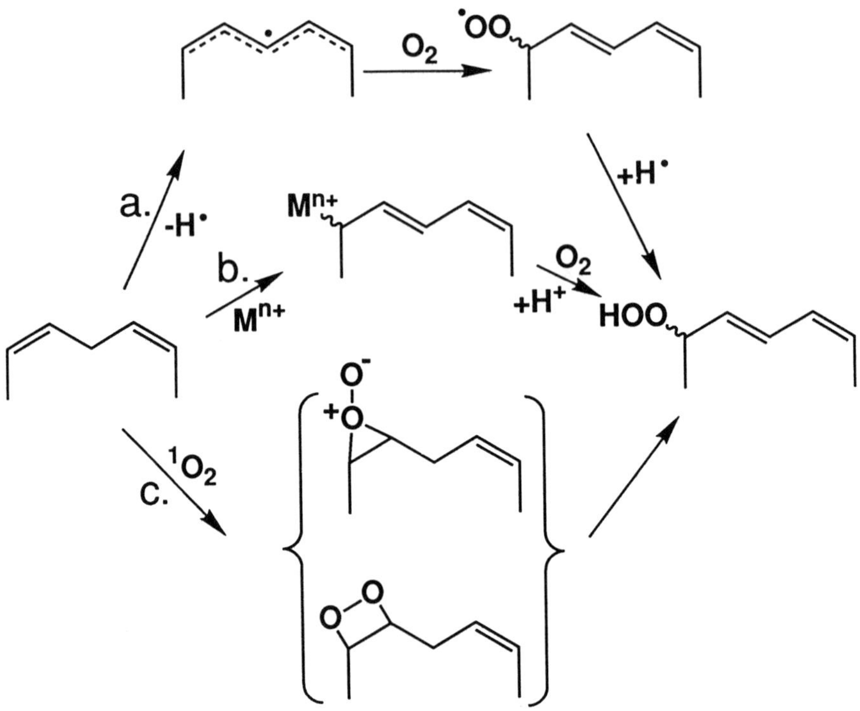

Figure 5.2. Precedents for formation of allylic hydroperoxides. *a*) Autoxidation. *b*) Stabilized carbanion intermediate. *c*) Singlet dioxygen–ene reaction.

Mechanistic Studies of the Lipoxygenase Reaction

Structure and Reactivity of the Iron Site

Although controversial, the body of evidence suggests that the active form of lipoxygenase contains Fe^{3+} (13–15). Unfortunately, no crystal structure of ferric lipoxygenase has been published, but data from extended X-ray absorption fine structure (EXAFS) spectroscopy combined with extrapolations of the crystal structure of ferrous lipoxygenase suggest that the structure of the ferric ion site is as shown in Figure 5.3 (16–18). A cavity leads from the surface of the protein to the face of the iron occupied by the oxygen-donor amide, a weak ligand that should be easily displaced. This may be a route for coordination of dioxygen to Fe^{2+}, should a ferrous intermediate be present in the mechanism (17). A second potentially important feature is the hydroxide ligand that points toward the putative substrate-binding cavity. It is very attractive to think that oxidation of LA in the active site proceeds by abstraction of a hydrogen atom by the ferric–hydroxide unit, to yield a ferrous–water complex. It has recently been reported that organic radicals react with ferrous–water units very rapidly to produce σ-organoiron complexes (19). That suggests a hybrid mechanism, in which a fatty acid radical intermediate reacts with the $[Fe–OH_2]^{2+}$ product of substrate oxidation to form a σ-organometallic complex that inserts dioxygen into the Fe–C bond.

The reactions of ferric lipoxygenase with catechols suggest that the iron has a reduction potential in excess of 0.5 V (vs. NHE [20]), the highest yet reported for a nonheme iron enzyme, supporting the radical mechanism. More directly telling for the mechanism is that partial reaction of lipoxygenase, the treatment of ferric lipoxygenase with LA in the absence of dioxygen, leads to reduction of the ferric ion at a rate comparable to the rate of the dioxygenation reaction (3,21). Thus oxidation of LA by the nonheme ferric ion in the enzyme is kinetically able to be a step in the dioxygenation mechanism. The ultimate organic product of the anaerobic reaction of LA with ferric lipoxygenase is a fatty acid pentadienyl radical that has been trapped and identified after dissociation from the enzyme (22,23).

The Structure of the Fatty Acid Intermediate

The data cited previously do not demonstrate that a pentadienyl radical is an intermediate of the reaction. For example, one could conceive of a competition between dioxygen insertion and homolysis of the Fe–C bond in mechanism 2. At low concentrations of dioxygen the latter would generate Fe^{2+} and pentadienyl radical as a side

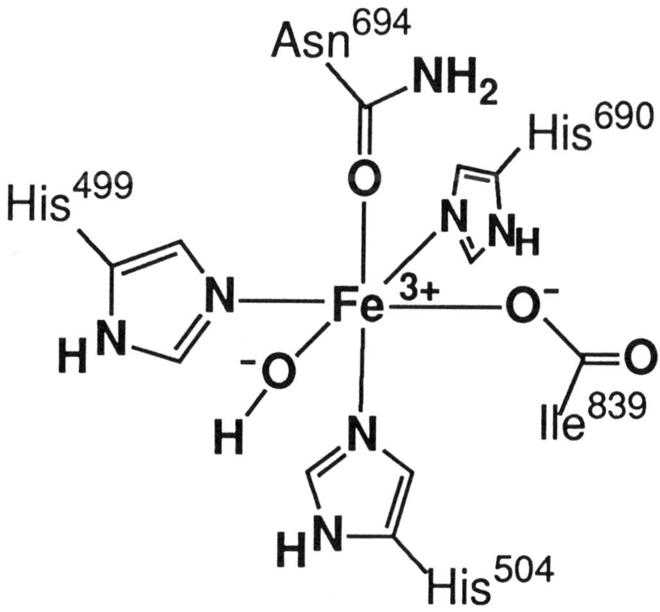

Figure 5.3. Model for the structure of the metal site in ferric (active) lipoxygenase.

product of the dioxygenation reaction. On the other hand, a true fatty acid radical intermediate in the active site might be prevented sterically from attaining the planar pentadienyl radical structure, and exist as a twisted, vinyl allyl radical instead. Only upon release into solution would it transform into the more stable conformer, the planar pentadienyl radical.

An important experiment was carried out by Funk et al. (24). They showed that LA analogs with altered double-bond stereochemistry could be substrates for lipoxygenase. In particular, 9,12(*E,Z*)-octadecadienoic acid is a slow substrate, yielding predominantly 13-hydroperoxy-9,11(*Z,E*)-octadecadienoic acid with 91:9 (*S:R*) stereoselectivity in the hydroperoxide. The stereoselectivity of dioxygen addition to C-13 suggests that this product is made via the normal enzymatic route. The isomerization of the Δ-9 bond suggests that this double bond is sufficiently weakened during the reaction to allow rotation about the 9,10 bond. Thus the Δ-9 bond in LA appears to be delocalized at some step in the mechanism. This is potentially the most damaging evidence against mechanism 2; if H^+ abstraction and Fe^{3+} addition are simultaneous (as expected on thermodynamic grounds) there would be no discrete carbanion intermediate to delocalize over the Δ-9 bond.

Purple Lipoxygenase

The solutions change color when product, 13(*S*)-HPOD, or substrates, LA and dioxygen, are added to ferric SBL-1 (25). The native ferric enzyme is faintly yellow, the result of imidazole to iron charge-transfer transitions (26). After addition of substrates or product the enzyme solutions turn purple. This purple species is metastable, but can be preserved for many minutes at 4°C. The color is the result of a new absorption band at 585 nm (27). At cryogenic temperatures these samples also show a new electron paramagnetic resonance (EPR) signal at $g = 4.3$, arising from high-spin Fe^{3+} in a rhombic electronic environment (25). The intensity of the EPR signal correlates with the intensity of the 585 nm absorption over a broad range of conditions (28), suggesting that the chromophore contains the active-site Fe^{3+}. Recently photolysis of the chromophore has been shown to lead to loss of the rhombic ferric ion and production of radicals derived from 13(*S*)-HPOD (29). This suggests that the chromophore is a Fe^{3+}-OOR complex between the active-site ferric ion and the enzyme product. In support of this, a synthetic nonheme iron–alkyl peroxide complex exhibits a low-energy peroxide to iron charge-transfer band at 515 nm (30).

When generated using LA and dioxygen, the purple species appears at a rate consistent with the enzyme-catalyzed dioxygenation reaction (21). Thus, the purple complex is also kinetically able to be an intermediate on the dioxygenation pathway. Together with the structural work discussed previously, this suggests that the purple chromophore is the penultimate mechanistic intermediate, an Fe^{3+}-13(*S*)-HPOD complex. Based on this we formulated the following hypotheses:

1. Purple lipoxygenase is a mechanistically relevant species.
2. Its lifetime is sufficient for it to partition out among other mechanistically relevant species.
3. Consequently, solutions of purple lipoxygenase are mixtures of species, including the purple chromophore; species with structures that will be informative about the dioxygenation mechanism.

As fatty acid radicals are frequently invoked reaction intermediates, we began by studying samples of purple lipoxygenase by EPR spectroscopy.

Fatty Acid Radicals in Solutions of Purple Lipoxygenase

Electron paramagnetic resonance spectra of frozen samples of purple lipoxygenase show evidence for fatty acid radicals in the solutions (31,32). These radicals are not present in samples of native ferric lipoxygenase, and the intensity of the radical spectra correlate with the intensity of the 585 nm absorption band over a range of enzyme concentrations.

Fatty Acid Peroxyl Radicals

When the samples are prepared in the presence of dioxygen the spectra include contributions from nearly axial species ($g_{1,2,3}$ = 2.035, 2.008, 2.002) shown in Figure 5.4. Use of ^{17}O-enriched dioxygen leads to spectral features corresponding to ^{17}O-hyperfine splittings from two oxygen atoms (31). That and the g-values allow us to assign this spectrum to a peroxyl radical. We have characterized these peroxyl radical spectra from samples prepared using [12,13-$^2H_2$]-13(*S*)-HPOD, [11-^2H]-13(*S*)-HPOD, [9,10-$^2H_2$]-13(*S*)-HPOD, [8,8-$^2H_2$]-13(*S*)-HPOD, and [per-^2H]-13(*S*)-HPOD (33). The spectra and their simulations are shown in Figure 5.5. In order to get adequate simulations it was necessary to assume unresolved hyperfine splittings from the protons or deuterons at C-8 and C-9. The magnitude of the coupling to the proton at C-9 tells us that the peroxyl radical is a 9-peroxyl radical derived from 13(*S*)-HPOD. This is not an intermediate on the pathway to 13(*S*)-HPOD, but it may be on the path to the minor lipoxygenase products, 9(*R,S*)-HPOD.

Fatty Acid Alkyl Radicals

The EPR spectra of purple lipoxygenase prepared with 13(*S*)-HPOD under anoxic conditions show an isotropic signal at g = 2. The quartet structure of the signal indicates that the radical contains three protons with partially resolved hyperfine splittings (32). Repeating the experiment in ^2H_2O had no effect on the signal, so none of these protons is exchangeable. However, using the set of ^2H-labeled fatty acid hydroperoxides to generate the purple lipoxygenase samples did result in a set of triplet EPR spectra (Fig. 5.6 [33]) demonstrating that the three strongly coupled protons are one on C-11, one of the two on C-9 and C-10, and one of the two on C-8 of the fatty acid.

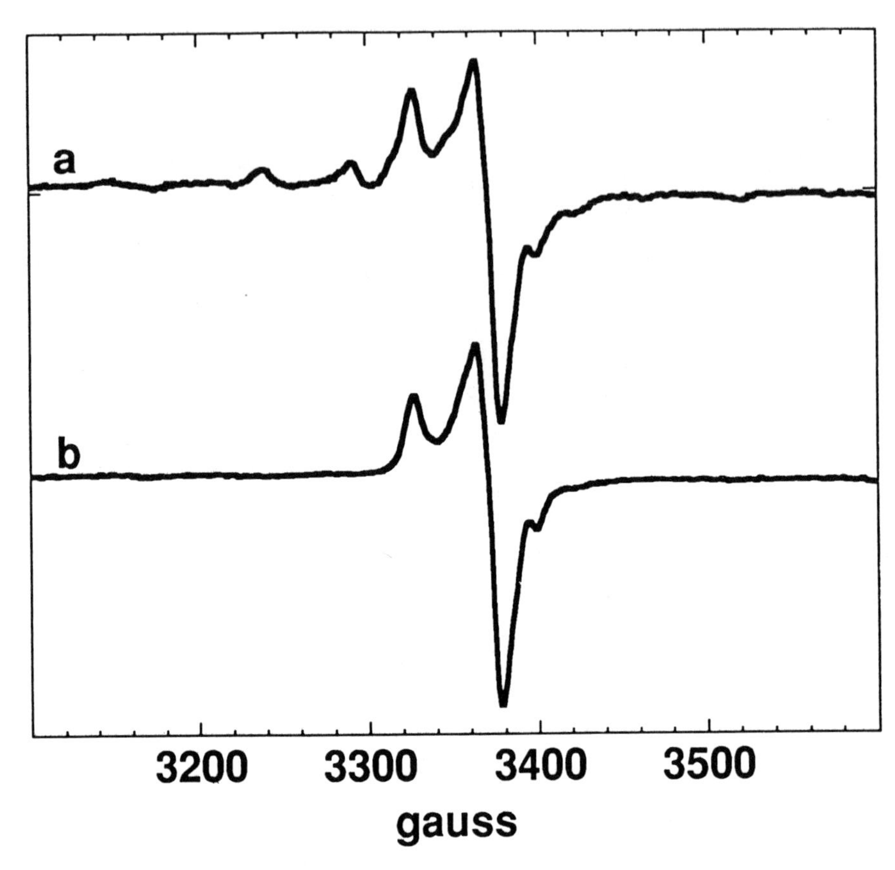

Figure 5.4. EPR spectra of peroxyl radicals in frozen solutions of purple lipoxygenase. *a*) Prepared using LA and 25% ^{17}O-enriched dioxygen. *b*) Prepared using LA and dioxygen.

This demonstrates that the alkyl radical is a [9,10,11]-allyl radical derived from 13(*S*)-HPOD.

Interestingly, there is no evidence for unresolved hyperfine splitting by the protons at C-12 or C-13 in this radical. The lack of coupling by the proton at C-13 tells us that there is no 12,13–double bond conjugated to the unpaired spin of the allyl radical. This can be so if the 12,13 bond is a single bond in the radical, or if there is a rotation about the 11,12–single bond sufficient to eliminate the overlap of the 12,13 and 9,10,11 π-orbitals. The latter is ruled out because a rotation about the 11,12 bond sufficiently large to eliminate the π-orbital overlap should bring the proton at C-12 into position for significant coupling to the allyl radical. Failure to observe splitting from the C-12 proton argues against rotation of the 11,12 bond in the species we observe.

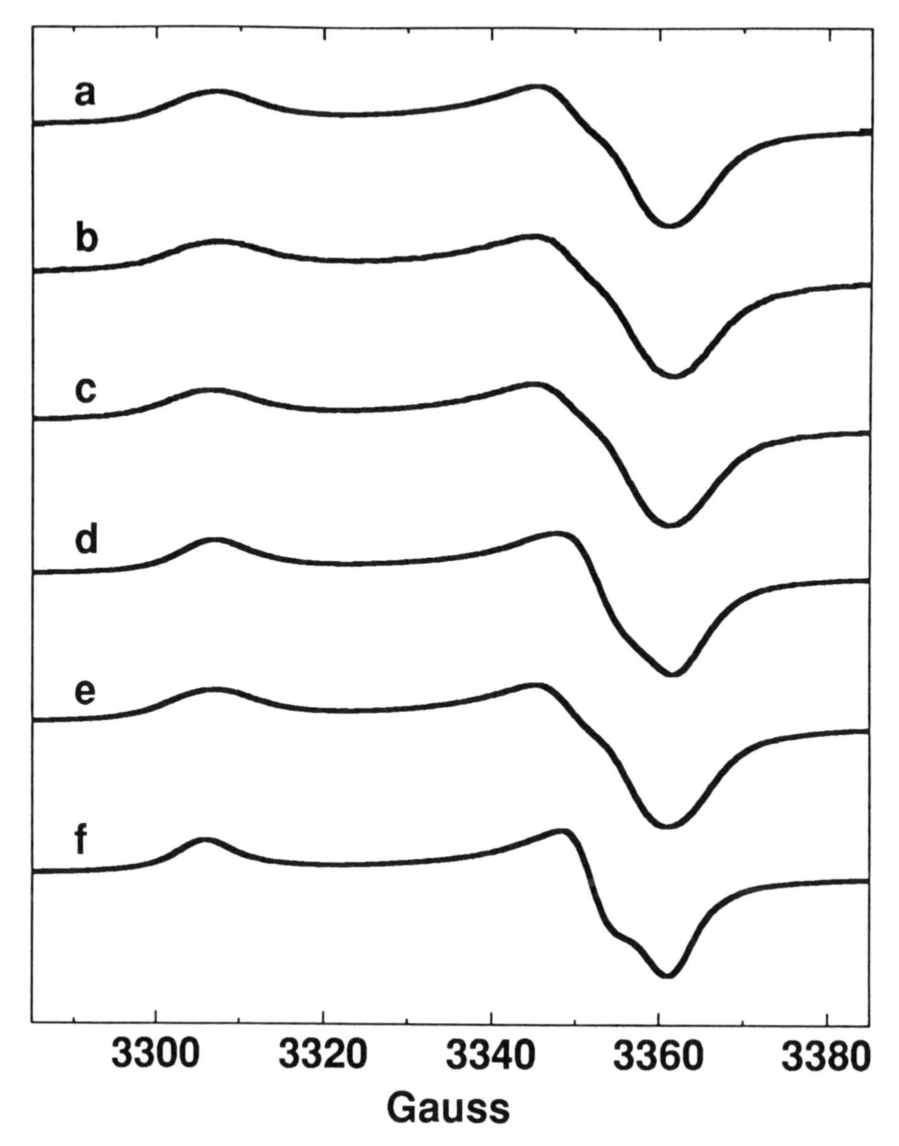

Figure 5.5. EPR spectra of peroxyl radicals in frozen solutions of purple lipoxygenase. Prepared using isotopically labeled LA and dioxygen. *a)* Natural abundance LA. *b)* [12,13-^{2}H]. *c)* [11,11-^{2}H]. *d)* [9,10-^{2}H]. *e)* [8,8-^{2}H]. *f)* [per-^{2}H].

The data do not allow any more specific analysis of the structure at C-12 and C-13 in the fatty acid allyl radical. In particular, attempts to observe the presence of ^{17}O gave equivocal results, even when the comparison was between radicals generated

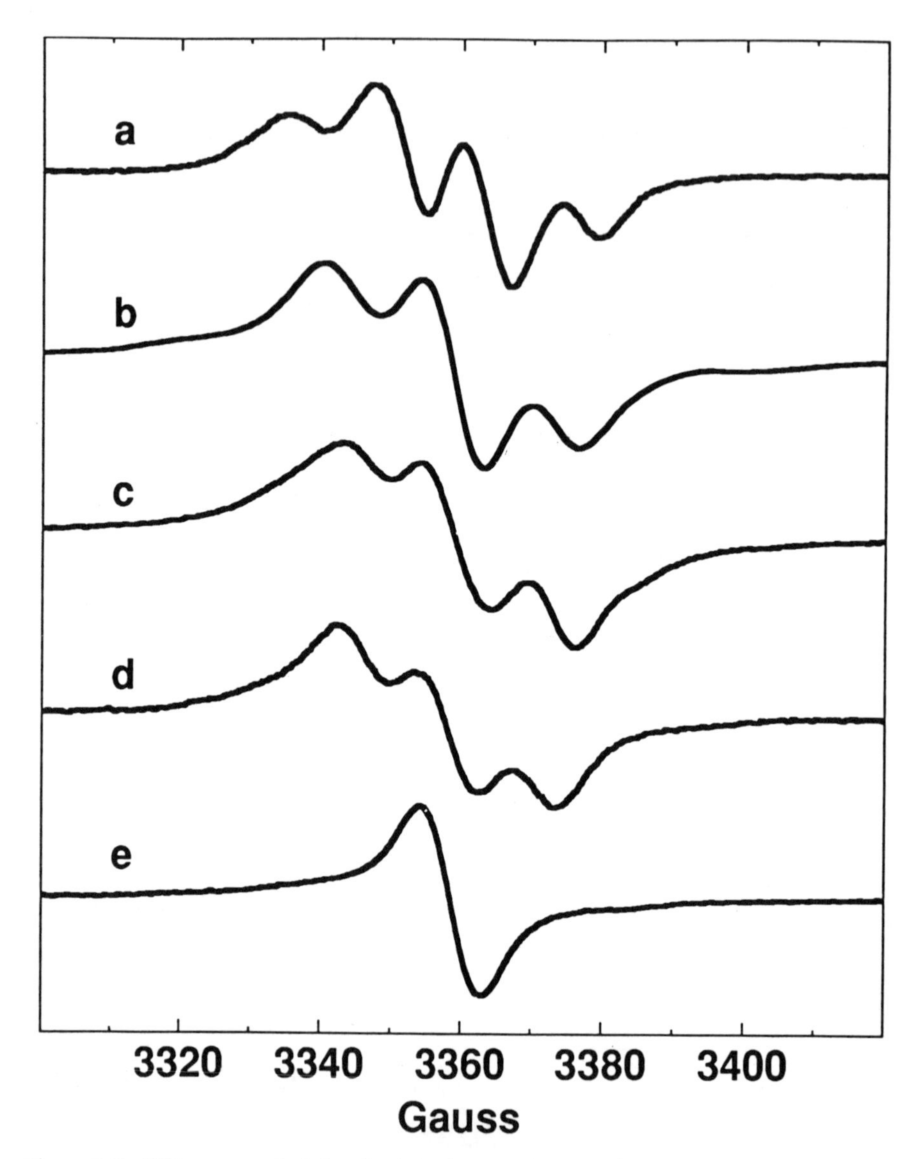

Figure 5.6. EPR spectra of allyl radicals in frozen solutions of purple lipoxygenase. Prepared using isotopically labeled 13(*S*)-HPOD under argon. *a*) Natural abundance 13(*S*)-HPOD. *b*) [11-^{2}H]. *c*) [9,10-^{2}H]. *d*) [8,8-^{2}H]. *e*) [per-^{2}H].

from [per-^{2}H]13(*S*)-HPOD (which gives a very narrow signal, Fig. 5.6) and ^{17}O-enriched [per-^{2}H]13(*S*)-HPOD (33).

If one is careful to maintain the temperature of the sample below about 5°C, the purple lipoxygenase EPR samples may be thawed and refrozen without losing their purple color. A sample that was prepared in the absence of dioxygen and showed the allyl radical EPR signal was thawed, equilibrated with dioxygen, and refrozen. The EPR spectrum of that sample revealed the fatty acid peroxyl radical. Rethawing the sample, this time under argon, followed by equilibration and refreezing led to a sample that once again showed the allyl radical. Consequently the allyl and peroxyl radicals appear to be in a dioxygen-dependent equilibrium (Fig. 5.7).

Location of the Fatty Acid Radicals in Purple Lipoxygenase

One test of the mechanistic relevance of these radicals is to determine their location. There are many reports in the literature of fatty acid radicals produced by lipoxygenase that have been trapped in solution, off the enzyme surface (23,34–40). The EPR spectra of the fatty acid allyl and peroxyl radicals produced in this chapter are very suggestive that the radicals are bound to the enzyme rather than free in solution.

The fatty acid allyl radical may show splittings from as many as five protons—the C-9 and C-11 protons are strongly coupled by a through-bond mechanism, and the C-12 and two C-8 protons are coupled via a through-space mechanism. The through-space coupling for protons α to the radical may be strong or weak; the magnitude varies as $\cos^2\theta$, where θ is the angle between the C–H bond and the π-bond that contains the unpaired spin (41). Consequently the magnitude of the coupling is a sensitive reporter of the structure of the radical. There is no observable coupling from the C-12 proton (substitution with ^2H leads to no observable change in the EPR spectrum) and only one of the protons at C-8 gives resolvable splitting. By itself this indicates that the 8,9 and 11,12 bonds of the radical each exist as one dominant rotomer.

If the radical were free in solution at the time of freezing, one would expect a distribution of rotomers, leading to species with a variety of through-space couplings from these three protons. The resulting EPR spectrum would be the sum of the spectra of all of these individual conformers, and thus would show poor resolution. To illustrate, when one generates the pentadienyl radical of LA in a neat frozen sample, where multiple conformations are present, the resulting EPR spectrum shows no resolved splittings (32). Consequently, observation of significant resolution in the EPR spectrum of the fatty acid allyl radical we observe strongly suggests that it exists in one dominant conformation in the vicinity of the unpaired spin. The 11,12 bond is in one

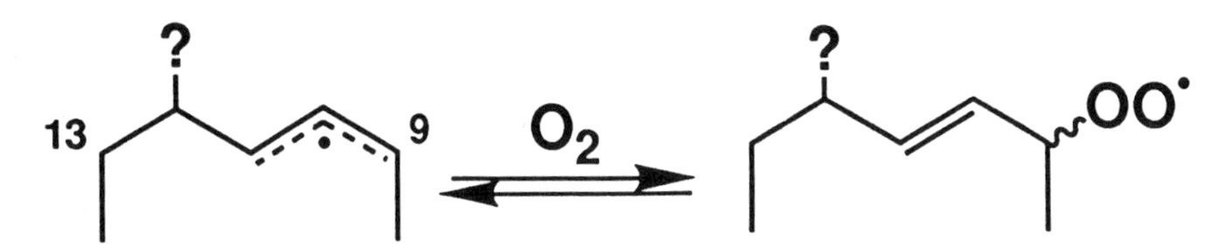

Figure 5.7. The fatty acid allyl radical is in a dioxygen-dependent equilibrium with the fatty acid peroxyl radical.

rotomer in which the C–H bond has poor overlap with the radical π-orbital (and hence small coupling), and the 8,9 bond is in a rotomer in which one C–H bond has good overlap and one C–H bond has poor overlap. The existence of one dominant conformation of the radical derives from steric constraints resulting from its being bound to the enzyme.

A similar argument holds for the fatty acid peroxyl radical, although the effect is much more subtle. The most stable structure for a secondary peroxyl radical in solution has through-space isotropic hyperfine coupling to the α-proton of 4.5 gauss (42). The effective isotropic hyperfine coupling from the 9-proton of the fatty acid peroxyl radical, obtained from simulating the EPR spectra of the natural abundance and [9,10-$^{2}H_2$]-peroxyl radicals, is 6 gauss, a substantially different value. This suggests that the fatty acid peroxyl radical is not in its most stable conformer, again probably because it experiences steric constraints by being bound to the enzyme.

Finally, both fatty acid radicals have shorter relaxation times than normally expected for organic radicals. We have not measured these relaxation times directly; however, the radicals are relatively difficult to saturate. The allyl radical has a $P_{1/2}$[1] of 1 mW, and the peroxyl radical a $P_{1/2}$ of 8 mW at 30 K (32). In contrast, the pentadienyl radical formed in neat LA has a $P_{1/2}$ of 0.008 mW, and the mixture of peroxyl radicals formed from that pentadienyl radical has a $P_{1/2}$ of 1 mW. This suggests that an additional relaxation mechanism is available to the radicals in the purple lipoxygenase samples; the most likely candidate is dipolar relaxation from the nonheme iron ion in the protein. This mechanism weakens as the distance between the iron and the radical increases, so effective relaxation requires the proximity of the two species.

Mechanistic Relevance of Fatty Acid Radicals in Purple Lipoxygenase

Is there any? It is perfectly plausible that formation of these radicals is a side reaction of the formation of purple lipoxygenase, and that their relevance to the mechanism lies only in the confirmation of the ability of the enzyme to carry out radical chemistry. We have argued in detail elsewhere that it is unlikely that these radicals are products of peroxidase chemistry (33). Nevertheless, we have recently shown that the same or a very similar allyl radical is formed stoichiometrically upon photolysis of the purple chromophore (29). The photolysis reaction is thermally reversible, suggesting that if the allyl radical were formed in the course of the dioxygenation reaction it would proceed to the ferric–fatty acid peroxide complex and thence to product.

A hypothetical mechanism for the dioxygenation of LA that includes a [9,10,11]-allyl radical is shown in Figure 5.8 (33). The steps in this mechanism are as follows.

---

[1]Microwave power at which the intensity of the EPR signal is one-half that of an unsaturated sample. $P_{1/2} \alpha T_1T_2$.

1. The fatty acid substrate binds in the active site.
2. The substrate is oxidized and a proton at C-11 is removed. This could be two steps, or a single hydrogen atom transfer. There is sufficient rotation about the 11,12 single bond that the resulting radical is delocalized only over C-9 through C-11, and the 12,13–double bond is left intact. Note that this allows interconversion of stereoisomers of the Δ-9 bond (24).
3. Dioxygen binds to the ferrous ion, quenching the triplet spin state of molecular oxygen. This is a necessary step in the dioxygenation mechanism because the reaction between triplet oxygen and the singlet state of the double bond is very slow.
4. The ferrous-dioxygen complex attacks the 12,13–double bond, yielding an intermediate analogous to either a dioxetane or perepoxide. These structures are precedented as intermediates in the reaction of singlet oxygen with double bonds (11). This is the allyl radical we observe.
5. The cyclic intermediate opens, producing the Fe^{3+}-13(*S*)-HPOD complex that is the purple chromophore.
6. The peroxide is protonated, yielding the product.

The allyl radical formed in the attack of the 12,13–double bond could react with dioxygen at C-9 instead of opening the ring. The resulting 9-peroxyl radical is the species we observe by EPR in samples prepared under dioxygen. It would not be able to carry out the ring-opening reaction; instead it would lose the dioxygen across the 12,13 bond, leaving the 9-peroxyl and Fe^{2+}. Reduction of the peroxyl by the ferrous ion would lead to 9-HPOD.

One attractive feature of this mechanism is that it predicts the observed stereochemistry of the soybean lipoxygenase reaction. The 13-HPOD formed in this mechanism is made stereospecifically, controlled by the relative position of the iron and the 12,13–double bond. In contrast, relatively equal amounts of 9(*R*)- and 9(*S*)-HPOD are made as the result of dioxygen addition to a planar radical. Nevertheless, it must be pointed out that there is no precedent for a metal–dioxygen complex adding to a double bond in this fashion. Further, an allyl radical is expected to be 7–10 kcal/mol less stable than a pentadienyl radical (43), reducing its attractiveness as a mechanistic intermediate.

The proposal in Figure 5.8 is eminently testable. It has been shown that the soybean lipoxygenase reaction exhibits a β secondary kinetic isotope effect: substitution of all of the vinylic protons of AA leads to ${}^{1}k_{H}/{}^{3}k_{H} = 1.16$, consistent with the delocalization of one or both of the double bonds involved in the reaction at or before the rate-limiting step (44). The mechanism proposed in Figure 5.8 requires that this isotope effect be fully expressed upon substitution of only the C-9 and C-10 protons, and not at all upon substitution of the C-12 and C-13 protons. In contrast, a pentadienyl radical intermediate would lead to partial expression of the secondary isotope effect upon substitution of each set of vinylic protons.

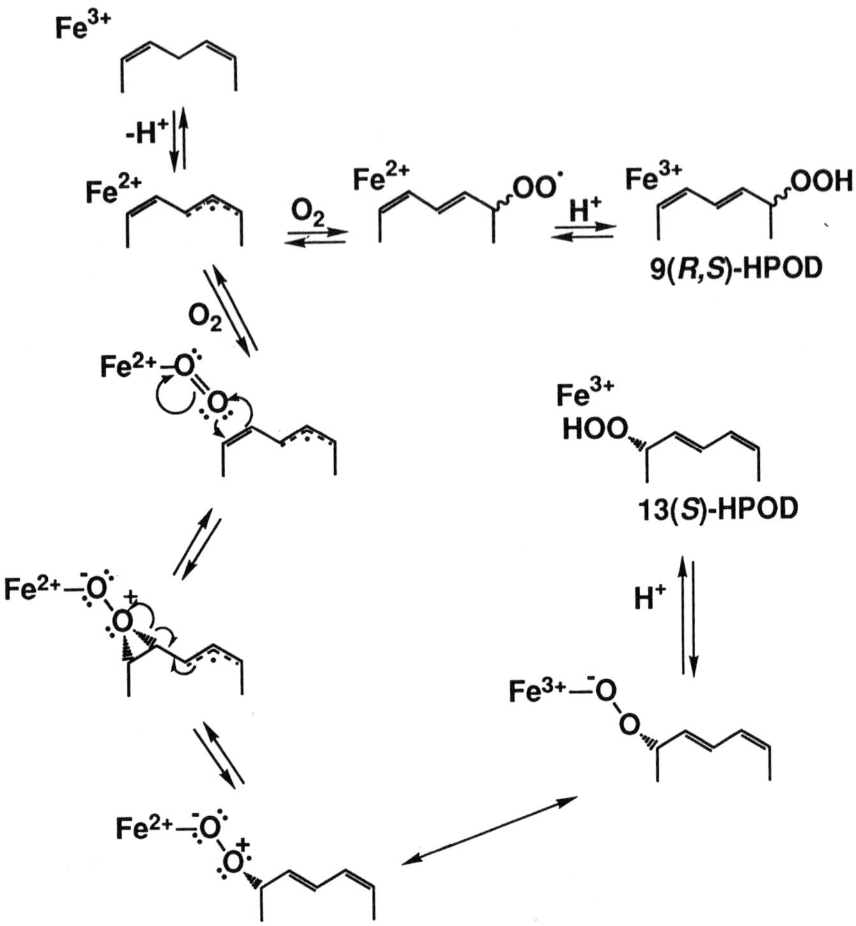

Fe3+
-H+
Fe2+
O2
Fe2+
OO•
H+
Fe3+
OOH
9(R,S)-HPOD
O2
Fe2+
Fe3+
HOO
13(S)-HPOD
Fe2+
H+
Fe3+
Fe2+

Figure 5.8. A hypothetical mechanism for the lipoxygenase reaction that includes a fatty acid allyl radical as an intermediate.

Conclusion

Frozen samples of purple lipoxygenase generated with either LA and dioxygen or 13(*S*)-HPOD show [9,10,11]-allyl or 9-peroxyl fatty acid radicals. These may result from decomposition of an Fe^{3+}-13(*S*)-HPOD complex. If they are mechanistic intermediates, they suggest a novel hypothesis for the mechanism of the lipoxygenase dioxygenase reaction. This hypothesis is directly testable by a secondary kinetic-isotope-effect experiment.

Acknowledgment

I gratefully acknowledge a long-standing, intellectually stimulating collaboration with Steven P. Seitz.

References

1. Gardner, H.W. (1991) *Biochim. Biophys. Acta 1084,* 221–239.
2. Siedow, J.N. (1991) *Annu. Rev. Plant Physiol. Plant Mol. Biol. 42,* 145–188.
3. de Groot, J.J.M.C., Veldink, G.A., Vliegenthart, J.F.G., Boldingh, J., Wever, R., and Van Gelder, B.F. (1975) *Biochim. Biophys. Acta 377,* 71–79.
4. Corey, E.J., and Nagata, R. (1987) *J. Am. Chem. Soc. 109,* 8107–8108.
5. Porter, N.A. (1986) *Acc. Chem. Res. 19,* 262–268.
6. Porter, N.A., Lehman, L.S., Weber, B.A., and Smith, K.J. (1981) *J. Am. Chem. Soc. 103,* 6447–6455.
7. Hamburg, M., (1971) *Anal. Biochem. 43,* 515–526.
8. Nikolaev, V., Reddanna, P., Whelan, J., Hildenbrandt, G., and Reddy, C.C. (1990) *Biochem. Biophys. Res. Commun. 170,* 491–496.
9. Swern, D. (1979) in *Comprehensive Organic Chemistry,* Stoddart, J.F., Pergamon Press, New York, pp. 909–940.
10. Arasasingham, R.D., Balch, A.L., Cornman, C.R., and Latos, G.L. (1989) *J. Am. Chem. Soc. 111,* 4357–4363.
11. Foote, C.S., and Clennan, E. (1995) in *Reactive Oxygen Species in Chemistry,* Foote, C.S., Valentine, J.S., Liebman, J.F., and Greenberg, A., Chapman & Hall, New York, in press.
12. Gollnik, K., and Kuhn, H.J. (1979) in *Singlet Oxygen,* Wasserman, H.H., and Murray, R.W., Academic Press, New York, pp. 287–429.
13. Wang, Z.X., Killilea, S.D., and Srivastava, D.K. (1993) *Biochemistry 32,* 1500–1509.
14. Schilstra, M.J., Veldink, G.A., and Vliegenthart, J.F.G. (1993) *Biochemistry 32,* 7686–7691.
15. Schilstra, M.J., Veldink, G.A., and Vliegenthart, J.F.G. (1994) *Biochemistry 33,* 3974–3979.
16. Scarrow, R.C., Trimitsis, M.G., Buck, C.P., Grove, G.N., Cowling, R.A., and Nelson, M.J. (1994) *Biochemistry 33,* 15023–15035.
17. Boyington, J.C., Gaffney, B.J., and Amzel, L.M. (1993) *Science (Washington, DC) 260,* 1482–1486.
18. Minor, W., Steczko, J., Bolin, J.T., Otwinowski, Z., and Axelrod, B. (1993) *Biochemistry 32,* 6320–6323.
19. van Eldik, R., Cohen, H., and Meyerstein, D. (1994) *Inorg. Chem. 33,* 1566–1568.
20. Nelson, M.J. (1988) *Biochemistry 27,* 4273–4278.
21. Egmond, M.R., Fasella, P.M., Veldink, G.A., Vliegenthart, J.F.G., and Boldingh, J. (1977) *Eur. J. Biochem. 76,* 469–479.
22. Garssen, G.J., Vliegenthart, J.F.G., and Boldingh, J. (1972) *Biochem. J. 130,* 435–442.
23. de Groot, J.J.M.C., Garssen, G.J., Vliegenthart, J.F.G., and Boldingh, J. (1973) *Biochim. Biophys. Acta 326,* 279–284.
24. Funk, M.O., Jr., Andre, J.C., and Otsuki, T. (1987) *Biochemistry 26,* 6880–6884.
25. de Groot, J.J.M.C., Garssen, G.J., Veldink, G.A., Vliegenthart, J.F.G., and Boldingh, J. (1975) *FEBS Letters 56,* 50–54.

26. Zhang, Y., Gebhard, M.S., and Solomon, E.I. (1991) *J. Am. Chem. Soc. 113,* 5162–5175.
27. Spaapen, L.J.M., Veldink, G.A., Liefkins, T.J., Vliegenthart, J.F.G., and Kay, C.M. (1979) *Biochim. Biophys. Acta 574,* 301–311.
28. Slappendel, S., Veldink, G.A., Vliegenthart, J.F.G., Åasa, R., and Malmström, B.G. (1983) *Biochim. Biophys. Acta 747,* 32–36.
29. Nelson, M.J., Chase, D.B., and Seitz, S.P. (1995) *Biochemistry 34,* 6159–6163.
30. Zang, Y., Elgren, T.E., Dong, Y.H., and Que, L. (1993) *J. Am. Chem. Soc. 115,* 811–813.
31. Nelson, M.J., and Cowling, R.A. (1990) *J. Am. Chem. Soc. 112,* 2820–2821.
32. Nelson, M.J., Seitz, S.P., and Cowling, R.A. (1990) *Biochemistry 29,* 6897–6903.
33. Nelson, M.J., Cowling, R.A., and Seitz, S.P. (1994) *Biochemistry 33,* 4966–4973.
34. Connor, H.D., Fischer, V., and Mason, R.P. (1986) *Biochem. Biophys. Res. Commun. 141,* 614–621.
35. Chamulitrat, W., and Mason, R.P. (1989) *J. Biol. Chem. 264,* 20968–20973.
36. Chamulitrat, W., and Mason, R.P. (1990) *Arch. Biochem. Biophys. 282,* 65–69.
37. Chamulitrat, W., Hughes, M.F., Eling, T.E., and Mason, R.P. (1991) *Arch. Biochem. Biophys. 290,* 153–159.
38. Iwahashi, H., Albro, P.W., McGown, S.R., Tomer, K.B., and Mason, R.P. (1991) *Arch. Biochem. Biophys. 285,* 172–180.
39. Iwahashi, H., Parker, C.E., Mason, R.P., and Tomer, K.B. (1991) *Biochem. J. 276,* 447–453.
40. Albro, P.W., Knecht, K.T., Schroeder, J.L., Corbett, J.T., Marbury, D., Collins, B.J., and Charles, J. (1992) *Chem. Biol. Interact. 82,* 73–89.
41. Morton, J.R. (1964) *Chem. Rev. 64,* 453.
42. Bennett, J.E., and Summers, R. (1973) *J. Chem. Soc., Faraday Trans 2,* 1043–1049.
43. MacInnes, I., and Walton, J.C. (1985) *J. Chem. Soc. Perkin Trans. II,* 1073–1076.
44. Wiseman, J.S. (1989) *Biochemistry 28,* 2106–2111.

Chapter 6

Isotopic Probes of the Soybean Lipoxygenase-1 Mechanism

Michael H. Glickman[a] and Judith P. Klinman[b]

[a]Department of Cell Biology, Harvard Medical School, Boston MA 02115; [b]Department of Chemistry, University of California, Berkeley CA 94720, USA.

Introduction

This chapter describes experiments designed to unravel the mechanism of soybean lipoxygenase-1 (SBL-1). First we will discuss the kinetic mechanism of lipoxygenase using isotopic probes to elucidate the sequence of events, especially in relation to the C–H bond cleavage. Then we will attempt to draw conclusions regarding the role of dioxygen in the lipoxygenase reaction.

Previous results have shown that the kinetic isotope effect measured by comparing the reaction rate of the protonated substrate, linoleic acid (LA), to that of deuterated linoleic acid (D-LA), (k_H/k_D) for the lipoxygenase reaction at room temperature is in the range of 30 for both the first- and second-order rate constants at saturated substrate concentrations (kcat and kcat/Km, respectively [1,2]). The magnitude of this observed isotope effect is much larger than expected on a single step after accounting for the differences in zero-point energies of the two isotopes in a thermally activated process (3). Because of the complexity of enzymatic reactions, isotope effects on single steps are often not fully expressed, complicating the analysis of the isotope effects measured. In order to make optimal use of kinetic isotope effects exhibited in enzyme systems, it is necessary to know the intrinsic effect on the isotopically sensitive step. As demonstrated in this chapter, the combination of kinetic studies as a function of temperature and viscosity allows an estimation of the magnitude of the intrinsic isotope effect and of the different rate-limiting steps in SBL-1. These results prove to be a powerful tool in dissecting the different steps in lipoxygenase catalysis and their relative rate limitation.

The reaction with LA, the standard substrate for SBL-1, preferentially produces 13(*S*)-hydroperoxy-9(*Z*),11(*E*), octadecadienoic acid (LOOH). The most widely accepted mechanism for lipoxygenase proposes that hydrogen is abstracted from the C-11 position on the substrate to yield a delocalized carbon-based radical that reacts with dioxygen and produces a peroxyl-radical. Electron paramagnetic resonance spectroscopy (EPR) has detected both a linoleyl- and a peroxyl-based radical, supporting the idea that C–H bond cleavage produces an organic radical that reacts with oxygen (4). The larger than unity secondary hydrogen isotope effects are also in agreement with a radical mechanism (1). At what stage oxygen enters the reaction has not yet been determined. Other questions remain: Does oxygen bind to the enzyme, and in what form? Is it activated in the absence of the fatty acid substrate? What is the precise sequence of events? Is there an order to the reaction-binding sequence or is there random binding? Or is it different altogether? The work presented in this chapter attempts to answer some of these questions.

The Nature of the Rate-Limiting Steps

Kinetic isotope effects at different reaction conditions were performed as noncompetitive experiments by comparing the absolute rates of lipoxygenase (SBL-1) with LA and D-LA under different conditions. Initial rates for SBL-1 were measured spectrophotometrically by monitoring product production at 234 nm via methods similar to those described previously (1,5). All the results presented in this chapter were performed in 0.1 M borate buffer pH 9, using SBL-1 purified from soybeans (6).

Temperature Dependence

Arrhenius plots of the kinetic isotope effect parameters, ^{D}(kcat) and ^{D}(kcat/Km), are shown in Figure 6.1*a*. The reaction of lipoxygenase on both LA and D-LA between 0 and 50°C shows an interesting temperature dependence. The primary k_H/k_D isotope effect on kcat increases from about 18 at 0°C, to a maximum of about 60, and then decreases with further increase in temperature, as expected for a single minute step (7).

The increase in ^{D}(kcat/Km) is more pronounced than for ^{D}(kcat) up to 32°C, at which point a break in behavior occurs that leads the isotope effect to only a small dependence on temperature. A similar change in behavior at 32°C is noticed for the absolute rates on kcat and kcat/Km for proteo-linoleic acid (Figs. 6.1*b*, 6.1*c*). The same figures show that even though a slight break in the behavior of kcat and kcat/Km is evident with D-LA, it is much less apparent than with H-LA. This last point is important in establishing the nature of the rate-limiting steps. Due to the extremely large k_H/k_D primary isotope effect in this reaction, the expectation is that the isotopically sensitive step with the deuterated substrate would be close to fully rate limiting under most conditions. For this reason the behavior of the reaction rate with D-LA as the substrate is monotonic with temperature.

The simplest explanation for this type of behavior is a change in the rate-limiting step. Below the transition temperature of 32°C, the isotopically sensitive step, in this case the C–H bond cleavage, is not fully rate limiting and therefore the isotope effect is not fully expressed. Above 32°C, hydrogen abstraction limits the overall reaction rate, therefore the observed behavior of the isotope effect represents the behavior of the small, isotopically sensitive step.

Viscosity Dependence

If the C–H bond cleavage is only partially limiting the rate of reaction at room temperature, what are other potential rate-limiting steps? Soybean lipoxygenase-1 is a fast enzyme with a second-order rate constant of about 3 x 10^7 $M^{-1}sec^{-1}$ at room tempera-

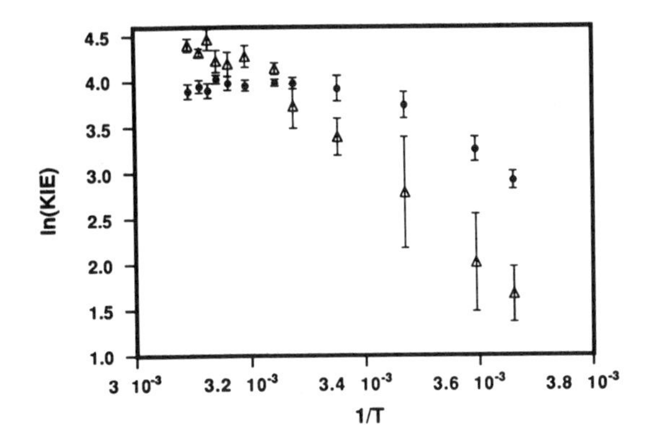

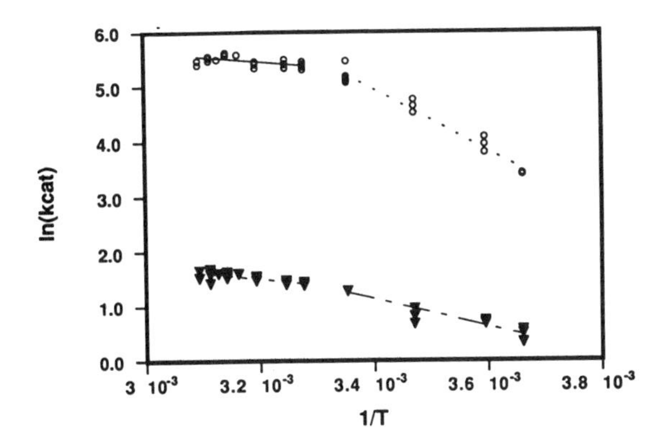

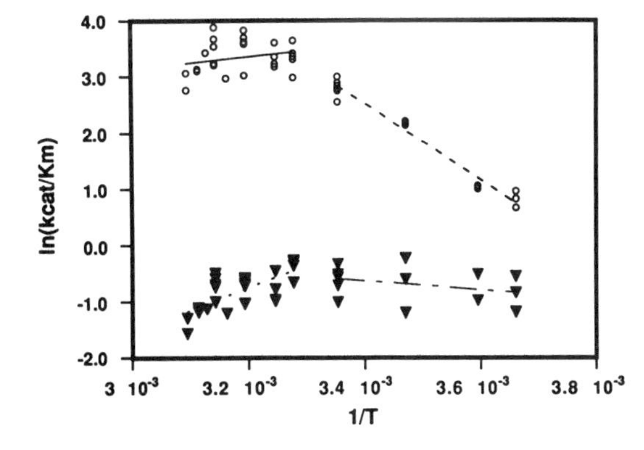

Figure 6.1. Temperature dependence of kinetic parameters. *a)* The observed k_H/k_D isotope effect; natural log of ^{D}(kcat/Km) (Δ), and of ^{D}(kcat) (•) in the temperature range 0–50°C, pH 9. *b)* Natural log of kcat for H-LA (o), and D-LA (▼). *c)* Natural log of kcat/Km for H-LA (o) and D-LA (▼).

ture (results not shown). This rate is nearing the upper limit for a bimolecular reaction, due to the diffusional encounter of the two substrates (8).

Viscosity studies can be used to test whether substrate binding to enzyme is partially rate limiting (9). Figure 6.2 shows the reciprocal of the relative rate constant of lipoxygenase, (kcat/Km)°/(kcat/Km), plotted against the relative viscosity of the reaction solution, h/h°, attained by increasing amounts of dissolved glucose. A bimolecular reaction that is fully diffusion controlled would have a slope of unity in such a plot (9).

From Figure 6.2, it can be seen that lipoxygenase is close to 50% diffusion controlled at room temperature as its slope is 0.48. When reacted with D-LA however, the rate of lipoxygenase catalysis is much slower and insensitive to viscosity (Fig. 6.2). The results with D-LA serve as a double control; lipoxygenase is not inhibited by glucose, since the reaction rate with D-LA is unaffected by addition of glucose to the buffered solution, and the C–D bond cleavage can be assumed to be fully rate limiting. Since these are identical substrates, their different rates should be due solely to the differences in their chemical step. The absence of viscosity effects on kcat/Km for the deuterated substrate (Fig. 6.2), or on kcat for either substrate (results not shown), demonstrates that the viscosity effect on kcat/Km for proteo-linoleic acid arises from substrate binding to enzyme. The step sensitive to viscosity is therefore distinct from the step that is sensitive to isotopic substitution. Since at room temperature and pH 9,

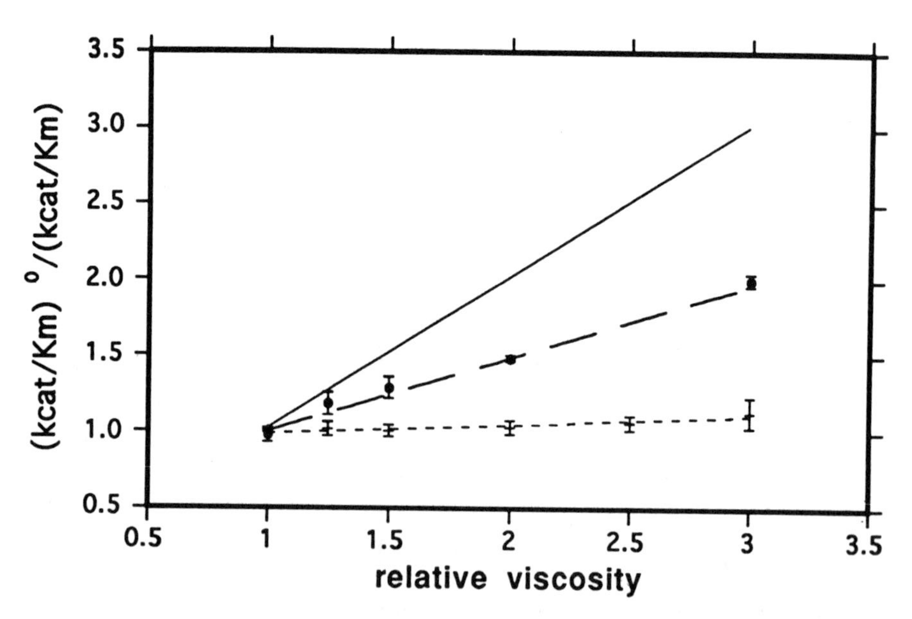

Figure 6.2. Effect of relative viscosity (h/h°) on normalized values of kcat/Km at 20°C, pH 9. Slope of the line is 0.48 ± 0.03 for H-LA (•), and 0.05 ± 0.02 for D-LA (+). The thin line is the theoretical behavior (slope = 1) of a fully diffusion-controlled bimolecular reaction. The viscosity of the reaction solution was obtained by increasing the amount of dissolved glucose, relative viscosity at the specific temperature was measured using a viscosimeter.

the reaction of SBL-1 is close to 50% diffusion controlled, the abnormally large kinetic isotope effect (KIE) of about 30 seen on kcat/Km under these conditions (Fig. 6.1*a*, and published results [1,2]) can not be fully expressed. This conclusion is in agreement with the temperature dependence in Figure 6.1.

Viscosity Studies at 37 and 5°C

From Figure 6.1, it can be concluded that rate limitation by the C–H bond cleavage step increases with temperature, so that only above 32°C is it fully rate limiting. The observation described in Figure 6.2—substrate binding is partially rate limiting at room temperature—is in agreement with the preceding statement. Consequently substrate binding should be less rate limiting at higher temperatures than at room temperature. At 37°C, the reaction of lipoxygenase with H-LA is basically insensitive to viscosity as can be seen in Figure 6.3. While the reaction is 48% diffusion controlled at 20°C (Fig. 6.2), it is virtually insensitive to viscosity at 37°C (Fig. 6.3). The assumption that the C–H bond cleavage is more rate limiting above 32°C is therefore corroborated by the concomitant decrease in reaction sensitivity to the viscosity of solution.

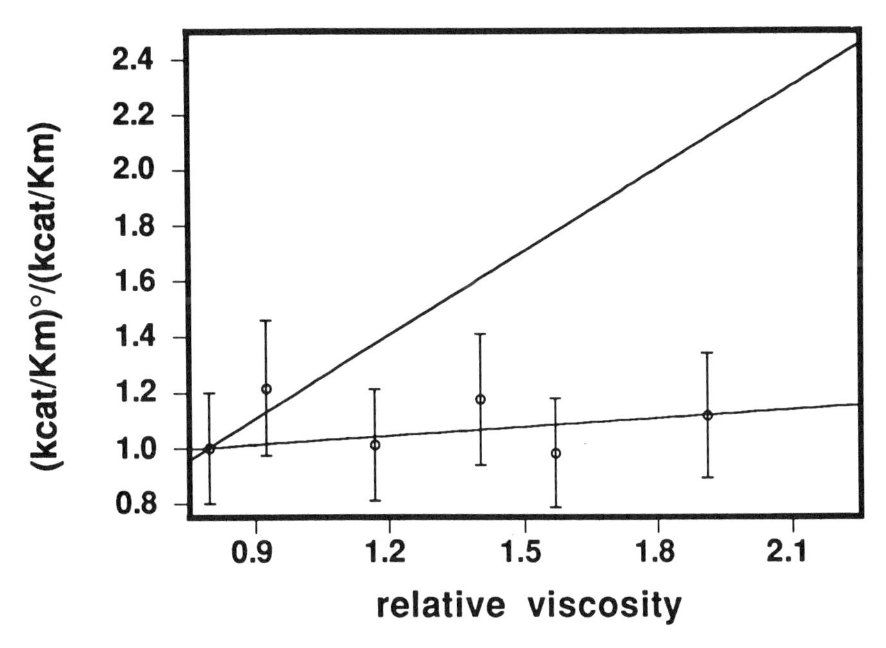

Figure 6.3. Effect of viscosity on normalized values of kcat/Km for the reaction with H-LA at 37°C, pH 9. Within experimental error, the reaction of lipoxygenase with H-LA is insensitive to the viscosity of the solution. The thin line is the theoretical behavior of a fully diffusion-controlled reaction.

We could predict that the reaction rate should be more diffusion controlled at 5°C, and therefore show a greater viscosity sensitivity, since the kinetic isotope effect at this temperature is small. However, the reaction is 24% diffusion controlled at 5°C, while it was 48% diffusion controlled at 20°C (Figs. 6.4 and 6.2, respectively).

The reaction of lipoxygenase is most sensitive to viscosity at room temperature, with substrate binding becoming less rate limiting as the temperature is either increased or decreased. The conclusion is that there must be another step that becomes increasingly rate limiting at lower temperatures. The reaction is insensitive to viscosity when D-LA substrate is used (results not shown), once again indicating that the step exhibiting the kinetic isotope effect (hydrogen abstraction) is distinct from that exhibiting the viscosity effect (substrate binding).

Solvent-Isotope Effects

Another probe of kinetic steps is solvent-isotope effects (SIE), where reaction rates in aqueous solutions are compared with those in which H_2O is replaced with D_2O. At 25°C a significant SIE is observed for kcat/Km, and a smaller SIE is also noticed for kcat (Table 6.1); values of both increase at 5°C. The magnitude of the observed SIE is rather large and could arise from a solvent-dependent hydrogen transfer, or from

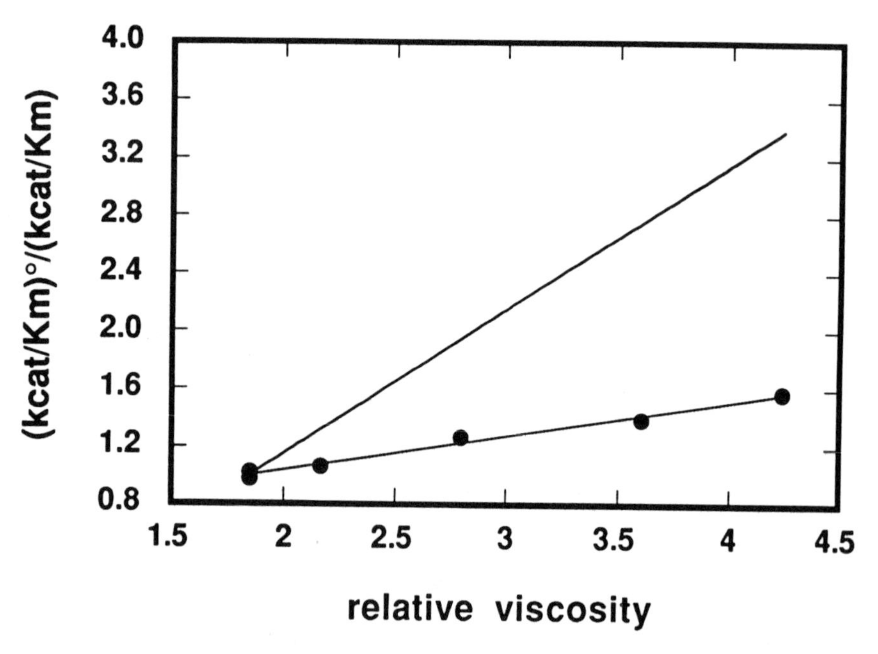

Figure 6.4. Effect of viscosity on the reaction with H-LA at 5°C, pH 9. The slope of the line is 0.24 (±0.04). The unmarked line is the theoretical behavior of a fully diffusion-controlled reaction.

multiple hydrogen-bond rearrangements—possibly an enzymatic conformational change upon substrate binding or displacement of water by substrate in the active site. Evidence for bound water in the active site has been published (10).

The fact that the value of the SIE is greater at lower temperatures, while the values of the KIE and viscosity effects decrease at lower temperatures, point to the conclusion that the SIE is on a separate mechanistic step. This point is further emphasized since the SIE at 25°C is obliterated when D-LA is used as substrate, showing that the C–H bond cleavage and steps sensitive to solvent deuteration are separate. To conclude, a step exhibiting an SIE is partially rate limiting the reaction at room temperature, and becomes more so at 5°C.

The Nature of the Rate-Limiting Steps

The combination of the changes of the rate and KIE with temperature, viscosity, and solvent clearly characterize three distinct steps: binding of substrate, a solvent-dependent step, and C–H bond cleavage. All three are partially expressed at room temperature, yet none is fully rate limiting. At higher temperatures, hydrogen abstraction increasingly limits the reaction, while at lower temperatures the solvent-sensitive steps dominate. Focusing on behavior above 32°C, where C–H bond cleavage is rate limiting, could be useful in understanding the nature of this step. A schematic presentation of the relative rate limitation of these three steps as a function of temperature is displayed in Figure 6.5.

At room temperature lipoxygenase is close to the diffusion-controlled rate for an enzymatic reaction, yet no one step is fully rate limiting. This type of behavior has been termed as a "perfect enzyme," since there is no evolutionary pressure to perfect the efficiency of single steps more than the rate dictated by the diffusional encounter of enzyme and substrate (11,12).

Oxygen Chemistry

A major question regarding the utilization of molecular oxygen in chemical and biological reactions is that of oxygen activation. While oxidation by dioxygen is often thermodynamically favorable, molecular oxygen is kinetically unreactive with most

TABLE 6.1 Solvent-Isotope Effect for LA and D-LA

	H-LA[a]		D-LA[a]	
Temp. (°C)	D2O(kcat)	D2O(kcat/Km)	D2O(kcat)	D2O(kcat/Km)
25	1.2(±0.1)[b]	2.5(±0.7)	0.9(±0.1)	1.2(±0.1)
5	2.2(±0.2)	3.7(±0.9)	1.12(±0.06)	1.5(±0.2)

[a]Ratio of rates when comparing reactions in H_2O and D_2O for either substrate.
[b]Values are solvent-isotope effects (unitless); numbers in parentheses represent errors as determined from replicate determinations.

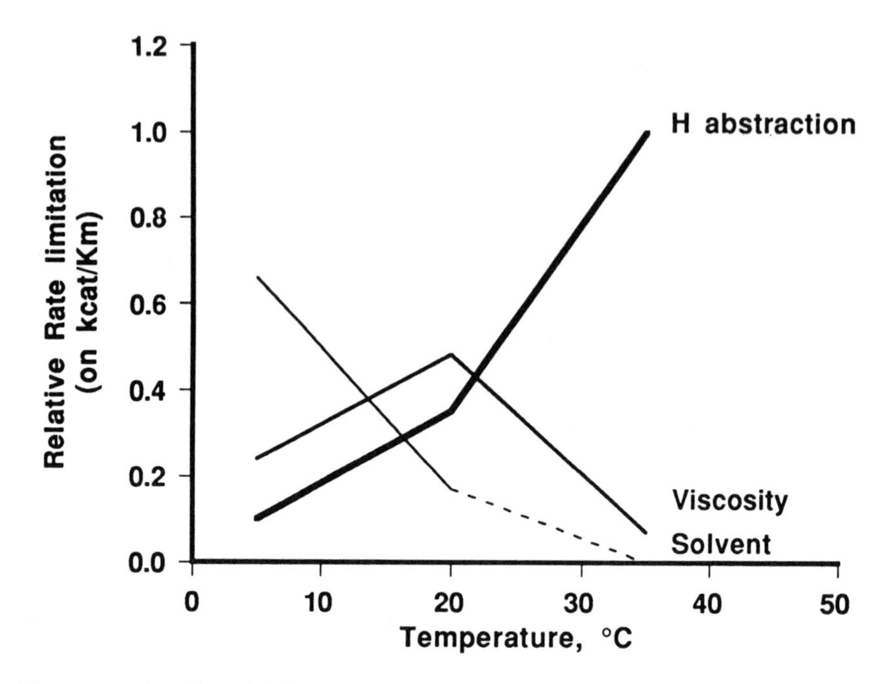

Figure 6.5. The effect of different steps on the relative rate limitation. Schematic presentation of the importance of C–H bond cleavage (thick line), steps affected by viscosity (substrate binding), and a step sensitive to replacing H_2O with D_2O, as function of temperature. At room temperature soybean lipoxygenase is a "perfect enzyme"; almost diffusion controlled, with no one step being fully rate limiting. At lower temperatures, a solvent-dependent step is dominant, whereas at higher temperatures C–H bond cleavage limits the reaction rate.

biological compounds. The limitations on the spontaneous reaction of atmospheric oxygen can be explained by spin conservation rules. O_2 is a diradical with two electrons in their highest energetic states and can exist in two electronic configurations, singlet (electron spins paired), and triplet (unpaired); however, all molecular oxygen in the atmosphere is comprised of triplet oxygen. Covalent diatomic bonds are singlet and can react only with singlet oxygen.

For atmospheric triplet dioxygen to react with a covalent bond it usually has to be "activated" by changing the electronic configuration to singlet. This is the mechanism by which most oxygen-utilizing enzymes catalyze reactions of molecular oxygen with biological substrates; a major role for enzymes utilizing dioxygen is activating oxygen to overcome the barrier described by spin conservation rules (13,14).

Another pathway for inducing oxygen chemistry is by first homolytically cleaving a covalent bond. The organic radicals thus produced are doublets, as are their products after the reaction with molecular oxygen, hence spin conservation is maintained. Radicals are therefore reactive species with triplet dioxygen.

Order of Substrate Binding

The primary task in unraveling the path by which dioxygen chemistry takes place is to differentiate between a mechanism that involves a ternary enzyme–substrate–O_2 complex necessary for catalysis, and one where the hydrogen abstraction occurs prior to the binding of dioxygen. In the case of a ternary complex we can further distinguish between a truly random mechanism where either substrate can bind to a free enzyme (Scheme 6.1*a*), an ordered binding of dioxygen followed by binding of LA (Scheme 6.1*b*), or an ordered binding of LA prior to O_2 (Scheme 6.1*c*). A distinct possibility, where O_2 binds after C–H bond cleavage, is shown in Scheme 6.1*d*.

Kinetic isotope effects measured at varying concentrations of either substrate can yield information necessary to exclude one or more of the possibilities. Examples and derivations of isotope effect expressions for reactions involving ternary complexes have been published by Klinman and co-workers (15,16). We have adapted these derivations to lipoxygenase, and summarized them in Table 6.2. In Table 6.3 we have

A

$$E+LA \underset{k_2}{\overset{k_1[LA]}{\rightleftharpoons}} ELA \underset{k_4}{\overset{k_3[O_2]}{\rightleftharpoons}} ELAO_2 \overset{k_5}{\dashrightarrow} \dashrightarrow E+P$$

$$E+O_2 \underset{k'_2}{\overset{k'_1[O_2]}{\rightleftharpoons}} EO_2 \underset{k'_4}{\overset{k'_3[LA]}{\rightleftharpoons}} ELAO_2$$

B

$$E+O_2 \underset{k_2}{\overset{k_1[O_2]}{\rightleftharpoons}} EO_2 \underset{k_4}{\overset{k_3[LA]}{\rightleftharpoons}} ELAO_2 \overset{k_5}{\dashrightarrow} \dashrightarrow E+P$$

C

$$E+LA \underset{k_2}{\overset{k_1[LA]}{\rightleftharpoons}} ELA \underset{k_4}{\overset{k_3[O_2]}{\rightleftharpoons}} ELAO_2 \overset{k_5}{\dashrightarrow} \dashrightarrow E+P$$

D

$$E+LA \rightleftharpoons ELA \rightleftharpoons EL\cdot \overset{[O_2]}{\rightleftharpoons} ELOO\cdot \dashrightarrow$$

Scheme 6.1. Possible mechanisms for the lipoxygenase reaction. Scheme 6.1*a* represents a random binding mechanism where either substrate can bind to free enzyme. Scheme 6.1*b* is an ordered binding mechanism where only dioxygen can bind to free enzyme. Scheme 6.1*c*, describes ordered binding of LA as a prerequisite for oxygen binding. A mechanism that involves hydrogen abstraction from the substrate prior to oxygen binding to the enzyme is shown in Scheme 6.1*d*.

derived the isotope-effect equations for the unique mechanism described in Scheme 6.1*d* in an effort to predict the behavior of a mechanism where O_2 binds after C–H bond cleavage.

Noncompetitive isotope effects were determined by directly comparing the reaction rates of SBL-1 with LA to SBL-1 with D-LA. Rates were determined by measuring

TABLE 6.2 Isotope Effect Expressions for Reactions Described in Scheme 6.1a, b, and c

		Random Binding	Ordered Binding	
KIE	Limit	Scheme 1*a*	Scheme 1*b* (O_2 first)	Scheme 1*c* (LA first)
$^D(kcat)$	High LA High O_2	$\frac{^Dk_5 + k_5/k_6}{1 + k_5/k_6}$	$\frac{^Dk_5 + k_5/k_6}{1 + k_5/k_6}$	$\frac{^Dk_5 + k_5/k_6}{1 + k_5/k_6}$
$^D(kcat/Km)_{LA}$	Low LA High O_2	$\frac{^Dk_5 + k_5/k'_4}{1 + k_5/k'_4}$	$\frac{^Dk_5 + k_5/k_4}{1 + k_5/k_4}$	1
$^D(kcat/Km)_{O_2}$	High LA Low O_2	$\frac{^Dk_5 + k_5/k_4}{1 + k_5/k_4}$	1	$\frac{^Dk_5 + k_5/k_4}{1 + k_5/k_4}$
$^D(kcat/Km)_{LA[O_2 \to 0]}$ or $^D(kcat/Km)_{O_2[LA \to 0]}$	Low LA Low O_2	$\frac{^Dk_5 + k_5/(k'_4 + k_4)}{1 + k_5/(k'_4 + k_4)}$	$\frac{^Dk_5 + k_5/k_4}{1 + k_5/k_4}$	$\frac{^Dk_5 + k_5/k_4}{1 + k_5/k_4}$

Expressions were derived for the limiting cases at extreme high and low concentrations of each substrate, for the three different possibilities of reactions involving a ternary complex, [enzyme-LA-O_2], occurring before the C–H bond cleavage.

TABLE 6.3 Isotope Effect Expressions for the Mechanism Described in Scheme 6.1*d*

KIE	Limit	Expression	Experimental Value
$^D(kcat)$	High LA High O_2	$\frac{^Dk_2 + k_2/k_2 + (k_2/k_5)(k_{-4}/k_4) + k_2/k_4}{1 + k_2/k_5 + (k_2/k_5)(k_{-4}/k_4) + k_2/k_4}$	62±6 (Fig. 6.8*b*, circles) 64±5 (Fig. 6.8*a*, circles)
$^D(kcat/Km)_{LA}$	Low LA HighO_2	$\frac{^Dk_2 + k_2/k_{-1}}{1 + k_2/k_{-1}}$	59±18 (Fig. 6.8*b*, triangles) ~60 (Fig. 6.8*a*, circles)
$^D(kcatKm)_{ox}$	High LA Low O_2	$\frac{^Dk_{eq2} + k_2/k_2}{1 + k_2/k_{-2}}$	<5 (Fig. 6.8*a*, triangles) <2.5 (Fig. 6.8*b*, circles)
$^D(kcat/Km)_{ox,LA}$	Low LA Low O_2	$^Dk_{eq2}$	<4 (Fig. 6.8*a*, triangles)

Expressions were derived for limiting cases of concentrations of each substrate and compared to experimental values obtained from Figure 6.8.

oxygen consumption on a Clark oxygen monitor at different dioxygen concentrations. Reaction solutions were presaturated with different mixtures of N_2 and O_2 to obtain a specific dioxygen concentration prior to injection of SBL-1 via a gastight syringe to initiate reaction. The KIE was calculated from the ratio of the kinetic parameters for each substrate thus determined. First, we will present the experimental results, then we will discuss how a kinetic model for lipoxygenase can be obtained by comparing the primary k_H/k_D isotope effects at varying concentrations of dioxygen and LA.

The kinetic parameters of lipoxygenase were deduced by measuring rates as a function of LA or D-LA at a given concentration of O_2, as plotted in Figure 6.6. These experiments were replotted to obtain rates as a function of O_2 at different concentrations of LA or D-LA (Fig. 6.7). In Figure 6.6*a* it is evident that when LA is the substrate, both the maximal velocity at saturating substrate (kcat) and the second-order rate at low LA concentrations ($kcat/Km_{(LA)}$) decrease with the concentration of the second substrate, O_2. In contrast, when substituting LA with D-LA (Fig. 6.6*b*), the kinetic parameters are unaffected by dioxygen concentrations within the attainable experimental range (258 to 5 μM of O_2).

This characteristic is even clearer in Figure 6.7; both kcat(app), and kcat/Km(O_2) decrease as the LA concentration decreases (Fig. 6.7*a*), however when D-LA is the substrate, the reaction is saturated with O_2 under all experimental conditions (Fig. 6.7*b*). Since the observed enzymatic rate is independent of all oxygen concentrations measured when D-LA is the second substrate (Fig. 6.7*b*), we can conclude that the Km for O_2 is below the lowest oxygen concentration attainable by the experimental procedure, 5 μM. The upper limit for the value of Km(O_2) is therefore 2.5 μM, and Km is possibly much smaller. This feature makes it impossible to obtain an accurate value for the isotope effect ^{D}kcat/Km(O_2). However, an upper limit for the isotope effect ^{D}kcat/Km(O_2) can be determined from an estimate of the upper limit for Km(O_2) with deuterated substrate from the experimental results. It is important to note that, as evident from Figures 6.6 and 6.7, an accurate measurement of Km(O_2) has only been obtained for LA, and not D-LA, therefore the isotope effect $^{D}kcat/Km(O_2)_{app}$ overestimates the second-order rate constant kcat/Km(O_2) when D-LA is used as the second substrate and the true behavior of the isotope effects at very low [O_2] is not known.

Regretfully, kinetic measurements can never prove a mechanism, only rule specific ones out. In addition we are limited, since we do not have accurate measurements of kcat/Km at low O_2 concentrations. We will therefore use these results to narrow down the mechanistic possibilities. The simplest case is that of ordered binding where LA binds to the free enzyme first, as a requirement for O_2 binding (Scheme 6.1*c*); the prediction is that the isotope effect ^{D}kcat/Km(LA) will be unity under conditions where O_2 is saturating the enzyme (Table 6.2). Since ^{D}kcat/Km(LA) is in the range of 60 at high O_2 concentrations (Fig. 6.8*b*), this scenario can definitely be ruled out as a mechanism for lipoxygenase.

The prediction for an ordered binding sequence of dioxygen prior to LA (Scheme 6.1*b*) is that the kcat/Km isotope effect will be independent of the first substrate

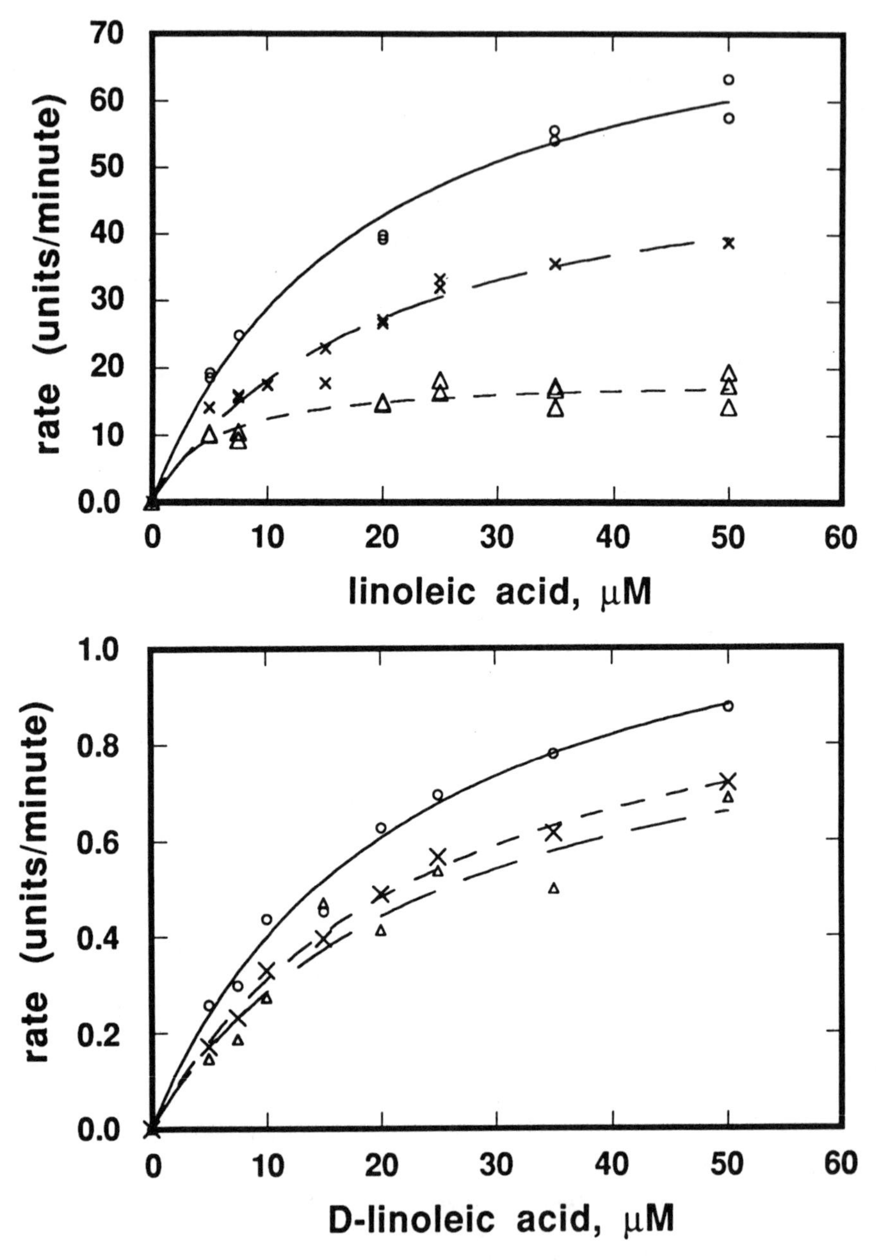

Figure 6.6. Reaction rates of lipoxygenase as a function of LA concentration (Panel A) or D-LA (Panel B). Rates of oxygen consumption were determined using a Clark Oxygen Electrode, and are defined as μM dioxygen consumed/min for every mg lipoxygenase injected to reaction chamber. Examples are shown for 258 μM O_2 (o), 77 μM O_2 (x), and 13 μM O_2 (Δ). Linoleic acid concentrations are in μM.

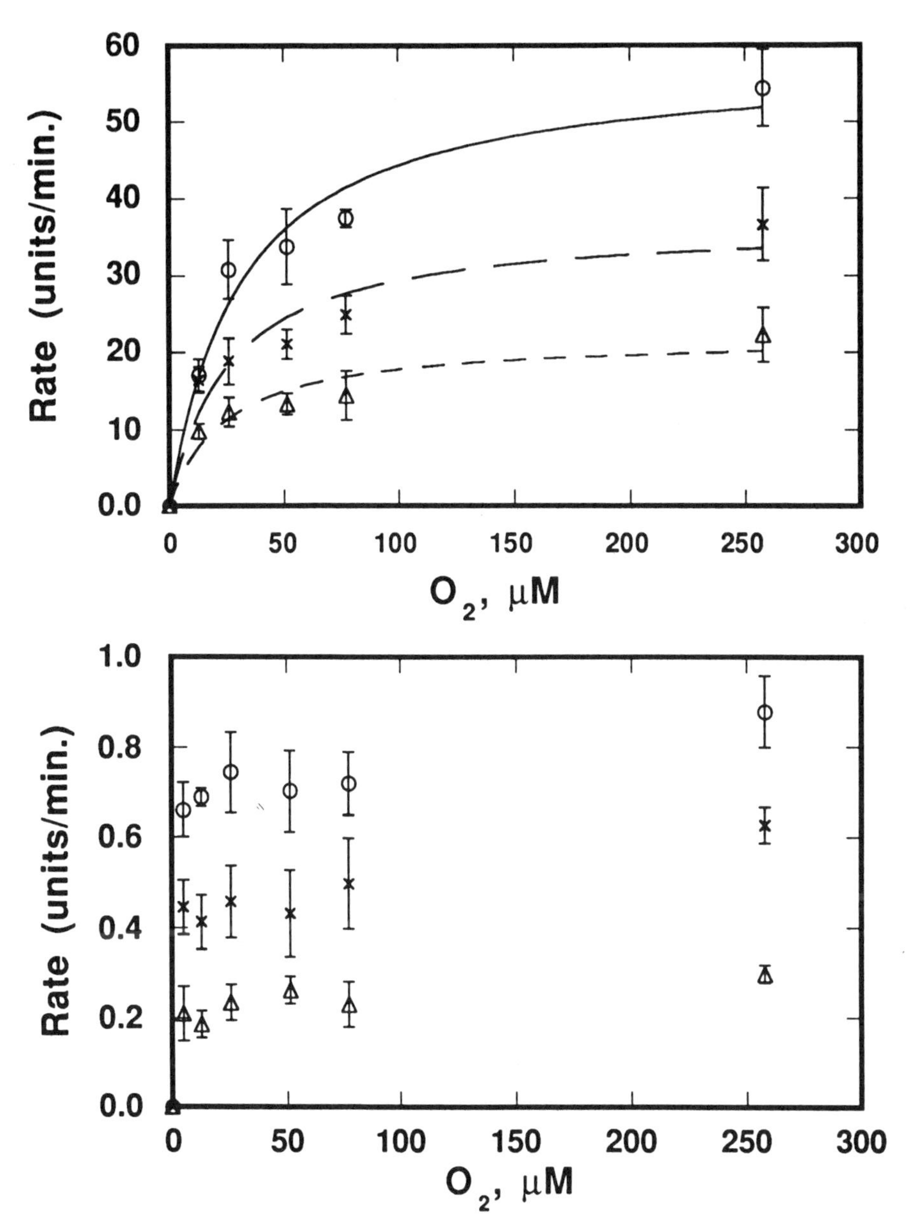

Figure 6.7. Reaction rates of lipoxygenase as a function of O_2 with LA as substrate (Panel A), or D-LA (Panel B). Rates of oxygen consumption were determined using a Clark Oxygen Electrode, and are defined as µM dioxygen consumed/min for every mg lipoxygenase injected to reaction chamber. Examples in Panel A are shown at 50 µM LA (o), 20 µM LA (x), and 5 µM LA (Δ). In Panel B rates are at 50 µM D-LA (o), 20 µM D-LA (x), and 5 µM D-LA (Δ).

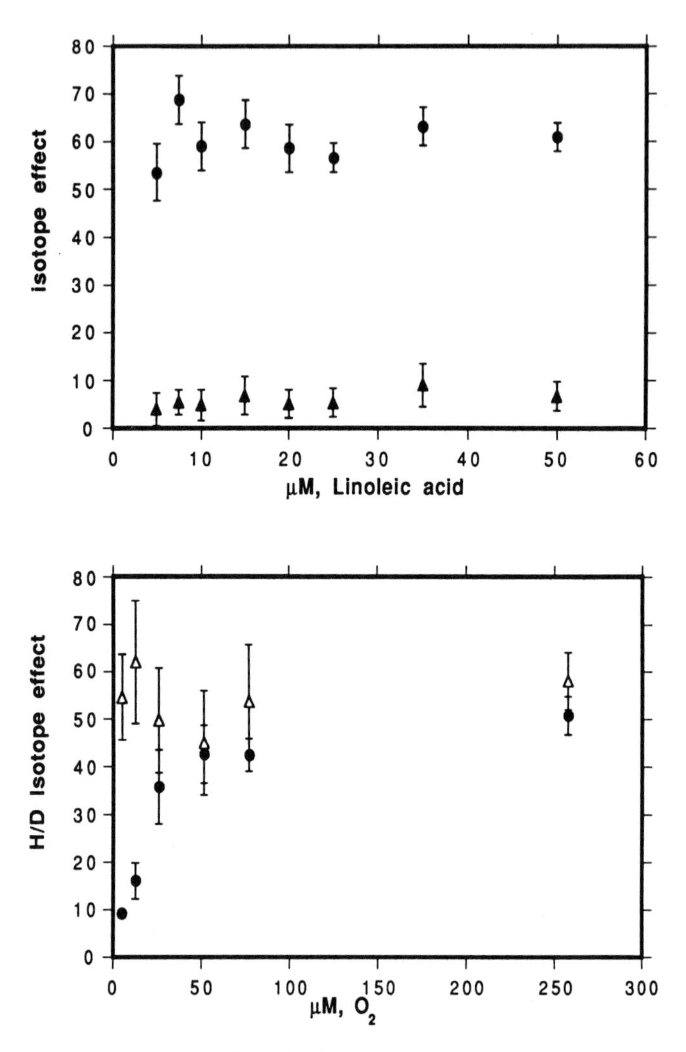

Figure 6.8. *a*) Kinetic isotope effects for the lipoxygenase reaction as a function of substrate concentration. The isotope effects $^{D}kcat_{app}$ (•) and $^{D}kcat/Km(O_2)_{app}$ (Δ) are displayed as a function of LA concentration. From this plot, a number of kinetic parameters can be extracted: $^{D}kcat = 64 \pm 5$ at saturating LA and O_2 concentrations (•); $^{D}kcat/Km(LA) = 60$ at saturating O_2 and low LA concentrations (•); $^{D}kcat/Km(O_2) < 5$ at low O_2 concentrations, (Δ). It is important to note that $^{D}kcat/Km(O_2)_{app}$ (Δ) is an upper limit and not a true value representing the isotope at kcat/Km conditions for O_2. *b*) Kinetic isotope effects for the lipoxygenase reaction as a function of substrate concentration. The isotope effects $^{D}kcat_{app}$ (•) and $^{D}kcat/Km(LA)_{app}$ (Δ) are plotted as a function of O_2 concentration. A number of kinetic parameters are extracted: $^{D}kcat = 62 \pm 6$ at saturating O_2 and LA concentrations (•); $^{D}kcat/Km(LA) = 59 \pm 18$ at saturating O_2 and low LA concentrations (Δ); $^{D}kcat/Km(O_2) < 2.5$ at low O_2 concentrations, (•).

concentration (Table 6.2). In the case of lipoxygenase, this means that ^{D}kcat/Km(LA) should be independent of O_2 concentration. Within the attainable experimental range of O_2 concentrations, this statement is valid (Fig. 6.8*b*). In addition, since we did not obtain accurate measurements of ^{D}kcat/Km(O_2) we cannot accurately determine whether the value of ^{D}kcat/Km(O_2) at high LA concentrations is unity. Therefore, the conclusion is that these experiments alone cannot rule out the possibility that O_2 binds to free enzyme as a prerequisite for LA binding. However, another characteristic of an ordered binding mechanism (as evident from examination of the expressions in Table 6.2) is that ^{D}kcat/Km(O_2) is expected to increase as the concentration of LA is decreased, until it equals the value of ^{D}kcat/Km(LA). From the upper limit obtained for ^{D}kcat/Km(O_2) (Fig. 6.8*b*), there is no evidence that it increases to approximate 60, the value of measured for ^{D}kcat/Km(LA). While we cannot rule out such a mechanism, we think it unlikely.

Finally, if we consider the general case of a random binding mechanism (Scheme 6.1*a*), the isotope effect on kcat/Km should alternate between two finite values for each of the two substrates. Considering a mechanism such as described in Scheme 6.1*a*, the isotope effect at low LA concentration or low O_2 concentration can be expressed as follows (similar to derivations in [15]):

$$^{D}(V/K)_{LA(O_2 \rightarrow \infty)} = \frac{^{D}k_5 + k_5/k_4'}{1 + k_5/k_4'} \qquad \text{Eqn. 1}$$

$$^{D}(V/K)_{O_2(LA \rightarrow \infty)} = \frac{^{D}k_5 + k_5/k_4}{1 + k_5/k_4} \qquad \text{Eqn. 2}$$

While an accurate experimental value for Eqn. 1 is presented in this work, only an upper limit value for Eqn. 2 can be estimated (Fig. 6.8). However, it is evident from the results that ^{D}kcat/Km(LA) >> ^{D}kcat/Km(O_2). For this result to be consistent with a random binding mechanism, the binding of O_2 to the ternary complex must be much tighter than that of LA. In other words, the rate of LA release from the ternary complex, k'_4, must be much faster than the rate of dioxygen release, k_4. This increased "stickiness" of oxygen will show up as a very large commitment to catalysis (k_5/k_4), decreasing the value of ^{D}kcat/Km(O_2) compared to ^{D}kcat/Km(LA). In a practical sense, the kinetic pathway is literally pushed to ordered binding of O_2 prior to LA (Scheme 6.1*b*). Due to the huge difference in magnitude between the isotope effects of ^{D}kcat/Km(O_2) and ^{D}kcat/Km(LA), the mechanism cannot be truly random binding, but rather it resembles an ordered mechanism. Previously, we discussed an ordered binding mechanism for lipoxygenase, so that oxygen binds prior to LA, is possible. Our conclusion is that truly random binding is unlikely to be the mechanism for the lipoxygenase reaction.

Another distinct mechanism that can explain the observed results is a reaction in which the isotopically sensitive chemical step occurs prior to oxygen binding (Scheme 6.1*d*). Since oxygen binding occurs after the isotopically sensitive step in this scenario, the KIE is not expected to be expressed in kcat/Km(O_2), which only includes steps after substrate binding. This last statement is accurate unless the reaction is fully reversible, in which an equilibrium isotope effect could show up in kcat/Km(O_2). The kinetic parameters, displayed in Table 6.3, demonstrate how the isotope effects are influenced by substrate concentrations. This is a rather unusual enzymatic mechanism; however, the behavior is similar to a mechanism described by Cook (17).

The derivations presented in Table 6.3 predict that ^{D}kcat/Km(O_2) should be insensitive to LA concentration, since the expressions do not include an intrinsic KIE, so that they are expected to be close to unity. The striking characteristic of this type of mechanism is that when dioxygen concentrations are low the isotope effect reflects the equilibrium isotope effect regardless of the concentration of LA. At high dioxygen concentrations, the isotope effects also reflect the KIE regardless of the LA concentration. Primary K_H/K_D equilibrium isotope effects are small compared to KIE and are usually close to unity (18). This is consistent with our observation that ^{D}kcat/Km(LA) is much greater than ^{D}kcat/Km(O_2), since only the former includes the KIE. Due to the abnormally large intrinsic KIE in the lipoxygenase reaction, the disparity between ^{D}kcat/Km(LA) and ^{D}kcat/Km(O_2) is made clear.

It is evident that this mechanism (Scheme 6.1*d*) can explain all the trends observed in the data presented. While we think that it is possible that the observed ^{D}kcat/Km(O_2) plotted in Figure 6.8*b* represents the intrinsic equilibrium isotope effect, we are not suggesting that the equilibrium isotope effect is in the observed range of 4–6, which would be abnormally large. As mentioned previously, the observed value of this isotope effect is an upper limit and could be significantly smaller since it includes an overestimation for Km(O_2) with D-LA as substrate. For future work it will be interesting to determine the precise equilibrium K_H/K_D isotope effect to see if it is abnormally large, and whether that correlates to the abnormally large KIE, or whether it is close to unity as expected.

To summarize these results, kinetic studies suggest that molecular oxygen enters the lipoxygenase-catalyzed reaction only after the substrate C–H bond is cleaved. However, the mechanistic possibility in which dioxygen binds to the free enzyme before LA has not been excluded completely.

Does Free Lipoxygenase Bind Dioxygen

The previously mentioned kinetic experiments suggest two mechanistic possibilities. Fortunately, they are radically different: either O_2 binds tightly to the free enzyme in the absence of LA (Scheme 6.1*b*), or dioxygen cannot bind at all to the free enzyme (Scheme 6.1*d*). It should be easy to distinguish between these possibilities by testing whether lipoxygenase can bind dioxygen. The presence of bound oxygen to SBL-1 was determined by analyzing concentrated enzyme solutions for released oxygen by degassing the solution into a vacuum line, separating dioxygen from other gases, and trapping the released oxygen in an oxygen trap. The partial pressure of total O_2 in a concentrated SBL-1 solution was determined after the oxygen was separated from other gases and converted to CO_2, in a method similar to an experimental procedure published for other oxygen-utilizing enzymes (19,20).

Lipoxygenase is known to exist in a number of forms, with one of the differences between them being the state of the iron in the active site. Free active enzyme is considered to be in the ferric state (21,22). Therefore, similar determinations were performed on solutions containing different forms of lipoxygenase: native (ferrous) enzyme, oxidized (ferric) enzyme, enzyme that was first oxidized and then reduced back to ferrous, and enzyme in the presence of a substrate analog. Soybean lipoxygenase was oxidized to ferric form by an excess of the product peroxide, 13(*S*)-hydroperoxy-9(*Z*),11(*E*)-octadecadienoic acid, produced enzymatically from LA. Lipoxygenase was reduced to the ferrous form by nordihydroguaiaretic acid (NDGA [23]). Vaccenic acid, in excess of enzyme concentration, was used as a substrate analog.

Our experimental results clearly showed that free ferric enzyme does not bind molecular oxygen. However, most oxygen-carrier proteins, such as hemoglobin, are ferrous and not ferric. We therefore analyzed a number of different enzyme forms to see if we could detect any form capable of binding dioxygen. Reduced ferrous lipoxygenase does not bind dioxygen, even though other nonheme iron-containing proteins, including other dioxygenases, have the capability to bind molecular oxygen in their ferrous state (13,14). Lipoxygenase that was reduced after it was first oxidized does not bind O_2 either, indicating that binding of O_2 is not dependent on a ligand change during redox reactions of iron in the active site. Another possibility is that oxygen binding is dependent on an enzymatic conformational change upon substrate binding. Obviously, we could not measure oxygen binding in the presence of the LA substrate, since it would then react with oxygen. It is probable that a similar enzymatic conformation would occur upon the binding of a substrate or product analog. The presence of vaccenic acid, a competitive inhibitor that is a substrate analog, does not produce any oxygen binding with the different enzyme forms.

To summarize these results, we could not detect a lipoxygenase form that binds dioxygen without the presence of an activated substrate. In conjunction with the models predicted by the KIE mentioned previously, it would seem that the only plausible mechanism for the lipoxygenase reaction is one in which the C–H bond cleavage takes place prior to the trapping of O_2 by the activated substrate–enzyme complex (Schemes 6.1*d* and 6.2).

Conclusion

Lipoxygenase is a unique enzyme in at least three aspects characterized in this work:

1. Catalysis is achieved via "substrate activation" instead of the more familiar "oxygen activation." Lipoxygenase is a unique iron-containing enzyme demonstrated to follow this mechanism.
2. The isotope effect measured for this reaction is the largest known observed primary k_H/k_D isotope effect in an enzymatic system, and it rivals the largest reported isotope effects for chemical reactions at room temperature.

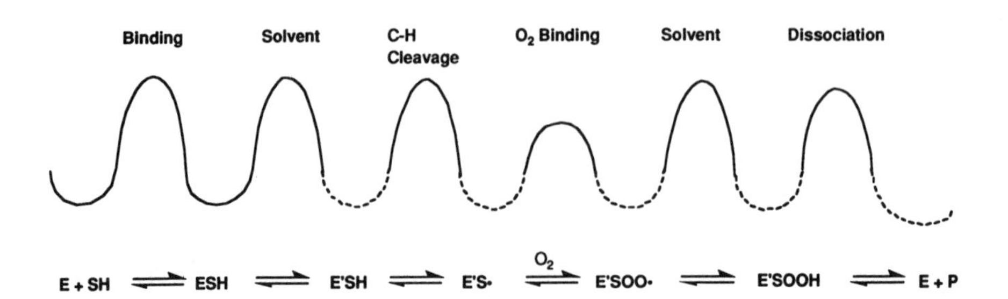

Scheme 6.2. Free energy profile of lipoxygenase at 25°C. A summary of reaction steps resolved by this work. Dotted lines represent energies of intermediates that have not been determined. Three kinetic barriers are of similar height, as the three steps are partially rate determining: substrate binding, an unidentified solvent-sensitive step, and substrate C–H bond cleavage. Molecular oxygen enters the reaction site only after hydrogen abstraction has occurred, and is trapped rapidly by activated substrate. The intermediate is reduced and reprotonated, and released as product peroxide.

3. The second-order rate constant of SBL-1 at room temperature and pH 9, is close to the diffusional limit for enzymatic reactions. Under these conditions, the free energy profile of SBL-1 involves a series of steps; all of which have similar energy barriers. Consequently, the hydrogen-abstraction step is not fully expressed at room temperature.

In Scheme 6.2, we proposed a mechanism for lipoxygenase based on the results presented in this chapter. Three distinct steps are clearly characterized in the lipoxygenase reaction prior to addition of dioxygen: binding of substrate, a solvent-dependent step, and the C–H bond cleavage. All three are partially expressed at room temperature, yet none is fully rate limiting. After substrate binding, some type of rearrangement is necessary for hydrogen abstraction from the bound substrate to occur, after which molecular oxygen enters the reaction. The "solvent-dependent step," exhibiting the SIE, possibly reflects an enzymatic conformational change involving hydrogen-bond rearrangements, or displacement of water from the enzyme active site. At higher temperatures (above 32°C) hydrogen abstraction increasingly limits the reaction, while at lower temperatures (5°C) the solvent-sensitive steps dominate (Fig. 6.5).

While the lipoxygenase reaction is close to the diffusion-controlled rate for an enzymatic reaction, surprisingly no one step is fully rate limiting. This type of behavior has been termed that of a "perfect enzyme" since there is no evolutionary pressure to perfect the efficiency of single steps more than the rate dictated by the diffusional encounter of enzyme and substrate (11,12). Maximal catalytic efficiency is obtained when increasing the rate of any one single step, does not significantly increase the overall enzymatic rate. Under this definition, SBL-1 is clearly optimized for catalysis at room temperature. It would be interesting to see if mammalian lipoxygenases are similarly optimized for catalysis at 37°C.

Molecular oxygen does not actually bind or react with free lipoxygenase, but rather is trapped by the activated enzyme–substrate complex. Linoleic acid binding, a diffusion-controlled and partially rate-limiting step (at pH 9 and 25°C), is followed by the isotopically sensitive "chemical" step. The C–H bond cleavage step is also partially rate limiting, probably producing a linoleyl-based radical that directly traps dioxygen in a very fast step. The peroxyl radical thus produced is reduced and protonated, leading to a product peroxide that is then released in another partially rate-limiting step controlling kcat.

A mechanism involving radical intermediates is supported by a number of published results; under anaerobic conditions lipoxygenase forms a carbon-centered linoleic free-radical that was detected by spin trapping (24). The radical was proposed to react with oxygen to give a peroxyl radical and a fatty acid hydroperoxide as the final product (25). Direct observation of the peroxyl radical was detected by EPR (26). The secondary k_H/k_D isotope effects are normal in magnitude and behavior, and are consistent with the formation of a delocalized allylic radical on the secondary positions—carbon 9, 10, 12, or 13 (1).

The reaction in its entirety is likely to be reversible, as both linoleyl and peroxyl radicals are shown by EPR to be in equilibrium (4), and the enzyme catalyzes the stereospecific exchange of molecular oxygen with the OOH group of the product peroxide (4,27). As yet, there is no direct evidence whether the C–H bond cleavage itself is reversible. Both the work by Nelson et al. (4), and Matthew and Chan (27), and the fact that crystal structure and magnetic susceptibility have not detected bound oxygen in the absence of the substrate (28,29), suggest that chemistry can be performed on LA in absence of O_2.

Our results produce the first direct kinetic proof that O_2 is not activated by lipoxygenase in the catalytic pathway under normal turnover conditions. Furthermore, molecular oxygen may not bind to the active site of lipoxygenase, but the activated enzyme–substrate complex traps it in a fast diffusion process. This last conclusion is based on the fact that the reaction rate is close to the diffusion-controlled rate, and that we could not detect any form of enzyme, either ferrous or ferric, having the capability to bind molecular oxygen in the absence of activated substrate. Interestingly, this mechanism is much closer to that of the autoxidation of lipids (30,31), than some model systems for a lipoxygenase-type reaction involving oxygen activation (32).

A striking feature of this almost-diffusion-limited enzymatic reaction is that it involves a series of steps, none of which is fully rate limiting, similar to what has been described for a "perfect enzyme" (11,12). Yet at the same time, unsaturated fatty acid peroxidation is a fast, spontaneous process that can take place as an autocatalytic reaction in solution. What then is the role of lipoxygenase as an enzyme? Obviously lipoxygenase accelerates the rate of lipid peroxidation. If the lipoxygenase mechanism is similar to that of autoxidation, the observed rate acceleration is probably not due to acceleration of the propagation step of autoxidation, since that is considered to be almost diffusion limited (31,33). Rather, the acceleration is linked to facilitating the initiation process and eliminating termination mechanisms.

We propose that part of the answer also lies in the reaction selectivity, and the stereospecificity of the products. Whereas the major product of lipoxygenase-mediated oxidation of LA is 13(*S*)-hydroperoxy-9(*Z*),11(*E*)-octadecadienoic acid, myriad isomers and stereoisomers are produced in autoxidation. This could be an example of what has been coined "Negative Catalysis" by Rétey, "Some enzymatic reactions involving highly reactive intermediates (such as radicals) can achieve reaction selectivity by preventing undesired reactions rather than facilitating the target pathway" (34).

References

1. Glickman, M.H., Wiseman, J., and Klinman, J.P. (1994) *J. Am. Chem. Soc. 116,* 793–794.
2. Hwang, C.C., and Grissom, C.B. (1994) *J. Am. Chem. Soc. 116,* 795–796.
3. Melander, L., and Saunders, W.H. (1987) *Reaction Rates of Isotopic Molecules.* R.E. Krieger Publishing, FL, Chapters 1–4.
4. Nelson, M.J., Seitz, S.P., and Cowling, R.A. (1990) *Biochem. 29,* 6897–6903.
5. Glickman, M.H. (1994) Mechanistic Studies of Soybean Lipoxygenase, Ph.D. Dissertation, University of California, Berkeley.
6. Axelrod, B., Cheesbrough, T.M., and Laakso, S. (1981) *Methods Enzymol. 71,* 441–451.
7. Bell, R.P. (1980) *The Tunnel Effect in Chemistry.* Chapman Hall, New York.
8. Solc, K., and Stockmayer, W.H. (1973) *Int. J. Chem. Kinet. 5,* 733–752.
9. Brouwer, A.C., and Kirsch, J.F. (1982) *Biochem. 21,* 1302–1307.
10. Nelson, M.J. (1988) *J. Am. Chem. Soc. 110,* 2985–2986.
11. Albery, W.J., and Knowles, J.R. (1976) *Biochem. 15,* 5631–5640.
12. Burbaum, J., Rains, R.T., Albery, W.J., and Knowles, J.R. (1989) *Biochem. 28,* 9293–9305.
13. Feig, A.L., and Lippard, S.J. (1994) *Chem. Rev. 94,* 759–805.
14. Que, L. (1980) *Struct. Bonding 40,* 39–72.
15. Klinman, J.P., Humphries, H., and Voet, J.G. (1980) *J. Biol. Chem. 255,* 11648–11651.
16. Ahn, N., and Klinman, J.P. (1983) *Biochem. 22,* 3096–3106.
17. Cook, P.F., Yoon, M.Y., Hara, S., and McClure, G.D. (1993) *Biochem. 32,* 1795–1802.
18. Klinman, J.P. (1978) *Adv. Enz. Relat. Areas Mol. Biol. 46,* 415–494.
19. Tian, G., and Klinman, J.P. (1993) *J. Am. Chem. Soc. 115,* 8891–8897.
20. Tian, G., Berry, J.A., and Klinman, J.P. (1994) *Biochem. 33,* 226–234.
21. Schilstra, M.J., Veldink, G.A., and Vliegenthart, J.F.G. (1994) *Biochem. 33,* 3974–3979.
22. deGroot, J.J.M.C., Garssen, G.J., Veldink, G.A., Vliegenthart, J.F.G., and Boldingh, J. (1975) *FEBS Letters 56,* 50–54.
23. Kemal, C.E.A. (1987) *Biochem. 26,* 7064–7072.
24. deGroot, J.J.M.C., Garssen, G.J., Vliegenthart, J.F.G., and Boldingh, J. (1973) *Biochim. Biophys. Acta 326,* 279–284.
25. deGroot, J.J.M.C., Veldink, G.A., Vliegenthart, J.F.G., Boldingh, J., Wever, R., and Van Gelder, B.F. (1975) *Biochim. Biophys. Acta 377,* 71–79.
26. Chamulitrat, W., and Mason, R.P. (1989) *J. Biol. Chem. 264,* 20968–20973.

27. Matthew, J.A., and Chan, H.W.S. (1983) *J. Food. Biochem. 7,* 1–6.
28. Boyington, J.C., Gaffney, B.J., and Amzel, L.A. (1993) *Science 260,* 1482–1486.
29. Petersson, L., Slappendel, S., and Vliegenthart, J.F.G. (1985) *Biochim. Biophys. Acta 828,* 81–85.
30. Porter, N.A., and Wujek, D.G. (1984) *J. Am. Chem. Soc. 106*, 2626–2629.
31. Porter, N.A. (1986) *Acc. Chem. Res. 19*, 262–268.
32. Rao, S.I., Wilks, A., Hamberg, M., and Ortiz de Montellano, P.R. (1994) *J. Biol. Chem. 269*, 7210–7216.
33. Gutteridge, J.M.C., and Halliwell, B. (1990) *TIBS 15*, 129–135.
34. Rétey, J. (1990) *Angew. Chem. Int. Ed. 29*, 355–361.

Chapter 7

Large Deuterium Kinetic Isotope Effects in Soybean Lipoxygenase

Chi-Ching Hwang[a] **and Charles B. Grissom**[b]

[a]Department of Biology, University of North Texas, Health Science Center at Fort Worth, 3500 Camp Bowie Blvd., Fort Worth, TX 76107; and [b]Department of Chemistry, University of Utah, Salt Lake City, UT 84112, USA.

Introduction

Lipoxygenase (EC.1.13.11.12) is a nonheme iron enzyme that catalyzes the dioxygenation of 1,4-diene units in polyunsaturated fatty acids to form conjugated diene hydroperoxides (Scheme 7.1). The kinetics and mechanism of soybean lipoxygenase have been extensively studied by physical and chemical methods (1–8). Soybean lipoxygenase catalyzes the regio- and stereospecific dioxygenation of 9(*Z*),12(*Z*)-octadecadienoic acid to form 13(*S*)-9(*Z*),11(*E*)-13-hydroperoxy-9,11-octadecadienoic acid (13(*S*)-HPOD) at pH 9.0 as the major product (9–10).

Lipoxygenase-catalyzed dioxygenation of the 1,4-pentadienyl moiety of fatty acids requires activation of either molecular oxygen or the unsaturated fatty acid. Molecular oxygen exists as a triplet ground state that is 23 kcal/mol lower in energy than the excited singlet. Therefore, triplet molecular oxygen cannot add directly across a double bond. Oxygen activation can occur by interaction with a transition metal ion to facilitate spin conversion and addition to an alkene. Dioxygen that is bound to a reduced transition metal ion can undergo a one-electron redox reaction in which activated dioxygen initiates the radical reaction (11). Evidence against this proposal has been presented by Vliegenthart and co-workers, who studied lipoxygenase by electron paramagnetic resonance (EPR) and magnetic susceptibility under aerobic and anaerobic conditions and found no evidence for O_2 binding to the native enzyme (12). This result suggests that the enzyme must activate the fatty acid, rather than O_2, to catalyze the hydroperoxidation of linoleate.

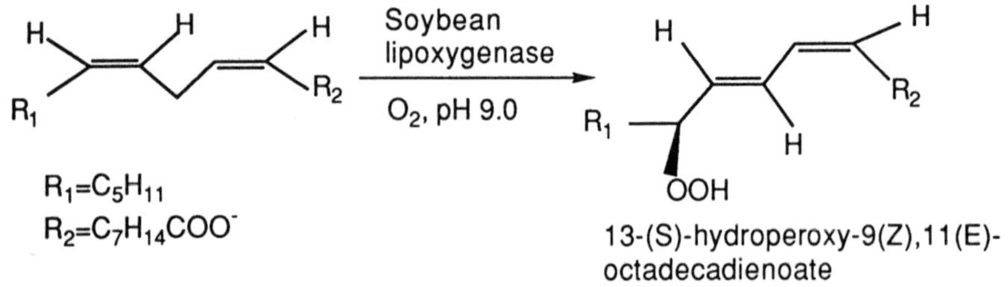

Soybean lipoxygenase

O_2, pH 9.0

$R_1 = C_5H_{11}$

$R_2 = C_7H_{14}COO^-$

13-(S)-hydroperoxy-9(Z),11(E)-octadecadienoate

Scheme 7.1. The reaction of soybean lipoxygenase with linoleate.

Differing catalytic mechanisms for lipoxygenase have been proposed with radical and organometallic species as viable intermediates (Scheme 7.2 [2,5,12]). In one proposed mechanism, the role of Fe^{III} is to oxidize the 1,4-diene moiety of the substrate to a pentadienyl radical. This radical species can either react with O_2 to form a peroxyl radical that reoxidizes Fe^{II} to yield the peroxide anion under aerobic conditions (Scheme 7.2*a*) or form a linoleate dimer under anaerobic conditions (1). Recently Nelson and co-workers have proposed that the radical intermediate in lipoxygenase is actually an allyl radical that is delocalized only over C-9, C-10, and C-11 (13). A nonradical mechanism has also been suggested with an organometallic intermediate to explain the stereospecificity of the reaction (Scheme 7.2*b* [14–16]). A concerted deprotonation by a basic group on the enzyme, followed by electrophilic addition of Fe^{III} to carbon, would give an organoiron intermediate. The allylic organoiron intermediate then would react with O_2, followed by protonation, to give product.

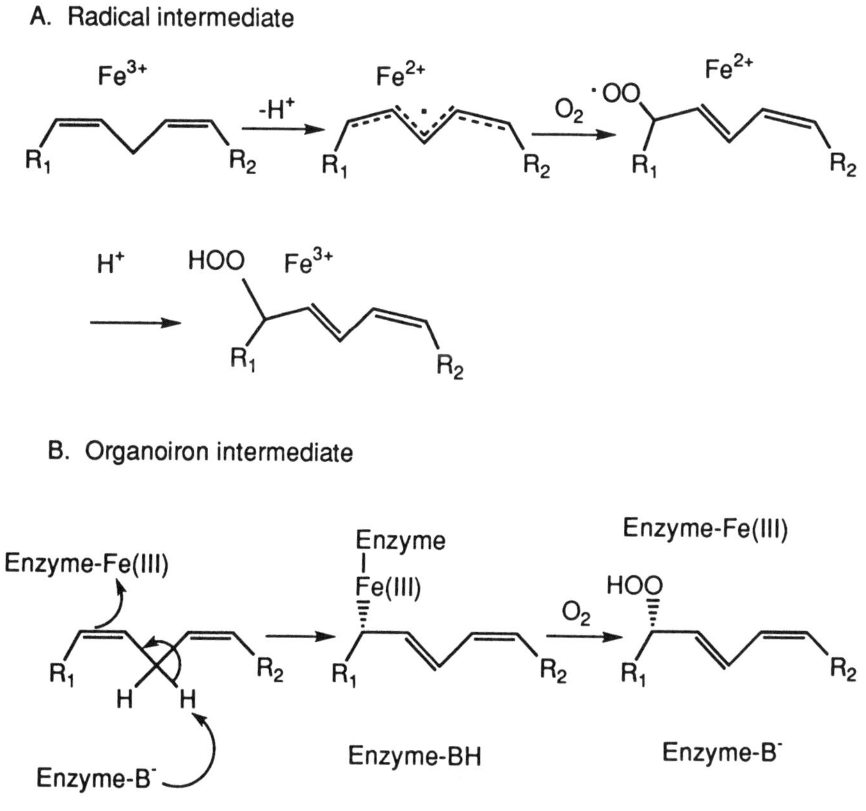

A. Radical intermediate

B. Organoiron intermediate

Scheme 7.2. Mechanism of soybean lipoxygenase-catalyzed reaction.

Stable radical intermediates that occur during catalysis can be detected by EPR based on the magnetic moment of the unpaired, spinning electron. The unpaired electron can be directly observed or trapped as a stable nitroxyl radical. Under anaerobic conditions, the pentadienyl radical has been trapped at C-9 and C-13 of linoleate by the radical scavenger 2-methyl-2-nitrosopropanol (17). Under aerobic conditions, peroxyl radicals from linoleate and arachidonate in the presence of lipoxygenase have been observed by EPR. These radical species have an electron spin resonance (ESR) signal at g = 2.014 (18). A spin-trapped linoleate-radical intermediate has also been observed using nitrosobenzene. The intermediate pentadienyl radical can be formed either by homolytic cleavage of the C–H bond or by heterolytic cleavage of the C–H bond, followed by oxidation of the resulting carbanion by the proximal Fe^{III}. The former mechanism produces the geminate carbon and hydrogen radical pair, whereas the latter produces the carbon radical and high-spin Fe^{II}.

Magnetic field effects on reaction rates have been used to provide unambiguous evidence for radical pair intermediates in some chemical reactions (20–24). Recently, Harkins and Grissom demonstrated the magnetic field dependence of the enzyme ethanolamine ammonia lyase by both steady-state and stopped-flow kinetic methods as proof of kinetically significant radical intermediates (25,26). In order to study the mechanism of soybean lipoxygenase–catalyzed oxygenation, and perhaps differentiate between radical and nonradical mechanisms, magnetic field effects and kinetic isotope effects (KIE) on the reaction of lipoxygenase with linoleate and arachidonate have been examined (27).

Recently, we reported an unusually large primary deuterium isotope effect of $^{D}V_{max} = 36 \pm 3$ and $^{D}(V_{max}/K_m) = 28 \pm 2$ on the soybean lipoxygenase–catalyzed oxygenation of [11,11]-$^2H_2$-linoleate (27). This observation was confirmed in a companion report from the Klinman laboratory (28); both reports were pre-dated by a similar observation of $^{D}V_{max} = 38$ for the lipoxygenase-catalyzed oxygenation of [13,13]-$^2H_2$-8,11,14-icosatrienoic acid (29). Another report estimated $^{D}V_{max} = 8.7$ and $^{D}(V_{max}/K_m) = 7.6$, as determined by comparing the kinetic parameters of unlabeled and [11,11-$^2H_2$]-linoleate of only 95% dideuterium isotopic purity (30). The lower fraction of deuteration in the latter experiment probably resulted in the lower estimation of KIE.

Experimental

Synthesis of [11,11-$^2H_2$] Linoleic Acid

[11,11-$^2H_2$] Linoleic acid (LA) was synthesized by a modification of the literature procedure (Scheme 7.3 [31]).

10-Undecynoic Acid (6)

Thirty-one milliliters of bromine was added to a stirred solution of 0.58 mol 10-undecynoic acid in 180 mL $CHCl_3$ in an ice–salt bath. The solution was stirred for 1 h and gradually warmed to room temperature and concentrated in vacuo to give a crude

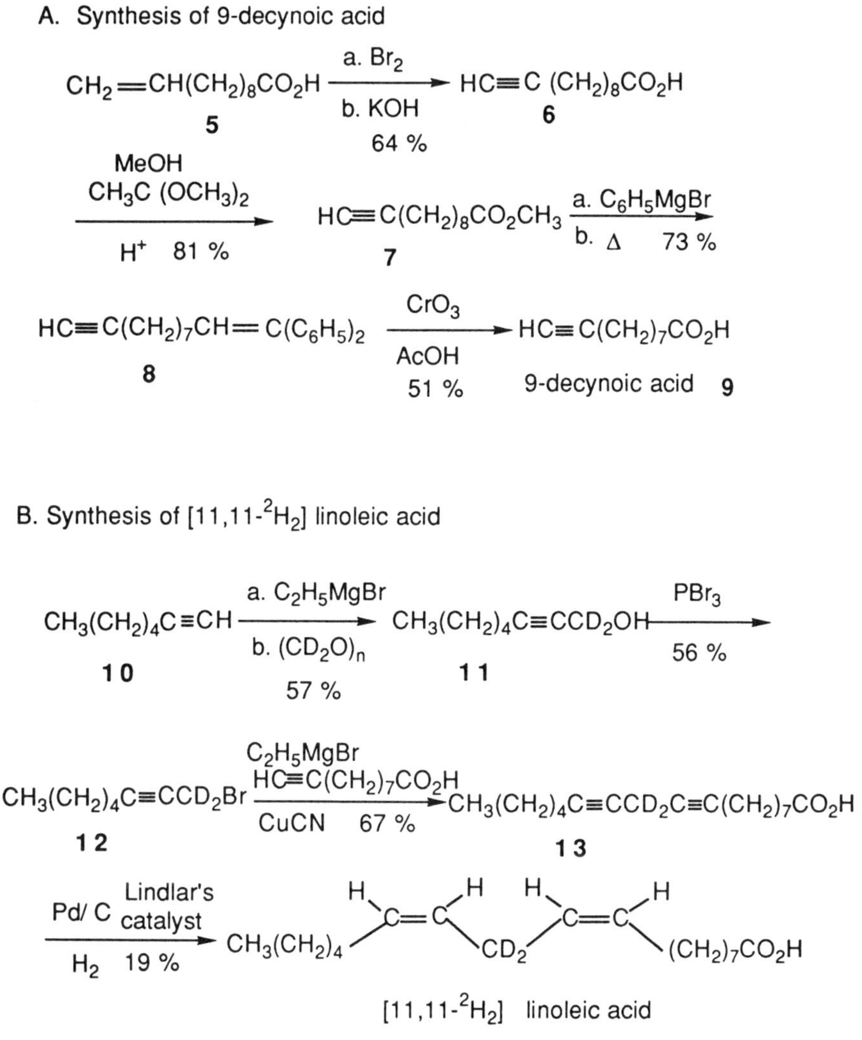

A. Synthesis of 9-decynoic acid

$$CH_2{=}CH(CH_2)_8CO_2H \ (\mathbf{5}) \xrightarrow[\text{b. KOH} \quad 64\ \%]{\text{a. } Br_2} HC{\equiv}C(CH_2)_8CO_2H \ (\mathbf{6})$$

$$\xrightarrow[H^+ \quad 81\ \%]{MeOH,\ CH_3C(OCH_3)_2} HC{\equiv}C(CH_2)_8CO_2CH_3 \ (\mathbf{7}) \xrightarrow[\text{b. } \Delta \quad 73\ \%]{\text{a. } C_6H_5MgBr}$$

$$HC{\equiv}C(CH_2)_7CH{=}C(C_6H_5)_2 \ (\mathbf{8}) \xrightarrow[AcOH \quad 51\ \%]{CrO_3} HC{\equiv}C(CH_2)_7CO_2H$$

9-decynoic acid **9**

B. Synthesis of [11,11-$^2H_2$] linoleic acid

$$CH_3(CH_2)_4C{\equiv}CH \ (\mathbf{10}) \xrightarrow[\text{b. } (CD_2O)_n \quad 57\ \%]{\text{a. } C_2H_5MgBr} CH_3(CH_2)_4C{\equiv}CCD_2OH \ (\mathbf{11}) \xrightarrow[56\ \%]{PBr_3}$$

$$CH_3(CH_2)_4C{\equiv}CCD_2Br \ (\mathbf{12}) \xrightarrow[CuCN \quad 67\ \%]{C_2H_5MgBr,\ HC{\equiv}C(CH_2)_7CO_2H} CH_3(CH_2)_4C{\equiv}CCD_2C{\equiv}C(CH_2)_7CO_2H \ (\mathbf{13})$$

$$\xrightarrow[H_2 \quad 19\ \%]{\text{Pd/C Lindlar's catalyst}} CH_3(CH_2)_4CH{=}CHCD_2CH{=}CH(CH_2)_7CO_2H$$

[11,11-$^2H_2$] linoleic acid

Scheme 7.3. Synthesis of [11,11-2H2] LA.

mixture of dibromide products. The dibromoacid was refluxed with 24.6 mmol KOH dissolved in 250 mL H_2O and kept in an oil bath at 150–160°C for 8 h, cooled, and left to stand overnight. The reaction mixture was dissolved in 1.5L H_2O, acidified to pH 4 with dilute H_2SO_4, and extracted with ethyl ether (4 x 100 mL). The ether layer was dried over $MgSO_4$ and concentrated in vacuo. The distillation product obtained was a pale yellow oil, and was allowed to solidify upon standing (67.5 g, 0.37 mol, 64%): 300 MHz ^1H NMR ($CDCl_3$) δ 2.34 (t, 2H), 2.16 (dt, 2H), 1.91 (t, 1H), 1.58–1.64 (m, 2H), 1.44–1.53 (m, 2H), 1.2–1.4 (m, 8H); IR (CCl_4) 3305, 1705 cm^{-1}.

Methyl 10-Undecynoate (7)

Ninety-three grams of 2,2-dimethoxypropane and 0.20 g toluene-*p*-sulfonic acid were added to a solution of 10-undecynoic acid in 20 mL MeOH. The solution was refluxed for 12 h and concentrated in vacuo to give a crude liquid that was distilled under vacuum to give methyl ester (60 g, 0.31 mol, 81%): 300 MHz ^{1}H NMR ($CDCl_3$) δ 3.62 (s, 3H), 2.3 (t, 2H), 2.16 (dt, 2H), 1.91 (t, 1H), 1.58–1.64 (m, 2H), 1.44–1.53 (m, 2H), 1.2–1.4 (m, 8H); IR (CCl_4) 3310, 1740 cm^{-1}.

1,1-Diphenylundec-1-en-10-yne (8)

A solution of 60 g methyl-10-undecynoate in 300 mL Et_2O was added dropwise to a phenyl magnesium bromide solution prepared by the addition of 1.0 mol bromobenzene in 1.0 mol Mg in 300 mL Et_2O. The reaction solution was refluxed for 2.5 h, then poured into an ice-water mixture, acidified by 2 N H_2SO_4, and extracted with ether. The ether layer was dried over $MgSO_4$ and evaporated. The crude alcohol was dehydrated in an oil bath at 220°C for 1 h. The product was obtained as a pale yellow oil (71 g, 0.24 mol, 73%): 300 MHZ ^{1}H NMR ($CDCl_3$) δ 7.2–7.4 (m, 10H), 6.1 (t, 1H), 2.1–2.3 (m, 4H), 2.0 (t, 1H), 1.2–1.6 (m, 10H).

9-Decynoic acid (9)

Twenty-seven grams of chromium trioxide in 30 mL H_2O was added to a stirred solution of 82.7 mmol 1,1-diphenylundec-1-en-10-yne. The solution was stirred at room temperature overnight and concentrated in vacuo. The residue was dissolved in 400 mL 2 N H_2SO_4, saturated with NaCl, and extracted with ether. The ether layer was extracted with dilute NaOH. The aqueous phase was acidified, extracted with ether, and concentrated in vacuo. The product was obtained as a colorless liquid (7.1 g, 42 mmol, 51%): 300 MHz ^{1}H NMR ($CDCl_3$) δ 2.35 (t, 2H), 2.18 (dt, 1H), 1.94 (t, 2H), 1.59–1.70 (m, 2H), 1.48–1.56 (m, 2H), 1.28–1.47 (m, 6H); 75 MHz ^{13}C NMR ($CDCl_3$) δ 180.40, 84.58, 68.20, 34.07, 28.90, 28.72, 28.50, 28.38, 24.58, 18.37.

1,1-Dideutero-2-octyn-1-ol (11)

Forty-two millimoles of ethyl magnesium bromide in ether was added to a stirred solution of 42 mmol 1-heptyne in 10 mL ether at 0°C. After stirring for an additional hour, 29 mmol paraformaldehyde-d_2 was added in one portion. The mixture was stirred at low temperature for 2 h, then 40 h at room temperature. The initial product was hydrolyzed by 20 mL 5% H_2SO_4, extracted with ether and evaporated. The product was obtained as a colorless liquid (1.93 g, 15 mmol, 52%): 300 MHz ^{1}H NMR ($CDCl_3$) δ 2.28 (s, 1H), 2.12 (t, 2H), 1.44 (m, 2H), 1.3 (m, 2H), 1.44 (m, 2H), 1.3 (m, 2H), 0.84 (t, 3H); 75 MHz ^{13}C NMR ($CDCl_3$) δ 86.4, 78.2, 31.0, 28.3, 22.2, 18.7, 13.9.

1,1-Dideutero-1-Bromo-2-Octyne (12)

Thirty-five hundredths of a milliliter of phosphorous tribromide was added to a stirred solution of 13 mmol 1,1dideutero-2-octyn-1-ol with 0.2 mL dry pyridine in an ice–salt bath for 0.5 h. The solution was refluxed for 3 h, then cooled to room temperature. The mixture was poured into 5 mL H_2O, extracted with Et_2O, dried over $MgSO_4$ and evaporated. The product obtained was a colorless liquid (1.39 g, 7.3 mmol, 56%) following chromatography on silica using hexane/ether (10:1, v/v): 300 MHz ^{1}H NMR ($CDCl_3$) δ 2.2 (t, 2H), 1.42–1.55 (m, 2H), 1.22–1.38 (m, 4H), 0.88 (t, 3H); 75 MHz ^{13}C NMR ($CDCl_3$) δ 88.3, 75.1, 31.0, 28.1, 22.2, 18.9, 14.0.

11,11-Dideuterooctadecadiynoic acid (13)

A solution of 0.59 mmol 9-decynoic acid in 10 mL THF was added dropwise to 1.2 mL of a 1 M THF solution of 1.18 mmol ethyl magnesium bromide under static N_2. The mixture was warmed to reflux for 1 h and cooled to room temperature. One milligram of copper(I) cyanide was added to the solution. After 15 min, a solution of 0.28 mmol 1,1-dideutero-1-bromo-2-octyne in 3 mL THF was added dropwise. The solution was refluxed for 24 h, then decomposed with cold water, acidified to pH 3 with dilute H_2SO_4, saturated with NaCl, and extracted with ether. The ether layer was dried over $MgSO_4$ and evaporated. The residue was purified by chromatography on silica using hexane/ethyl acetate (10:1). The product was obtained as white powder (55 mg, 0.20 mmol, 67%): 300 MHz ^{1}H NMR ($CDCl_3$) δ 10.8 (br, 1H), 2.32 (t, 2H), 2.13 (t, 4H), 1.54–1.68 (m, 2H), 1.42–1.54 (m, 4H), 1.2–1.4 (m, 10H), 0.87 (t, 3H); 75 MHz ^{13}C NMR ($CDCl_3$) δ 179.78, 80.56, 74.55, 74.39, 34.02, 31.09, 28.94, 28.76, 28.65 (2C), 28.48, 24.65, 22.23, 18.71 (2C), 13.99.

[11,11-$^2H_2$] Linoleic Acid

A solution of 0.20 mmol 11,11-dideuterooctadecadiynoic acid in 3 mL petroleum ether was treated with 0.5 mg Lindlar's catalyst and one drop of quinoline. The stirred solution was hydrogenated with a hydrogen-filled balloon for 6 h at room temperature. The reaction mixture was filtered and concentrated in vacuo. The residue was dissolved in 100 mL CH_3CN, and separated by C-18 reverse-phase HPLC on an Alltech 5-mm (25 x 0.46 cm) column with the solvent mixture composed of CH_3CN/H_2O (60/40, v/v) for 5 min, followed by a linear gradient to 100% CH_3CN over 15 min at a flow rate of 2.0 mL/min. The retention time of 11,11-dideuterolinoleic acid appeared at 16.3 min. The collected solution was concentrated in vacuo. The product was obtained as a colorless oil (10 mg, 35 mmol, 19%): 300 MHz ^{1}H NMR ($CDCl_3$) δ 5.3–5.42 (m, 4H), 2.34 (t, 2H), 1.9–2.1 (m, 4H), 1.6–1.7 (m, 2H), 1.2–1.4 (m, 14H), 0.88 (t, 3H); 75 MHz ^{13}C NMR ($CDCl_3$) δ 179.8, 130.25, 130.05, 127.96, 127.79, 34.0, 31.5, 29.6, 29.4, 29.1, 29.08, 29.04, 27.19 (2C), 24.7, 22.57, 14.07.

Enzyme

Soybean lipoxygenase was obtained from Sigma Chemical Co. as the $(NH_4)_2SO_4$ precipitate and desalted by dialysis and gel permeation chromatography on Sephadex G-10 in 0.1 M NaOAc pH 5.6 and 10% glucose. SDS-Polyacrylamide gel electrophoresis showed the enzyme to be greater than 80% pure with a minor contaminating band of higher electrophoretic mobility. Deuterated and unlabeled LA was purified by C-18 HPLC (CH_3CN 60% in H_2O initially, followed by a linear gradient to 100% CH_3CN), collected anaerobically, and stored at –20°C in the dark. Assay conditions were: 24.1°C, 100 mM Ches (2-[*N*-cyclohexylamino] ethanesulfonic acid) or borate (as indicated) pH 9.00, and 1 cm pathlength cuvette. The lipoxygenase reaction with arachidonate was studied in 100 mM borate at pH 10.0 to prevent further oxidation by lipoxygenase (32). The initial rates were calculated by taking the tangent of the recorded trace on the chart paper and fitting the velocity (tangent) vs [S] data to the Michaelis–Menten equation to obtain the parameters V_{max} and V_{max}/K_m.

Steady-State Kinetic Studies

The lipoxygenase-catalyzed oxygenation reaction was monitored spectrophotometrically by following the formation of the conjugated dienylperoxide at 234 nm for linoleate product and 238 nm for arachidonate product at different external magnetic field strengths in the range of 0 to 0.2 T. The concentration of linoleate and arachidonate was determined by spectrophotometric endpoint assay with $\varepsilon = 25000\ M^{-1}\ cm^{-1}$ (32). The initial rates were measured and fitted to the Michaelis–Menten equation. Magnetic field effects and magnetic isotope effects on the lipoxygenase-catalyzed reaction were determined by comparing the kinetic parameters V_{max}, V_{max}/K_m, $^{D}V_{max}$, and $^{D}(V_{max}/K_m)$ as a function of magnetic field strength.

Isotope Effect Measurement by Internal Competition

Unlabeled and [11,11-$^2H_2$] linoleate were mixed to give an 80:20 mixture. The isotopic mixture was incubated with enzyme in 100 mM borate at pH 9.00, at a final concentration of 48.1 μM total linoleate and quenched at 32% reaction by the addition of 2 N HCl. Linoleic acid was extracted three times with ethyl ether, and the solvent was removed by rotary evaporation. The resulting isotopic mixture of LA was purified by HPLC as described previously. Isotopic analysis was performed by EI-MS at 70 eV at a probe temperature of 70°C. The intensity at 282 amu (m + 2 relative to the unlabeled LA at m = 280 amu) was taken as a measure of dideuterium incorporation. The contribution to the m + 1 and m + 2 (and m + 3 and m + 4 for dideuterated LA) peaks from naturally occurring ^{13}C was accounted for by measuring the intensities of these peaks in the unlabeled and dideuterated LA prior to mixing. As a test of $^1H/^2H$ isotopic ratio quantification by mass spectrometry, samples containing known amounts of unlabeled and dideuterated LA were combined and the isotopic ratio was determined by mass spectrometry. A linear relationship of the known isotopic composition to the isotopic composition measured by MS was observed in the range from 0 to 60% dideuterated LA (Fig. 7.1). Calculations for determining the deuterium isotope effect, $^{D}(V_{max}/K_m)$ are shown in Eqn. 1 (33), and Table 7.1.

Magnetic Field Dependence of Kinetic Parameters

Figure 7.2 shows the experimental setup for the steady-state kinetic studies as a function of external magnetic field. In order to follow the rate of product formation spectro-

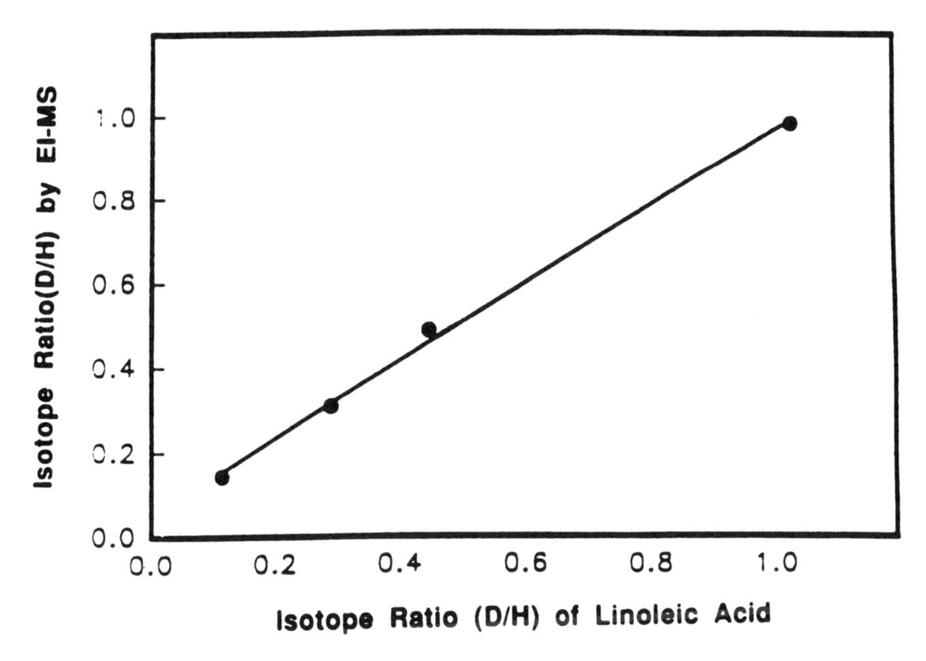

Figure 7.1. Linear relationship of the known isotopic composition to the isotopic composition of LA measured by mass spectrometer.

TABLE 7.1. Competitive Isotope Effects $^{D}(V_{max}/K_m)$ by Mass Spectrometric Isotope Quantification of Residual Substrate[a,b]

Isotope ratio (R_0)[c]	R	R/R_0	$1 - F_1$	$^{D}(V_{max}/K_m)$[d]
0.246				
0.243				
	0.397	1.630	0.605	33.9
	0.393	1.610	0.607	22.1

[a]Melander and Saunders (33).
[b]Calculations are shown in Eqns. 1 and 2.
[c]Average value: 0.245.
[d]Average value: $^{D}(V_{max}/K_m)$ is 28 ± 9.

photometrically as a function of magnetic field, a Beckman DU monochromator with solid state electronics was retrofitted with an electromagnet. The cell holder was a thermostatted brass block containing a Hall probe to measure the magnetic field flux. The magnetic field was homogeneous to ±2% within the area of the cuvette and the long-term stability was better than 0.1%. The photomultiplier tube was shielded from the applied magnetic field with mu metal and low-carbon high-iron steel. A Kepco constant-current power supply was controlled manually to adjust the magnetic field strength. A computer and a chart recorder were connected to the detector for data collection.

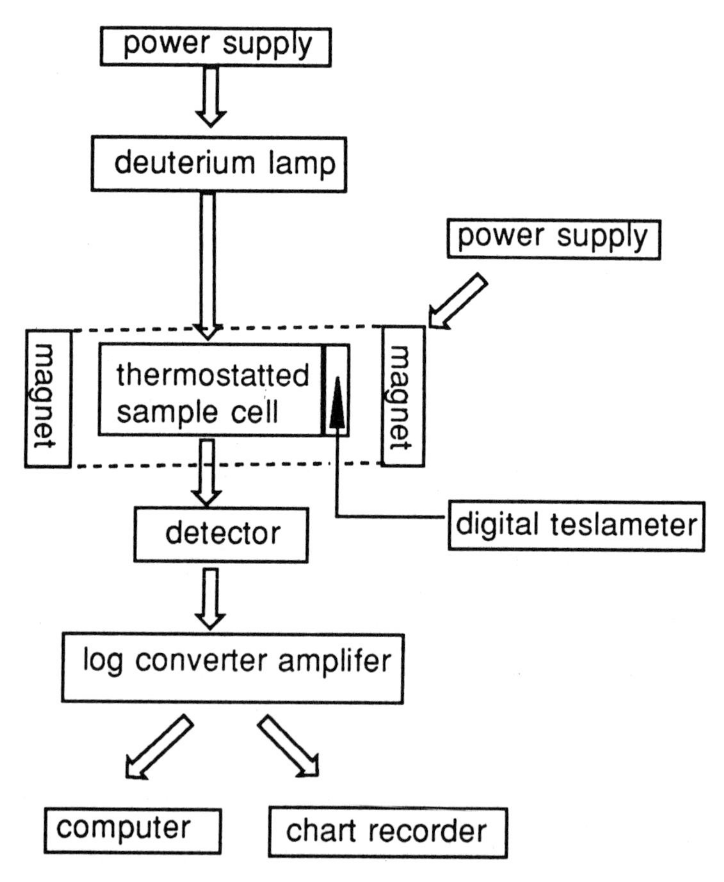

Figure 7.2. Spectrophotometer with electromagnet.

Results

Kinetic Studies

Smith and Lands have shown that a dilute enzyme solution undergoes a loss in specific activity over the 12–20 h required to perform the experiment (34). Attempts to stabilize commercial soybean lipoxygenase by adding bovine serum albumin, dithiothreitol, or glycerol to the enzyme-resuspending solution were not successful. Further studies on the stability of lipoxygenase were performed, including varying the pH and buffers (pH 9.0 in borate, 7.0 in Hepes, and 5.6 in acetate), adding polyhydroxy alcohols and detergents (glycerol, Ficoll, Tween 80, glucose, and sucrose), and adding chemical reagents (1 mM EDTA, 1 mM PMSF, 0.1 M NaCl, 1 mM DMSO, 1 mM KI, 1 mM $K_3Fe(CN)_6$ $3H_2O$, and 1 mM $K_4Fe(CN)_6$ $3H_2O$). Only a resuspending solution of 10% glucose and 0.1 M acetate, pH 5.6 under a nitrogen atmosphere stabilized against loss of enzyme activity for as long as 4 h. Storing the enzyme solution at –20°C until kinetic study use provided only a small amount of variability from vial to vial. Therefore, commercial soybean lipoxygenase was resuspended in a solution of 10% glucose, 0.1 M acetate, pH 5.6 in a nitrogen atmosphere, and the activity was normalized with a reaction rate at high linoleate concentration (27 μM) for routine kinetic studies. Soybean lipoxygenase–catalyzed oxygenation of linoleate and arachidonate obeys Michaelis–Menten kinetics with $K_m = 4.9 \pm 0.3$ μM, pH 9.0, and 4.7 ± 0.4 μM, pH 10.0, respectively. The kinetic parameter K_m is lower than the literature reports of 12 μM for linoleate (35), and 86 μM for arachidonate at pH 9.0 (32).

Kinetic Isotope Effects

The observed KIE was determined by either direct comparison of the kinetic parameters V_{max} and V_{max}/K_m for unlabeled and deuterated linoleate, or by internal competition by following residual substrate or product formation. A previous report estimated $^{D}V_{max} = 8.7 \pm 0.4$ and $^{D}(V_{max}/K_m) = 7.6$ at 25°C and pH 9.0, from direct comparison of the kinetic parameters with unlabeled and [11,11-$^2H_2$] linoleate of 95% dideuterium isotopic purity (30). Repetition of this measurement with [11,11-$^2H_2$] linoleate of 99.4% dideuterium isotopic purity gave a much larger KIE. In a direct comparison of the kinetic parameters determined by spectrophotometrically following 13(*S*)-HPOD product at 234 nm, $^{D}V_{max} = 36 \pm 3$ and $^{D}(V_{max}/K_m) = 28 \pm 2$.

In order to test for the presence of a potent inhibitor in the deuterated linoleate, assays with unlabeled linoleate were doped with increasing amounts of [11,11-$^2H_2$] linoleate. No significant decrease in the rate was observed until the amount of deuterated linoleate added was greater than 10%. In a similar experiment, biphasic kinetic traces were observed when deuterated and unlabeled linoleate were mixed in a 96:4 ratio (Fig. 7.3). In the early phase of the reaction, mostly unlabeled linoleate is consumed because of the large isotopic discrimination against deuterated linoleate. If the amount of unlabeled linoleate that is added is kept to less than 5%, a linear kinetic trace that approximates the rate with deuterated linoleate is obtained after unlabeled linoleate is depleted. These two experiments provide a qualitative indication of a large isotopic discrimination between ^1H and ^2H.

Internal Competition Isotope Effect Measurements

In order to quantify the KIE obtained by deuterated and unlabeled substrate competing in the same assay, the isotopic composition of residual substrate was determined by mass spectrometry. Under internal competition conditions, $^D(V_{max}/K_m) = 28 \pm 9$ at 24.2°C (Table 7.1). Analysis of these data are presented for Eqns. 1 and 2.

$$^D(V_{max}/K_m) = \ln(1 - F_1)/\ln[(1 - F_1)R/R_0] \quad (1)$$

where:

$R_0 = 0.245$ (ave.);
$a^0{}_1 = 36.2$;
$C_0 = a^0{}_1 + a^0{}_2 = 45\ \mu M$;
$a_1 = 30.6/(1 + R)$;
$C = a_1 + a_2 = 30.593\ \mu M$ (0.34 OD consumed);
$R = a_2/a_1$; and
$R_0 = a^0{}_2/a^0{}_1$

$$1 - F_1 = a_1/a^0{}_1 = C\,(1 + R_0)/(C_0\,(1 + R)) \quad (2)$$

where:

a_2 and $a^0{}_2$ = concentration of the labeled linoleate;
a_1 and $a^0{}_1$ = concentration of the unlabeled linoleate; and
F_1 = fractional amount of conversion of unlabeled linoleate

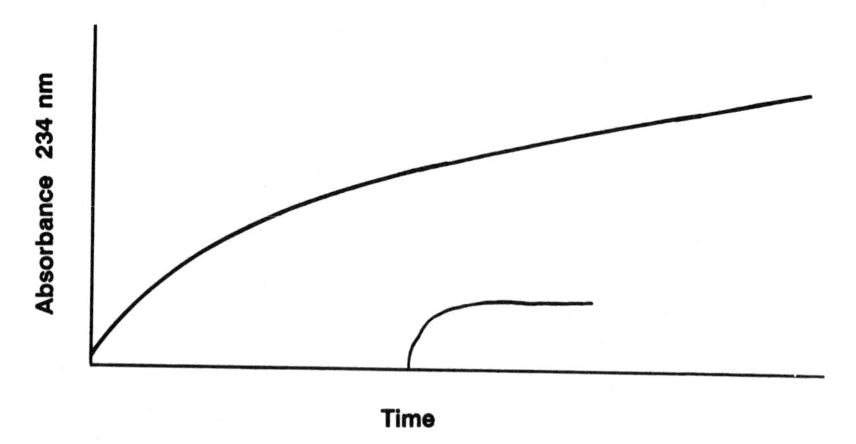

Figure 7.3. Kinetic traces of the mixture of [11,11-$^2H_2$] linoleate (24.6 μM) with unlabeled linoleate (1 μM) as the ratio of 96:4 (top), compared with the unlabeled linoleate (1 μM, bottom).

The large standard error of the isotope effect is due to the inherent unsuitability of determining large KIE by following residual substrate. The only interpretation possible from this measurement is that $^{D}(V_{max}/K_m)$ is indeed large and greater than 19. Further studies of the KIE of lipoxygenase-catalyzed oxygenation of the unlabeled and dideuterated linoleate by internal competition with mass spectrophotometric isotope quantification in the product were attempted in order to reduce the standard error (33). However, EI-MS gives a small molecular ion peak even after reduction with sodium borohydride and methylation with diazomethane. Chemical ionization with methane as the reagent gas was used to improve the determination of the isotopic ratio from the M + 1 peak. The spectrum for linoleate showed a major peak at M-17 that may result from loss of water and a small peak at M + 1 that caused difficulty in obtaining accurate and reproducible measurement of the isotopic ratio.

Magnetic Field Effect and Magnetic Isotope Effect Studies

The magnetic field dependence of the kinetic parameters V_{max} and V_{max}/K_m for soybean lipoxygenase–catalyzed oxygenation of unlabeled and [11,11-$^{2}H_2$] linoleate was studied at pH 9.0 and 24.2°C. The results are shown in Figure 7.4. Daily variations in enzymatic activity were corrected by normalizing all rates in a standard assay at 27 μM without magnetic field. No magnetic field dependence of the enzymatic reaction is observed with linoleate. An alternate substrate, arachidonate, was studied at pH 10.0 and also showed no magnetic field dependence on the reaction catalyzed by soybean lipoxygenase (Fig. 7.5). The absence of a magnetic field effect does not support the proposed radical mechanism, although it does not strictly rule it out either. Nelson observed both alkyl and peroxyl radicals in the purple form of lipoxygenase that is obtained by incubation of the native enzyme with an excess of product (36). Magnetic field effects on the enzymatic reaction were studied in the presence of product hydroperoxide (3 μM) to enhance any possible magnetic field effect. However, there is no magnetic field dependence to the kinetic parameters V_{max} and V_{max}/K_m in the presence of hydroperoxide either.

Figure 7.6 shows the absence of a magnetic field dependence on the isotope effects $^{D}V_{max}$ and $^{D}(V_{max}/K_m)$. The data in Figure 7.6 were obtained by completing kinetic isotherms with unlabeled and deuterated linoleate as quickly as possible at the indicated magnetic field. Since the data in Figure 7.6 are ratios, no normalization to a standard rate was required.

Discussion

Magnetic Field Effect Studies

There is no magnetic field dependence to the lipoxygenase-catalyzed reaction. The result is somewhat unexpected in light of the observed radical intermediates with linoleate and hydroperoxide product. A geminate radical pair may be formed from carbon and hydrogen radicals, or a carbon radical and high-spin Fe^{II}. A peroxyl

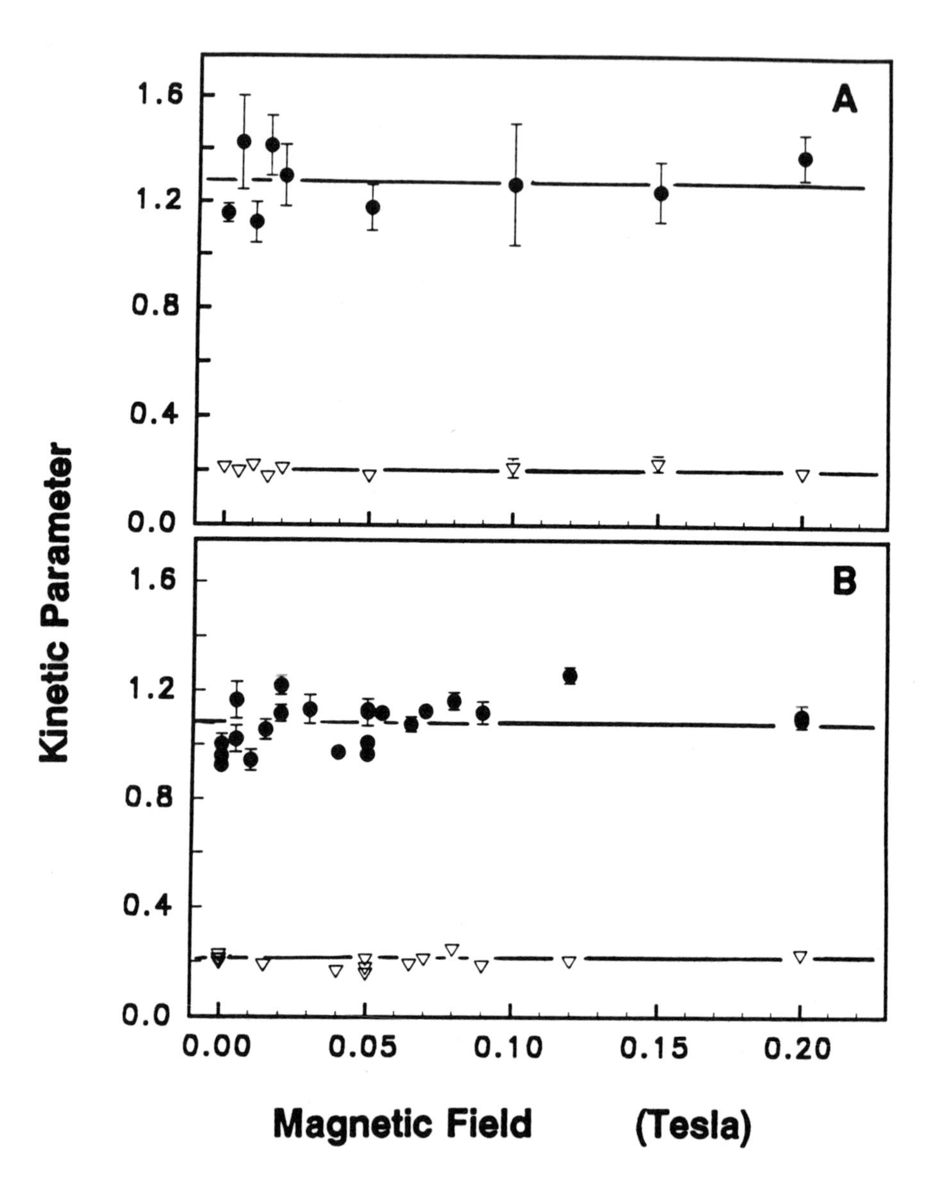

Figure 7.4. Magnetic field dependence of lipoxygenase kinetic parameters with linoleate as substrate • = V_{max} μM min^{-1}; ∇ = V_{max}/K_m min^{-1}. Assay conditions: 100 mM Ches pH 9.00, 24.2°C. *a)* Unlabeled linoleate: [Lipoxygenase] = 0.09 nM. *b)* Deuterated linoleate: [Lipoxygenase] = 2.7 nM in the assay. *Source:* Hwang and Grissom (27) by permission of the American Chemical Society, copyright 1994.

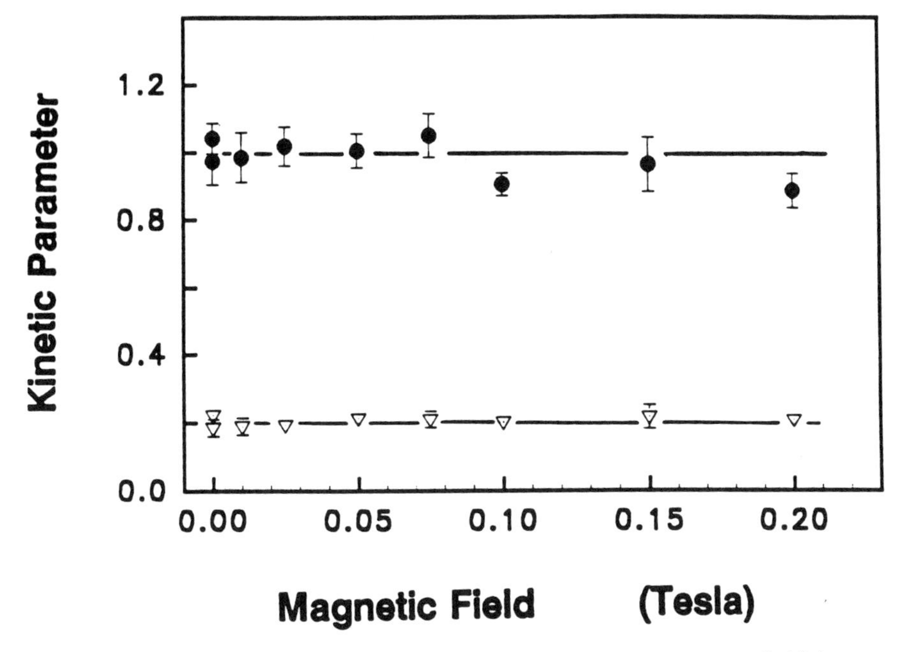

Figure 7.5. Magnetic field dependence of arachidonate kinetics. Magnetic field dependence of lipoxygenase kinetic parameters with arachidonate as substrate. Assay conditions: 100 μM borate pH 10.0, 24.1°C. Each point represents the indicated kinetic parameter obtained from fitting the substrate saturation isotherm to $d[P]/dt = V_{max}$ $[S]/K_m$ + [S] by nonlinear methods: • = V_{max} μM min^{-1}; ∇ = V_{max}/K_m min^{-1}. [Lipoxygenase] = 0.09 nm in the assay. The standard error for data points without plotted error bars is smaller than the symbol. *Source:* Hwang and Grissom (27) by permission of the American Chemical Society, copyright 1994.

radical may form a radical pair with the proximal Fe^{II} also. Recently Harkins and Grissom have shown the magnetic field dependence of the enzyme ethanolamine ammonia lyase by both steady-state and stopped-flow kinetic methods and attributed the magnetic field dependence to recombination of the radical pair formed by homolytic cleavage of the C–Co bond (25,26).

Grissom has considered the theoretical treatment of magnetic field effects in enzyme-catalyzed reactions with radical pair intermediates (20). In order to observe a magnetic field effect in an enzyme-catalyzed reaction, formation of the ES complex that contains the radical pair must be reversible. Considering the reversible formation of the ES complex in Scheme 7.4, k_4 would be the magnetic-sensitive step for the recombination of the [E• •S] radical pair, and k_7 would be the product-release step. Step k_7 is irreversible under initial rate conditions, since no product is present to support the reverse reaction. ES and EP are the initial binding complexes for E with S and P, and include any conformational change. The kinetic parameters, V_{max} and V_{max}/K_m,

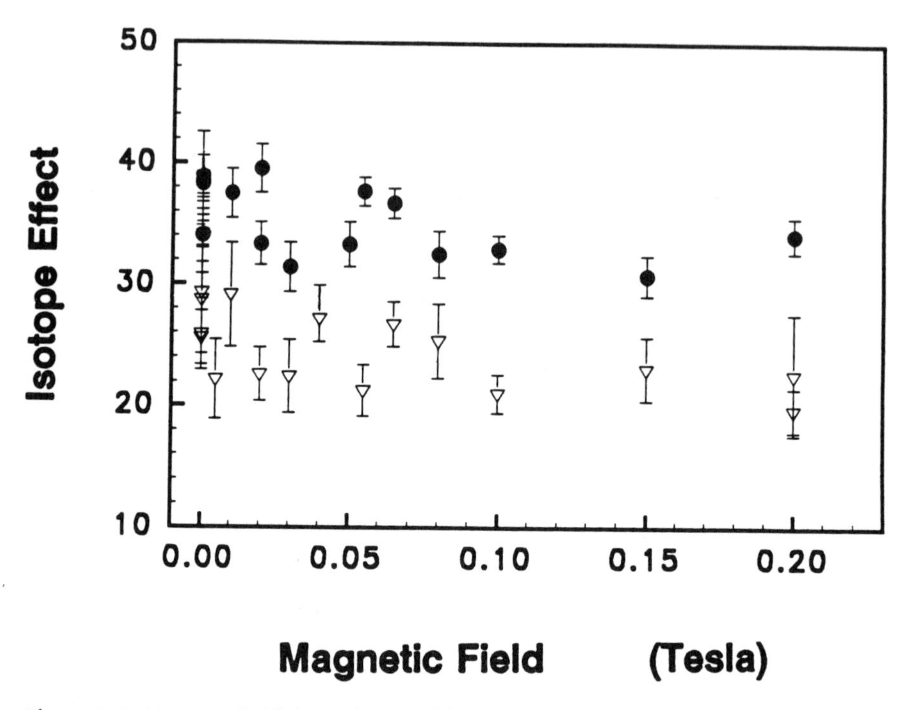

Figure 7.6. Magnetic field dependence of the deuterium KIE: • = $^{D}V_{max}$; ∇ = $^{D}(V_{max}/K_m)$. *Source:* Hwang and Grissom (27) by permission of the American Chemical Society, copyright 1994.

are expressed in Eqns. 4 and 5. The magnetic field-sensitive step, k_4, is present in both equations and suggests that magnetic field effects on the reversible enzymatic reaction can be observed in the kinetic parameters V_{max} and V_{max}/K_m.

Soybean lipoxygenase–catalyzed oxygenation of linoleate and arachidonate is carried out at saturated [O_2], around 220 μM at 24°C. The reported apparent K_m for dioxygen is 30 μM at 360 μM linoleate (37). The high concentration of dioxygen leads to an irreversible step between the formation of linoleate radical and peroxyl radical. In order to consider the possibility of magnetic field effects in the soybean lipoxygenase–catalyzed reaction, a kinetic scheme is proposed (Scheme 7.5) in which EA* and EP may include the radical pair intermediates so that k_4 and k_8 include magnetic field–sensitive steps for the recombination of the radical pair. Rate constant k_5 is the first irreversible step due to the saturating concentration of oxygen. Step k_9 represents the product (P)-release step that is irreversible when initial rates are measured (*vel*, Eqn. 6). The kinetic parameters V_{max} and V_{max}/K_m are given by Eqns. 7 and 8. The kinetic parameter V_{max} is dependent on the individual kinetic constants that relate different enzyme forms and reflect the slowest step during enzymatic turnover. Steps k_4 and k_8 are present in V_{max} and suggest a magnetic field dependence of V_{max} that may be diminished by kinetic complexity if k_4 and k_8 are not rate-limiting steps. The

$$E \underset{k_2}{\overset{k_1[S]}{\rightleftharpoons}} ES \underset{k_4}{\overset{k_3}{\rightleftharpoons}} [E^{\cdot}\ {}^{\cdot}S] \underset{k_6}{\overset{k_5}{\rightleftharpoons}} EP' \xrightarrow{k_7} E + P$$

$$V_{max} = \frac{k_3k_5k_7[E_t]}{k_3k_5 + k_3(k_6 + k_7) + k_4(k_6 + k_7) + k_5k_7} \quad (4)$$

$$V_{max}/K_m = \frac{k_1k_3k_5k_7[S][E_t]}{k_3k_5k_7 + k_2k_4\,(k_6 + k_7) + k_2k_5k_7} \quad (5)$$

Scheme 7.4. Magnetic field effects on the reversible reaction including the radical pair as an intermediate.

kinetic parameter V_{max}/K_m reflects the catalytic efficiency and is limited to events prior to the first irreversible step in the mechanism. If $k_5[B] >> k_4$, V_{max}/K_m is reduced to Eqn. 9, in which no magnetic field-sensitive step is present. This may be consistent with no magnetic field effect being observed on the kinetic parameter V_{max}/K_m in the soybean lipoxygenase–catalyzed oxygenation of linoleate. Another possibility may be due to spin-orbit coupling with the proximal iron that masks any magnetic field effect.

Kinetic Isotope Effect Studies

The primary and secondary KIE on the oxygenation of polyunsaturated fatty acids have been studied for the lipoxygenase reaction and for the corresponding nonenzymatic autoxidation reaction (Table 7.2). Egmond and co-workers observed a primary KIE of 8.7 ± 0.4 for [11,11-$^2H_2$] LA with soybean lipoxygenase at pH 9.0, 25°C by steady-state kinetic methods and suggested that the abstraction of hydrogen from the substrate is the rate-limiting step (30). Repeating this measurement with [11,11-$^2H_2$] linoleate of 99.4% dideuterium isotopic purity, we found $^DV_{max} = 36 \pm 3$ and $^D(V_{max}/K_m) = 28 \pm 2$ at 24.2°C (27). In order to quantify the KIE obtained by deuterated and unlabeled substrate competing in the same assay, the isotopic composition of residual substrate was determined by mass spectrometry. Under internal competition conditions, $^D(V_{max}/K_m) = 28 \pm 9$ at 24.2°C. The large standard error is due to the inherent unsuitability of determining large KIE by following residual substrate. The only interpretation possible from this measurement is that $^D(V_{max}/K_m)$ is indeed large and greater than 19. A complete summary of reported primary and secondary KIE with lipoxygenase is given in Table 7.2.

$$E \underset{k_2}{\overset{k_1[A]}{\rightleftharpoons}} EA \underset{k_4}{\overset{k_3}{\rightleftharpoons}} EA^{*} \xrightarrow{k_5[B]} EAB^{\bullet} \underset{k_8}{\overset{k_7}{\rightleftharpoons}} EP \xrightarrow{k_9} E + P$$

E: Soybean lipoxygenase
A: Linoleate or arachidonate
B: Dioxygen
P: Hydroperoxy product
$EA^{\bullet}$: Radical pair intermediate including linoleate radical
EP: Radical pair intermediate including peroxyl radical
E_t: Total concentration of lipoxygenase

$$vel = \frac{k_1k_3k_5k_7k_9[A][B][E_t]}{k_1k_3k_5k_7[A][B] + k_1k_3k_5(k_8 + k_9)[A][B] + k_1k_3k_7k_9[A] + k_1k_7k_9(k_4 + k_5[B])[A] + (k_2(k_5[B] + k_4) + k_3k_5[B])k_7k_9} \quad (6)$$

$$V_{max} = \frac{k_3k_5k_7k_9[B][E_t]}{k_7k_9(k_4 + k_5[B]) + k_3k_7k_9 + k_3k_5[B](k_8 + k_9) + k_3k_5k_7[B]} \quad (7)$$

$$V/K_A = \frac{k_1k_3k_5[B][E_t]}{k_3k_5[B] + k_2(k_4 + k_5[B])} \quad (8)$$

If $k_5[B] >> k_4$,

$$V/K_A = k_1k_3[E_t] / (k_2 + k_3) \quad (9)$$

Scheme 7.5. Magnetic field effects on the kinetics which the first irreversible step occurs before the releasing of product.

TABLE 7.2. Observed Kinetic Isotope Effects on the Oxygenation of Polyunsaturated Fatty Acids

Fatty acid	Primary KIE	Secondary KIE	T (°C)
[11,11-$^2H_2$] Linoleate[a]	$^DV_{max}$ = 8.7 ± 0.4		25
[11,11-$^2H_2$] Linoleate[b]	$^DV_{max}$ = 43.4 ± 1.9		25
	$^D(V/K)$ = 27.4 ± 3.8		
[11,11-$^2H_2$] Linoleate[c]	$^DV_{max}$ = 36 ± 3		25
	$^D(V/K)$ = 28 ± 2		
[9,10,12,13-$^2H_4$] Linoleate[a]		$^D(V/K)$ = 1.04 ± 0.03	25
6,9,12-[11,11-$^2H_2$, 1-^{14}C]	$^D(V/K)$ = 8.0[d]		0
Octadecatrienoate	^Dk = 4.5[e]		37
[10-DR-^3H] Arachdonate[f]		$^T(V/K)$ = 1.19	37
[5,6,8,9,11,12,14,15-^3H] Arachidonate[g]		$^T(V/K)$ = 1.16 ± 0.02	37
[9,10,12,13-^3H, 11,11-$^2H_2$] Linoleate[b]		$^T(V/K)$ = 1.16 ± 0.04	0

[a]Soybean lipoxygenase *Source:* Egmond et al. (30).
[b]Soybean lipoxygenase. *Source:* Glickman et al. (28).
[c]Soybean lipoxygenase. *Source:* Hwang and Grissom (27).
[d]Soybean lipoxygenase, and $^D(V/K)$ = 7.1 for platelet lipoxygenase at 37°C. *Source:* Hamberg (38).
[e]Autoxidation.
[f]Platelet lipoxygenase. *Source:* Brash et al. (39).
[g]Soybean lipoxygenase. *Source:* Wiseman (40).

There are at least six explanations for a deuterium KIE that exceeds the maximum value of about 7 for mass KIE (41).

1. Magnetic KIE on C-^1H and C-^2H bond homolysis and radical-pair recombination due to a difference in nuclear magnetic moment between ^1H and ^2H (42);
2. Quantum mechanical tunneling of ^1H relative to ^2H (43);
3. Branching to an alternate reaction pathway (44–46);
4. Multiplicative isotope effects on synchronous (coupled) bond scission at multiple isotopically labeled sites;
5. Autoxidation in a nonenzymatic chain process; and
6. Differential desolvation of unlabeled and deuterated linoleate in an isotope effect on binding to enzyme.

To address the first possibility of a magnetic isotope effect, we have determined the magnetic field independence of the lipoxygenase reaction. A magnetic isotope effect can only be a factor in radical-pair or biradical reactions. The nuclear magnetic moment of the relevant hydrogen isotopes is ^1H $\mu_N = 2.79 \times 10^5$ and ^2H $\mu_N = 0.86 \times 10^5$ (erg/T). In the early moments following C–H bond homolysis, intersystem crossing (ISC) due to hyperfine interactions in the C• •H radical pair will populate the three triplet spin states ($T_{0,\pm1}$) equally in the absence of a magnetic field. As the applied magnetic field is increased, the T_{+1} and T_{-1} levels are no longer degenerate with the singlet state (S_0), and ISC is decreased. The increased singlet population will enhance nonproductive radical-pair recombination. The greater nuclear magnetic moment of

^1H in the homolysis of a $C{-}^1H$ bond will lead to less nonproductive radical-pair recombination (relative to homolysis of a $C{-}^2H$ bond) and result in a greater isotopic discrimination against ^2H. A further increase in the overall observed KIE may result: any magnetic spin-induced isotope effect will be multiplied with a mass–isotope effect, such that the product of a mass–isotope effect of ≈9 and a magnetic spin enhancement of ≈4 could lead to an observed KIE in excess of 30. However, this explanation is precluded by the independence of the KIE with increasing magnetic field. The unusually large KIE cannot be due to differences in the nuclear magnetic moment of ^1H and ^2H.

A normal secondary tritium isotope effect $^T(V_{max}/K_m)$ of 1.19 was observed in the reaction of platelet lipoxygenase with [10-D_R-^3H] arachidonate (39). This result suggests that multiplicative isotope effects on synchronous (coupled) bond scission at multiple isotopically labeled sites is not the reason for the large observed KIE (a primary plus a secondary KIE). Similarly, autoxidation is also not the cause because the reaction catalyzed by soybean lipoxygenase is stereospecific.

The work of Glickman, Klinman, and Wiseman shows that the lipoxygenase reaction exhibits several other unusual features (28). First, the isotope effect increases with an increase in [linoleate] and it attains a value of $^D(V_{max}/K_m) = 80$ at 90 μM linoleate. Second, the observed isotope effect increases from $^D(V_{max}/K_m) = 6$ and $^DV_{max} = 17$ at 0°C to $^D(V_{max}/K_m) = 57$ and $^DV_{max} = 65$ at 35°C. This is the opposite trend expected from theoretical considerations of KIE. A quantum mechanical tunneling isotope effect or a mass-isotope effect should decrease as the temperature increases (41). This suggests quantum mechanical tunneling may not be the reason for the unusually large KIE.

In this last section, we will consider the possibility that the unusually large isotope effect may be caused by a branching reaction scheme that leads to the multiplication of KIE. In complex reactions with more than one isotope-sensitive step, each microscopic transformation exhibits only the isotopic discrimination afforded by the bond vibrational frequencies that are different in the unlabeled and deuterated reactants. In a sequential (linear) reaction, the observed isotope effect is largely determined by the isotopic discrimination on the rate-determining step. In no instance is the overall (observed) KIE described by the product of the isotope effects on individual microscopic rate constants.

Theory of Amplified KIE from Branching Reactions

In a multistep reaction with more than one isotope-sensitive step and a branch-point to a nonisotope-sensitive pathway that gives rise to a second product, the observed isotope effect on the minor product is increased and the isotope effect on the major product is decreased (33,44–47). Scheme 7.6 illustrates a branching reaction in which reactant A is converted to intermediate X (with rate constant k_1), then partitioned to products P and Q (with rate constants k_3 and k_5, respectively). With steady-state treatment of [X], the observed rate and KIE on the production of P is described by Eqns. 10 and 11, where k_{AP} is the observed rate for formation of P from starting material A. The isotope effect on each step is $k_1/k_1' = a_1$, $k_3/k_3' = a_3$, and $k_5/k_5' = a_5$. The ratio of product formation from intermediate X is b, which equals k_3/k_5. The corresponding rate constant with the heavy isotope is denoted with a prime (after Thibblin [47] and references therein; and Saunders and Melander [33]).

$$k_{AP} = k_1k_3/(k_3 + k_5) \tag{10}$$

$$\begin{aligned} k_{AP}/k_{AP}' &= k_1k_3(k_3' + k_5')/(k_1'k_3'(k_3 + k_5)) \\ &= a_1(b + (a_3/a_5))/(b + 1) \end{aligned} \tag{11}$$

If the ratio of isotope effect, $a_3/a_5 > 1$, the observed isotope effect on the formation of product P will be larger than a_1. If Q is the major product via a nonisotope-sensitive transformation ($b << 1$, $a_5 = 1$), Eqn. 11 simplifies to Eqn. 12 and a multiplicative isotope effect is observed.

$$k_{AP}/k_{AP}' = a_1a_3 \tag{12}$$

It is through Eqn. 12 that we can understand the unusually large isotope effects that can arise in a branching reaction. Large isotope effects caused by branching in chemical systems are known (Thibblin and Ahlberg [48] and references therein), and a revisionist analysis of large KIE previously ascribed to tunneling seems to be underway (44).

Branching in Soybean Lipoxygenase

Soybean lipoxygenase catalyzes the regio- and stereospecific dioxygenation of 9(*Z*),12(*Z*)-octadecadienoic acid to form 13(*S*)-HPOD at pH 9.0 as the major product with a small (3%) amount of 9(*S*)-10(*E*),12(*Z*)-9-hydroperoxy-10,12-octadecadienoic acid (9(*S*)-HPOD) (9,49,10). In order to observe multiple isotope effects on consecutive isotope sensitive steps in a branching mechanism, an alternate fast pathway from the intermediate must exist (Scheme 7.6). If branching from a high-energy intermediate occurs in lipoxygenase, this alternate product must meet the following criteria to observe an enhanced isotope effect:

1. It must not absorb at 234 nm;
2. It must be formed and released in a nonisotope-sensitive fashion; and
3. It must be a substrate for the enzyme.

The possibility of an alternate reaction pathway for lipoxygenase might involve a

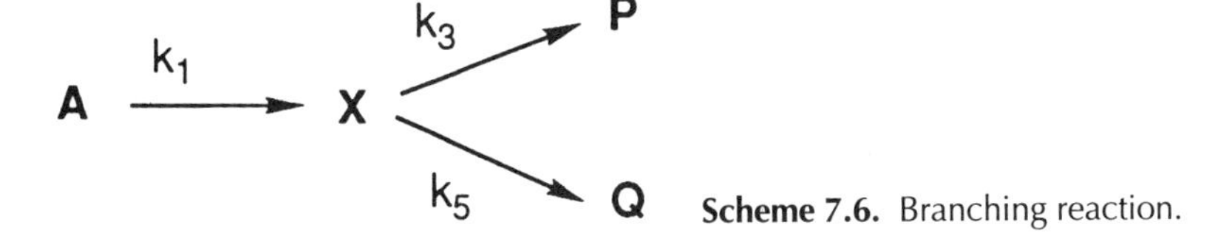

Scheme 7.6. Branching reaction.

linoleate radical intermediate that could dissociate from the enzyme into solution, followed by either hydrogen atom abstraction from solvent or isomerization to *cis,cis*-linoleate and back into the *cis,trans* isomer, which is an alternate substrate for lipoxygenase (50). To look for additional products, radioactive [1-^{14}C] linoleate was reacted with lipoxygenase. The reaction was quenched at 25% conversion, as determined spectrophotometrically. The reaction mixture was separated by reverse-phase HPLC and analyzed for radioactivity by scintillation counting. There were only two peaks in the radioactivity trace. Analyzing these fractions by NMR spectrometry, these two peaks were identified as LA and 13(*S*)-HPOD. No isomers of linoleate or additional products were observed. Attempts to trap the radical intermediate by D_2O and EtOH-d6, and determine the isotopic distribution of linoleate by MS were not successful after 25% conversion. No increase in the intensity of the M + 1 and M + 2 peaks (281 and 282 amu for mono- and dideuterated LA) was observed. This seems to preclude amplification of the primary deuterium KIE by branching to an alternate pathway that regenerates linoleate in a pathway that is not isotope sensitive.

Conclusion

Soybean lipoxygenase exhibits one of the largest deuterium KIE in an enzymatic reaction. The unusually large isotope effect cannot be attributed to a magnetic spin interaction involving a radical intermediate, or amplification of the isotope effect by dissociation of an intermediate through an isotope-insensitive pathway. An adequate explanation for the unusually large KIE awaits realization.

References

1. de Groot, J.J.M.C., Veldink, G.A., Vliegenthart, J.F.G., Boldingh, J., Wever, R., and Van Gelder, B.F. (1975) *Biochim. Biophys. Acta 377,* 71.
2. Veldink, G.A., Vliegenthart, J.F.K., and Boldingh, J. (1977) *Prog. Chem. Fats Other Lipids 15,* 131.
3. Gibian, M.J., and Galaway, R.A. (1977) in *Bioorganic Chemistry,* Van Tamelon, E.E., Academic Press, New York.
4. Vliegenthart, J.F.G., and Veldink, G.A. (1982) in *Free Radical in Biology,* Pryor, W.A., Vol. V, Chapter 2, Academic Press.
5. Corey, E.J. (1987) *Pure Appl. Chem. 59,* 269.
6. Boyington, J.C., Gaffney, B.J., and Amzel, L.M. (1993) *Science 260,* 1482.
7. Wang, Z.K., Killilea, S.D., and Srivastava, D.K. (1993) *Biochemistry 32,* 1500.
8. Schilstra, M.J., Veldink, G.A., and Vliegenthart, J.F.G. (1994) *Biochemistry 33,* 3974.
9. Gardner, H.W. (1989) *Biochim. Biophys. Acta 1001,* 274.
10. Egmond, R., Veldink, G.A., Vliegenthart, J.F.G., and Boldingh, J. (1972) *Biochem. Biophys. Res. Commun. 48,* 1055.
11. Aust, S.D., and Svingen, B.A. (1982) in *Free Radical in Biology,* Pryor, W.A., Vol. V, Academic Press, New York.
12. Feiters, M.C., Assa, R., Malmstrom, B.G., Slappendel, S., Veldink, G.A., and Vliegenthart, J.F.G. (1985) *Biochim. Biophys. Acta 831,* 302.

13. Nelson, M.J., Cowling, R.A., and Seitz, S.P. (1994) *Biochemistry 33,* 4966.
14. Corey, E.J., and Nagata, R. (1987) *J. Am. Chem. Soc. 109,* 8107.
15. Corey, E.J., and Walker, J.C. (1987) *J. Am. Chem. Soc. 109,* 8108.
16. Corey, E.J., Wright, S.W., and Matsuda, S.P.T. (1989) *J. Am. Chem. Soc. 111,* 1452.
17. de Groot, J.J.M.C., Garssen, G.J., Vliegenthart, J.F.G., and Boldingh, J. (1973) *Biochim. Biophys. Acta 326,* 279.
18. Chamulitrat, W., and Mason, R.P. (1989) *J. Biol. Chem. 264,* 20968.
19. Aoshima, H., Kajiwara, T., Hatanaka, A., and Hatano, H. (1977) *J. Biochem. 82,* 1559.
20. Grissom, C.B. (1995) *Chem. Rev. 95,* 3.
21. Hayashi, H. (1990) in *Photochemistry and Photophysics,* Rabek, J.F., Vol. I, Chapter 2, CRC Press, Boca Raton, FL.
22. Vanag, V.K., and Kuznetsov, A.N. (1988) *Izv. Akad. Nauk SSSR Ser. Biol. 2,* 215.
23. Turro, N.J. (1983) *Proc. Natl. Acad. Sci. USA 80,* 609.
24. Steiner, U.E., and Ulrich, T. (1989) *Chem. Rev. 263,* 958.
25. Harkins, T.T., and Grissom, C.B. (1994) *Science 263,* 958.
26. Harkins, T.T., and Grissom, C.B. (1995) *J. Am. Chem. Soc. 117,* 566.
27. Hwang, C.-C., and Grissom, C.B. (1994) *J. Am. Chem. Soc. 116,* 795.
28. Glickman, M.H., Wiseman, J.S., and Klinman, J.P. (1994) *J. Am. Chem. Soc. 116,* 793.
29. Takeguchi, C.A. (1974) *The Biosynthesis of Prostaglandins,* The University of Wisconsin, Ph.D. Thesis.
30. Egmond, R., Veldink, G.A., Vliegenthart, J.F.G., and Boldingh, J. (1973) *Biochem. Biophys. Res. Commun. 54,* 1178.
31. Tucker, W.P., Tove, S.B., and Kepler, C.R. (1971) *J. Labelled Compd. 7,* 11.
32. Bild, G.S., Ramadoss, C.S., and Axelrod, B. (1977) *Arch. Biochem. Biophys. 184,* 36.
33. Melander, L., and Saunders, W.H. (1987) *Reaction Rates of Isotopic Molecules,* Krieger Publishing, Florida.
34. Smith, W.C., and Lands, W.E.M. (1970) *Biochem. Biophys. Res. Commun. 41,* 846.
35. Axelrod, R., Cheesbrough, T.M., and Laakso, S. (1981) *Methods Enzymol. 71,* 441.
36. Nelson, M.J., Seitz, S.P., and Cowling, R.A. (1990) *Biochemistry 29,* 6897.
37. Tappel, A.C., Boyer, P.D., and Lundberg, W.O. (1952) *J. Biol. Chem. 199,* 267.
38. Hamberg, M. (1984) *Biochem. Biophys. Res. Commun. 117,* 593.
39. Brash, A.R., Ingram, C.D., and Mass, R.L. (1986) *Biochim. Biophys. Acta 875,* 256.
40. Wiseman, J. (1989) *Biochemistry 28,* 2106.
41. Bell, R.P. (1980) *The Tunnel Effect in Chemistry,* Chapter 5, Chapman Hall, New York.
42. Turro, N.J., and Kraeutler, B. (1984) in *Isotopes in Organic Chemistry,* Buncel, E., and Lee, C.C., Vol. 6, Chapter 3, Elsevier, Amsterdam.
43. Cha, Y., Murray, C.J., and Klinman, J.P. (1989) *Science 243,* 1325.
44. Osterheld, T.H., and Brauman, J.I. (1992) *J. Am. Chem. Soc. 114,* 7158.
45. Korzekwa, K.R., Trager, W.F., and Gillete, J.R. (1989) *Biochemistry 28,* 9012.
46. Thibblin, A., and Ahlberg, P. (1977) *J. Am. Chem. Soc. 99,* 7926.
47. Thibblin, A. (1988) *J. Phys. Org. Chem. 1,* 161.
48. Thibblin, A., and Ahlberg, P. (1989) *Chem. Soc. Rev. 18,* 209.
49. Hamberg, M. (1971) *Anal. Biochem. 43,* 515.
50. Funk, M.O., Andre, J.C., and Otsuki, T. (1987) *Biochemistry 26,* 6880.

Chapter 8

Phytooxylipins: The Peroxygenase Pathway

Elizabeth Blée

Institut de Biologie Moléculaire des Plantes (CNRS UPR 406), Département d'Enzymologie Cellulaire et Moléculaire, 28 rue Goethe, 67083 Strasbourg Cedex, France.

Introduction

Over the last 20 years, a massive body of information revealed the fundamental role that the recently named oxylipins (1), oxygenated metabolites of polyunsaturated fatty acids (especially those derived from arachidonic acid), played in mammalian inflammatory responses and more generally in stress conditions: infection, allergy, exposure to food, drug and/or environmental xenobiotics. By comparison, very little is known of plant responses to similar compounds. Recently, however, interest in phytooxylipins has increased, since these compounds were suggested to be involved in plant defense mechanisms.

Oxylipins are quite common metabolites in plants that are mainly derived from LA and LNA. Lipoxygenases (EC 1.13.11.12) produced from the C18 unsaturated fatty acids either 9- or 13-hydroperoxyoctadecadi(tri)enoic acids, or a mixture of both, depending on the source of the enzyme; these highly reactive aliphatic molecules are then rapidly metabolized by plant cells into a variety of physiologically active derivatives (Fig. 8.1). Two well-characterized enzymes have been shown to degrade these hydroperoxides into protective compounds, for example antibacterial and wound-healing agents, or precursors of regulatory substances, such as jasmonic acid (2). The first is a lyase that cleaves the bond between C-12 and C-13 (α-scission from the *trans*-11,12-double bond) yielding C_6 aldehydes and 12-oxo-*trans*-9-dodecenoic acid, a precursor of the wound hormone traumatin (3). The second is allene oxide synthase, a hydroperoxide dehydrase and nonclassical cytochrome P450 (4), that catalyzes the dehydration of 13-hydroperoxylinolenic acid into a very unstable allene oxide, that is then either enzymatically cycled into 12-oxo-phytodienoic (5), the precursor of jasmonic acid, or hydrolyzed into α- and γ-ketols (2).

Recently, we have described a new fate for fatty acid hydroperoxides: the peroxygenase cascade, which also leads to products relevant to plant defense mechanisms, since some of them are cutin monomers (6,7), or possess antifungal properties (8,9). This new pathway involves two enzymes, a peroxygenase that catalyzes an intramolecular transfer of oxygen from hydroperoxides yielding epoxyalcohols or an intermolecular oxygen transfer (cooxidation reaction) resulting, for example, in the epoxidation of double bond of unsaturated fatty acids (10); and an epoxide hydrolase that preferentially hydrates the epoxides formed by the peroxygenase (11). This chapter deals with our present knowledge of the peroxygenase cascade and its involvement in the defense responses of plants against pathogens and xenobiotics.

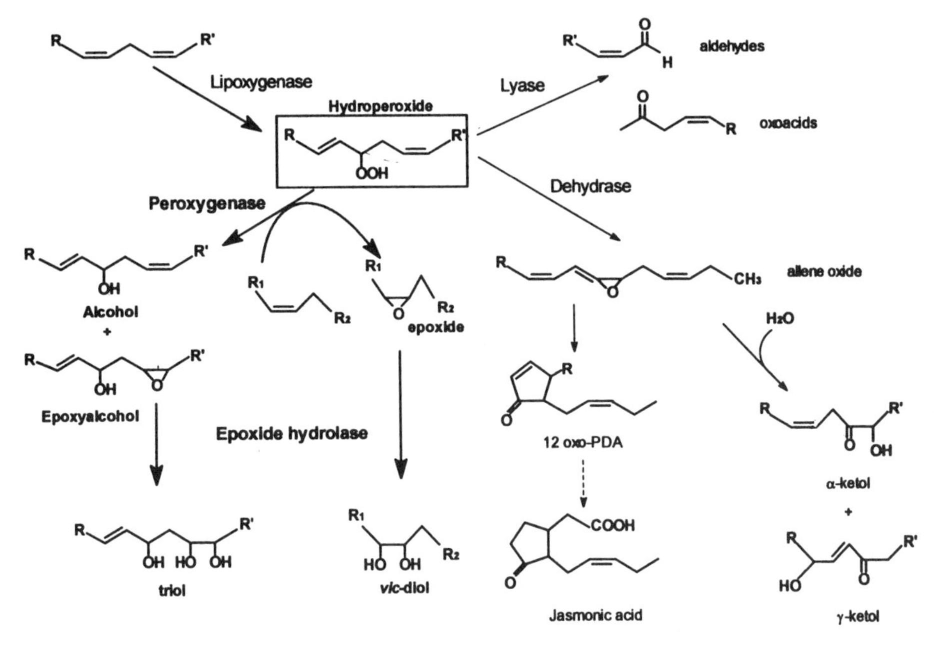

Figure 8.1. The lipoxygenase pathway involving the well-characterized hydroperoxide lyase and dehydrase pathways but also the newly described peroxygenase pathway (in boldface).

The Peroxygenase

In the late 1970s, besides the well-known cytochrome P450s (recently renamed "heme-thiolate protein"), a CO-binding hemoprotein, initially termed "H_2O_2-reducible hemoprotein," was described in microsomal fractions of peas by Ishimaru and Yamazaki (12). In the presence of hydroperoxides this new protein was shown to catalyze the hydroxylation of aromatic compounds, such as indole, phenol, 1-naphtol, and aniline into indoxyl, hydroquinone, naphtohydroquinone, and phenylhydroxylamine, respectively. Furthermore, experiments conducted in the presence of ^{18}O-labeled LA hydroperoxide demonstrated that the oxygen atom introduced into the products was derived from the hydroperoxide (13). Since, during the reaction the hydroperoxide was reduced to its corresponding alcohol, the trivial name of "peroxygenase" was given to this new oxygenase activity.

Almost ten years later, we found the occurrence of an oxidase in soybean seedlings that catalyzed the sulfoxidation of methiocarb, an insecticide, that did not need any cofactor, such as NADPH or FAD, for its activity, but required hydroperoxides as oxidants (14). Later, we showed that this enzyme was also able to actively epoxidize double bonds of unsaturated fatty acids (15). Hamberg and Hamberg confirmed the presence of this epoxygenase in the bean *Vicia faba* (16). Since then, we have demonstrated that all of these activities, hydroxylation, sulfoxidation, and epoxidation (or epoxygenation), are in fact co-oxidation reactions catalyzed by a single protein or closely related isozymes (10), therefore, we keep the name of "peroxygenase" originally given to this rather unique oxidase.

Characteristics of the Peroxygenase

In all plants examined so far, peroxygenase activity was predominantly found associated with microsomal fractions. It should be mentioned, however, that in such crude systems hydroperoxides can also serve as substrates for other oxidative reactions, which can be enzymatic, such as co-oxidations catalyzed by lipoxygenases (17,18), peroxidases (19), and heme-thiolate protein (20), or nonenzymatic, that is initiated by transition metals or heme complexes (21).

In order to characterize the peroxygenase, it was of great importance to have access to enzyme preparations that largely obliterate the complicating factors. This was achieved by carefully washing the membrane preparation to eliminate soluble peroxidases and lipoxygenases, then by solubilizing the peroxygenase with a nonionic detergent and purifying the enzyme on a cation-exchange chromatography column (22). This purification step gave a clear-cut separation of the peroxygenase from the solubilized membrane-bound lipoxygenases (active at pH 9.0 or 5.5 [15]). The dithionite-reduced peroxygenase fraction exhibited an absorbance at 412 nm in the presence of CO, but no peak at 450 nm or 420 nm, revealing the absence of active or denatured cytochrome P450. In the presence of pyridine, this reduced fraction showed a band at 556 nm characteristic of a pyridine-ferrohemoprotein complex of cytochrome *b* type protoheme (23). Indeed, optical spectra of the native peroxygenase indicated maximal absorption peaks at 630, 500, and 408 nm, which are the α, β, and Soret bands typical of high-spin (out-of-plane) Fe(III) heme (22,24). We have demonstrated that this hemoprotein supports peroxygenase-dependent oxidations; a perfect correlation could be observed between kinetics of inactivation of the peroxygenase activity and the destruction of the heme (measured as the disappearance of the Soret band) in the presence of cumene hydroperoxide (15).

Molecular Mechanism of the Peroxygenase

In order to study the molecular mechanism of peroxygenase, it was important to distinguish this enzyme from other enzymes, such as peroxidases and lipoxygenases, that were found to catalyze similar reactions. Peroxidases catalyze the oxidation of substrates using hydrogen peroxide as the electron donor. The product can be a dehydrogenated (Eqn. 1) or an oxygenated (Eqn. 2) compound.

$$SH_2 + H_2O_2 \rightarrow S + 2H_2O \qquad (1)$$

$$S + H_2O_2 \rightarrow SO + H_2O \qquad (2)$$

However, peroxygenase was demonstrated to be unable to catalyze "classical peroxidase" reactions, such as the oxidation of guaiacol (22). So far, only oxygenated compounds resulting from peroxygenase-mediated oxidation of the substrates have been found using hydrogen peroxide, cumene, or fatty acid hydroperoxides as the oxidant. It should also be pointed out that in plant systems similar oxidation reactions to the one catalyzed by peroxygenase have been shown to be catalyzed by peroxidases (19), or lipoxygenases (17). These reactions, which have been named co-oxidation reactions, have a very distinct mechanism, that is they involve free peroxy and or alkoxy radicals generated during the enzymatic process. In this case, the oxygen introduced into the products are derived from the radicals and thus originate from molecular oxygen. In contrast, experiments conducted in the presence of ^{18}O-labeled LA hydroperoxide revealed that in hydroxylation, sulfoxidation, or epoxidation reactions peroxygenase catalyzed the transfer of an oxygen atom from the hydroperoxides to the substrates (13,15,16,25, Eqn. 3), ruling out co-oxidative processes involving free radical species.

$$RO^*O^*H + S \rightarrow SO^* + RO^*H \quad (3)$$

Reactions catalyzed by peroxygenase imply the oxidation of the substrate inside the active site of the enzyme by some oxidized species of the hemoprotein, deriving from two possible modes of cleavage of the O–O bond of hydroperoxides by the enzyme (Eqn. 4). While a heterolytic cleavage will produce an Fe(V)=O species, a homolytic cleavage yields an RO· radical and an Fe(IV)-OH species susceptible to co-oxidize compounds. For example, both homolytic and heterolytic cleavages of the O–O bond were shown to take place during the reaction of cytochrome P450 enzymes with hydroperoxides and can occur as competing processes (26–29).

$$\mathrm{Fe(III) + ROH} \begin{cases} \xrightarrow{\text{homolytic}} \mathrm{Fe(IV)\text{-}OH + RO\bullet} \\ \xrightarrow{\text{heterolytic}} \mathrm{Fe\ (V){=}O\ + ROH} \end{cases} \quad (4)$$

Ishimaru and Yamazaki reported that during the peroxygenase hydroxylation of aromatic compounds, hydroperoxides used as co-oxidants were mostly converted to their corresponding alcohols and to some unidentified by-products (13). When studying the epoxidation of oleic acid, preliminary experiments allowed us to suggest that these minor products were isomers of epoxidized hydro(pero)xylinoleic acid, that were competing with oleic acid as substrates of the peroxygenase-oxidizing complex occurring during the reaction (15). These by-products were finally unambiguously characterized as 9,10-epoxy-13(*S*)-hydroxy-11(*E*)-octadecenoic acid (9,10-EHOD) by Hamberg and Hamberg (16). According to these authors, two pathways were taking place during LA oxygenation by microsomal fractions of *Vicia faba.* 13-HPOD, the hydroperoxide resulting from oxygenation of this fatty acid by a lipoxygenase, allowed a so-called epoxygenase (peroxygenase) catalyzed oxidation of LA into a 1:1 mixture of 9,10-epoxy-12-octadecenoic acid (9,10-EOD) and 12,13-epoxy-9-octadecenoic acid (12,13-EOD) in a major pathway (Eqn. 5). In a minor pathway, the hydroperoxide was transformed to an epoxyalcohol via intramolecular and intermolecular epoxidation (1:1) reactions (Eqn. 6 [16]).

$$13\text{-HPOD} + \text{LA} \rightarrow 13\text{-HOD} + 9{,}10 \text{ and } 12{,}13\text{-EOD} \quad (5)$$

$$13\text{-HPOD} + 13\text{-HOD} \rightarrow 13\text{-HOD} + 9{,}10\text{-EHOD} \quad (6)$$

The reduction of the hydroperoxide into 13-HOD, its corresponding alcohol, was consistent with a heterolytic cleavage of the peroxy O–O bond catalyzed by the peroxygenase, but the formation of epoxy–hydroxy fatty acids could also result from homolytic cleavage (30). Using 13-hydroperoxyoctadeca-9(*Z*),11(*E*),15(*Z*)-trienoic acid as a mechanistic probe, we have proposed the reaction scheme illustrated in Figure 8.2 (10). Accordingly, the peroxygenase mediated the exclusive heterolytic cleavage of this hydroperoxide, yielding the corresponding alcohol with a concomitant two-electron oxidation of the enzyme. This resulted in the formation of a ferryl-oxo complex (compound I-like). This intermediate then epoxidizes the more reactive nonconjugated *cis*-double bond (i.e., at position 15,16 or 9,10 of the 13-hydroxy derivatives of LNA and LA, respectively) either before it diffuses out of the active site (caged reaction, intramolecular mechanism), or after recomplexation within the active site of the molecule (co-oxidation reaction).

The proportion of intermolecular versus intramolecular oxygen transfer depends on the nature of the hydroxylated compound resulting from the reduction of the hydroperoxide and on its relative affinity for the peroxygenase, in other words, the ease with which it engages in an intramolecular oxidative process. That means that a hydroperoxide generating a good substrate for the iron-oxo intermediate should support little co-oxidation by the peroxygenase because of the incidence of high levels of intramolecular oxygen transfer. Conversely, a hydroperoxide that has no readily oxidizable functions after reduction to the corresponding alcohol, such as cumene hydroperoxide, is an excellent co-oxidant because of the lack of an intramolecular oxygen transfer pathway.

An immediate consequence is the extent of the mechanism-based inactivation of the enzyme that accompanies catalysis. In the absence of acceptor substrates, peroxygenase operating through unstabilized ferryl oxygen species in direct contact with the substrate appears extremely fragile during the catalytic cycle. But this enzyme is less susceptible to such degradation when reacting with hydroperoxides that have products that are easily oxidized. For example, cumene hydroperoxide is much more destructive than LA hydroperoxide.

Finally, reactions catalyzed by peroxygenase could proceed through a concerted or a multistep mechanism. We had addressed the problem of whether compound I- or II-like species act as the ultimate oxygen donors for the sulfide substrates by studying peroxygenase oxidation of para-substituted thioanisoles and using linear free-energy relationships (22). Studying such sulfoxidation reactions (presumably epoxidation and hydroxylation reactions occur via a comparable mechanism), we showed that co-oxidation reactions proceed via a nonconcerted mechanism (Eqns. 7–9) involving a

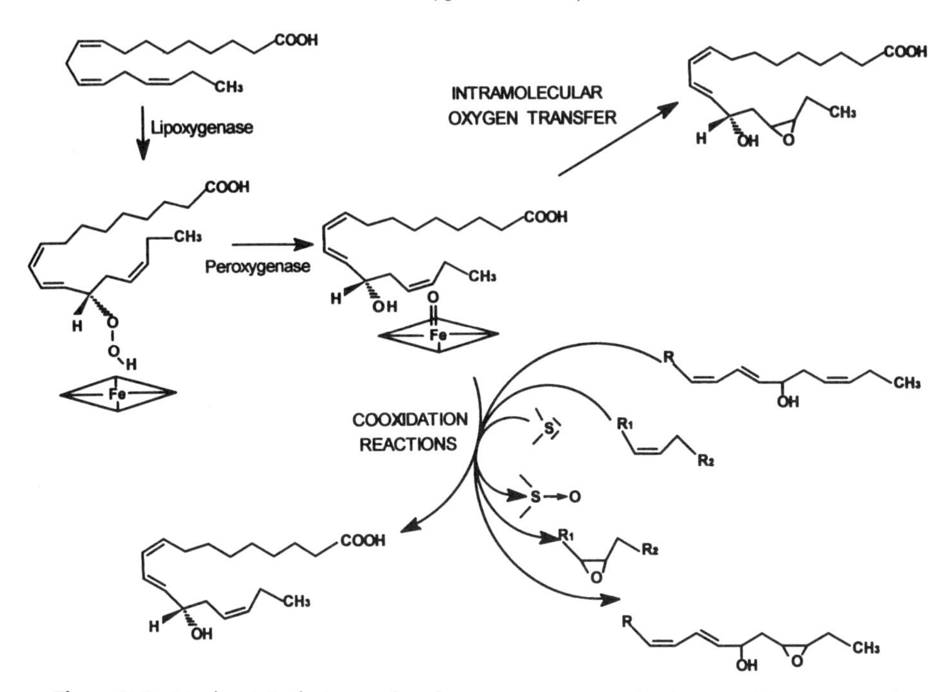

Figure 8.2. Mechanistic features of soybean peroxygenase. Hydroperoxides generated from unsaturated fatty acids by lipoxygenases react with peroxygenase according to intramolecular and intermolecular pathways. Co-oxidation reactions catalyzed by peroxygenase occur via the intermolecular mechanism.

heterolytic cleavage of the hydroperoxide O–O bond to yield a compound I-type oxo-heme that reacts with the substrate to form a radical-cation intermediate in the rate-limiting step and a rapid and stereoselective oxygen transfer leading to the sulfoxide.

$$Fe^{3+} + ROOH \rightarrow (Fe{=}O)^{3+} + ROH \quad (7)$$

$$(Fe{=}O)^{3+} + S \rightarrow (Fe{=}O)^{2+} + S^{\cdot+} \quad (8)$$

$$(Fe{=}O)^{2+} + S^{\cdot+} \rightarrow Fe^{3+} + SO \quad (9)$$

The resulting hydroxyl (ROH; Eqn. 7) undergoes intra- or intermolecular oxidation depending on its nature. In the overall oxidative process, the rate-limiting step is the transfer of the oxygen atom from the iron-oxo intermediate to the substrates. The important stereoselectivity observed in the oxygen-transfer step indicates that the putative intermediate must be very reactive. Moreover, in this hypothesis of a one-electron transfer process in the peroxygenase-catalyzed reaction, we must envision a tight coupling of the oxygenation step (Eqn. 9) with the electron abstraction (Eqn. 8), that is the trapping of the radical intermediate before it diffuses out of the active site. This would be in agreement with the lack of *S*-alkylation observed in the sulfoxidation reaction, with the inability of the peroxygenase to support a one-electron peroxidation of guaiacol and with the observation that a compound such as cyclopropyl phenyl sulfide is an excellent substrate for the peroxygenase and not a mechanism-based inhibitor.

However, one cannot exclude that we are dealing with a borderline reaction in which it becomes difficult to distinguish a two-step (via a radical-cation intermediate) from a one-step (direct insertion) oxygen-transfer mechanism. But recently, more direct evidence of the real existence of the radical intermediate was given during the formation of 9,10-epoxystearate by peroxygenase; a sizeable amount of *trans* epoxide (<16%) was formed during the enzymatic epoxidation of oleic acid, suggesting some isomerization of the radical cation while still in the active site (Blée, unpublished results).

Specificity, Regio- and Stereoselectivity of the Peroxygenase

Peroxygenase in presence of hydroperoxides uses co-oxidation to catalyze hydroxylation of aromatic compounds (13), sulfoxidation of different substituted thioanisoles (22), bioactivation of the herbicide EPTC (*S*-ethyl-*N,N,*-dipropylthiocarbamate [31]), and the insecticide methiocarb (14), but also oxidation of mono- and polyunsaturated fatty acids, such as oleic acid, LA, and LNA, into mono- and diepoxides (15,16,32). Thus peroxygenase seems to accept a rather broad range of structures that have hydrophobicity in common as cosubstrates. For example, while glutathione is not oxidized by peroxygenase, 2-mercaptoethanol is a powerful competitor and can be used as an inhibitor of peroxygenase-catalyzed co-oxidations (33), providing a convenient tool to distinguish the peroxygenasecatalyzed reactions from other oxidative enzymes.

Peroxygenase can also exhibit a strong stereoselectivity, since octadeca(ce)noic acids with double bonds in the *trans* configuration were not epoxidized at detectable rates (15,32). It seems that in this case, the double bonds of the fatty acids cannot be positioned correctly within the active site of the enzyme, since shortening the carbon chain of elaidic acid to 12 carbons allows the epoxidation of the *trans* double bond at 9,10-position of 9(*E*)-dodecenoic acid (Blée, unpublished results). This gives an interesting insight into the active site of the enzyme. From this data, we think that it has a funnel-shaped topology and that fatty acids must be able to adopt a hairpin configuration to enter it.

Another point is that the oleic acid methyl ester derivative is a much better substrate than the free fatty acid, but its C18-hydroxylated derivative is poorly oxidized by the peroxygenase. This suggests an important role of the unsaturated fatty acid's terminal methyl group for positioning into the active site of the enzyme. Peroxygenase also presents an important regioselectivity with LA and LNA: it predominantly catalyzes the epoxidation of the 9,10-double bond (15,34). The rate of epoxidation is affected not only by the position of the double bond, but also by the chain length of the substrate. The preferred monounsaturated fatty acids by soybean peroxygenase are as follows C18 > C16 > C14.

One characteristic of the peroxygenase-mediated co-oxidations is the important stereochemistry of the products examined so far. During enzymatic oxygenation of methyl *p*-tolyl sulfide, the *S*-sulfoxide was produced up to 90% enantiomeric excess (e.e.) in the presence of 13(*S*)-hydroperoxylinoleic acid as the oxidant (22). A similar enantiomer was formed by mammalian cytochrome-P450 dependent oxidases (35), while no chirality was induced by horse radish peroxidase (36). An opposite chirality, that is *R* enantiomer, is observed in the formation of the alkyl aryl sulfoxides catalyzed by a FAD-dependent mono-oxygenase purified from dog liver microsomes (90% e.e.[35]); and chloroperoxidase (12.7% e.e. [36]). The high stereoselectivity of the peroxygenase-catalyzed oxo-transfer, in addition to the incorporation of oxygen from the hydroperoxide into the substrate, may indicate direct access of the sulfide to the ferryl oxygen group, as has been suggested in the reaction catalyzed by chloroperoxidase and cytochrome-P450 dependent enzymes (37). In contrast, the lack of stereospecificity in sulfoxidation catalyzed by horseradish peroxidase might be the result of a reaction occurring at the heme edge.

Peroxygenase also presents an important enantiofacial selectivity during epoxidation reactions. All the epoxides from mono- or polyunsaturated fatty acids mainly exhibited the (*R*),(*S*) configuration with high optical purities, especially at the 9,10 position of naturally occurring plant fatty acids (16,32,34). However, one exception has been reported concerning the stereospecificity of the 12,13-epoxyoctadeca(ce)noic acids that mostly presents the configuration (*S*),(*R*) when produced by a microsomal fraction from *Vicia faba* (32). This could result from the preferential transformation of the 12(*R*),13(*S*)-epoxyoctadeca(ce)noate into the diepoxide (32), but it could also depend on the degree of purity of the peroxygenase used. For instance, epoxide hydrolase, which hydrates epoxides into diols, was mostly found in soluble fractions from beans (32,38). However, it can contaminate microsomes to a variable extent depending on the membrane-washing procedures (38). Moreover, a form of the hydrolase that was tightly bound to the membrane was also detected (38). These soluble and membrane-associated epoxide hydrolases present a very strong enantioselectivity in favor of the 12(*R*),13(*S*) enantiomer of 12,13-epoxyoctadecenoic acid (39). Thus the 12,13-monoepoxides obtained by epoxidation of LA by partially purified soybean peroxygenase or by soybean microsomes show opposite stereochemistries, that is the peroxygenase yields a 12(*R*),13(*S*)-isomer (30% e.e.), whereas crude microsomes gave mostly a 12(*S*),13(*R*)-isomer, the slowest substrate for the epoxide hydrolase (39).

In conclusion, peroxygenase is one of the rare microsomal heme-enzymes that catalyze the exclusive heterolytic cleavage of hydroperoxides. Hydroperoxide dehydrase, which appears to be the predominant cytochrome-P450 enzyme in many plant species (40), also utilizes organic peroxide substrates. Like peroxygenase, it is a ferric, high-spin hemoprotein, but it dehydrates the hydroperoxide into an intermediate allene-oxide, the precursor of "prostaglandin-like" compounds (jasmonic acid) and ketones. Compared to peroxygenase, it is likely that hydroperoxide dehydrase proceeds via a homolytic scission of the hydroperoxide, illustrating the importance of the apoprotein in determining the mechanism of hydroperoxide cleavage by hemoproteins. However, both enzymes exemplify a direct cooperation of lipoxygenase with another oxygenase (peroxygenase or hydroperoxide dehydrase) in different biosynthetic pathways. In that regard, it is noteworthy that in plants the lipoxygenase/peroxygenase couple plays the role devoted to the hydroperoxidase activity of the cyclooxygenase in animals, that is insertion of two atoms of oxygen into PUFA and reduction of the hydroperoxy group to the corresponding alcohol. Finally, from different aspects, peroxygenase could be considered to be the "missing-link" between peroxidases and cytochrome P450 in the plant kingdom.

The Epoxide Hydrolase

Epoxide hydrolases (EH; EC 3.3.2.3) catalyze the hydration of oxirane rings to the corresponding dihydrodiols. In mammals, microsomal and soluble forms of this enzyme are localized in the liver and have been well characterized (41,42).

The membrane-bound epoxide hydrolase (mEH) takes part in the detoxication mechanisms elaborated by organisms to transform liposoluble xenobiotics that accumulate in cellular membranes into hydrophilic molecules that are easier to conjugate and eliminate. Epoxides are electrophilic species that can react with nucleophilic residues present in proteins and nucleic acids. In the latter case this could lead to mutagenesis. Consequently, the epoxide hydrolysis by mEH has been considered to be a deactivation process. However, in recent years, the role of this enzyme in detoxication appeared more complex, since, in some cases, the resulting diols can be further transformed and become even more carcinogenic than the parent compounds. For example, the polycyclic hydrocarbon benzo[*a*]pyrene can be epoxidized by the mixed-function oxygenase system to epoxides that are hydrated by epoxide hydrolase. However, the resulting dihydrodiol 7,8-dihydro-7,8-dihydroxy-benzo[*a*]pyrene retains sufficient lipophilic character to serve as a substrate for a microsomal cytochrome P450 and gives rise to diolepoxide which has been shown to be poorly hydrated by mEH in vitro and highly mutagenic (43).

The soluble form of epoxide hydrolase (sEH) also hydrates some epoxidized xenobiotics. However, fatty acid epoxides appear to be the best substrates for this enzyme, suggesting that it is involved in the biotransformation of endogenous substrates. In addition, endogenous epoxides are hydrated by specific hydrolases, such as cholesterol epoxide hydrolase (44), leukotriene A_4 hydrolase (45), and hepoxilin hydrolase (46).

Compared with the wealth of information available on epoxide hydrolases in mammalian systems, this class of enzymes has been poorly investigated in higher plants until recently. The presence of such activity was suspected in seed extracts of *Vernolia anthelminta,* that converted vernolic acid (12,13-epoxy-9-octadecenoic acid) into an optically active form of the corresponding *vic*-diol (47), and in cultured bean (*Phaseolus vulgaris*) cells that hydrated *trans*-stilbene oxide (48). The only report on an epoxide hydrolase in plants was by Croteau and Kolattukudy concerning the hydration of 9,10-epoxy-18-hydroxystearic acid to *threo*-9,10,18-trihydroxystearic acid catalyzed by a *particulate* fraction prepared from skin of young apple fruits (49). This situation has changed recently, thus we have been able to purify and characterize a soluble form of an epoxide hydrolase from soybean that preferentially catalyzes the hydration of unsaturated fatty acid derived epoxides (11,38,39). The isolation and expression of cDNA coding for soluble epoxide hydrolases from two other plant systems have just been reported (50,51).

Some Characteristics and Specificities of Plant Epoxide Hydrolases

The occurrence of multiple forms of microsomal epoxide hydrolases has been demonstrated in rat and human livers. Although their molecular weights are very similar, they can be distinguished by their immunochemical properties, their amino acid compositions, and their activity toward different substrates.

It seems quite clear now that multiple forms of epoxide hydrolases also exist in plants, depending on the species and subcellular fractions examined. For example, subcellular fractions of soybean seedlings are capable of readily transforming *cis*-9,10-epoxystearate into 9,10-dihydroxystearate. The major part of this activity was observed in the 100,000 g supernatant; two forms could be separated from the cytosol by ammonium sulfate fractionation (Blée, unpublished results). We have purified to apparent homogeneity one of these isoforms that precipitated in a 40–60% saturated solution of ammonium sulfate (38); it was characterized as a dimer of 64 kDa that possesses a pI of 5.4 and an optimal pH of 7.4. In addition to the soluble forms, a minor fraction of epoxide hydrolase activity was found tightly associated with the microsomes and could therefore represent a membrane-bound form of this enzyme (38).

These hydrolases are different in their substrate specificity. Mammalian epoxide hydrolases are detoxifying enzymes and appear to have a wide substrate tolerance. The soybean epoxide hydrolase exhibits a quite narrow specificity for fatty acid epoxides that have physiological importance as precursors of cutin monomers and antifungal compounds. For example, styrene oxide and *trans*-stilbene oxide, good substrates for the mammalian microsomal and soluble epoxide hydrolases, were not found to compete for the hydration of 9,10-epoxystearate catalyzed by either the soluble or membrane-bound soybean enzyme (38). Similarly, the apple epoxide hydrolase described by Croteau and Kalatukkudy was reported to have a fairly stringent substrate specificity for fatty acid epoxides in favor of 18-hydroxylated derivatives (49), relatively poor substrates for soybean epoxide hydrolases. In this context, it is noteworthy that mouse liver microsomal and soluble epoxide hydrolases of similar substrate specificity were also reported (52), but there is no evidence that plant membrane-bound hydrolases possess any mammalian mEH-like biochemical and catalytic properties. In fact, compounds with *trans*-configuration, such as *trans* 9,10-epoxystearate, are poorly transformed by soybean hydrolases. In contrast, the recombinant hydrolases from potato and *Arabidopsis* actively catalyze the hydration of *trans*-stilbene oxide and are sensitive to mammalian soluble epoxide-hydrolase inhibitors (50,51). It should be noted that this latter enzyme is capable of hydrolyzing the *cis* and *trans* isomers of fatty acid epoxides equally well, and that hydrolysis of methyl trans 9,10-epoxystearate is slightly more rapid than the *cis*-isomer (53).

The amino acid sequences of the cloned plant enzymes exhibit about 35% homology with those of the mammalian soluble epoxide hydrolases but very little with mEH (50,51). Potato and *Arabidopsis* hydrolases also differ from the soybean hydrolase by their molecular weight and their pI. It is obvious that in the future, in order to compare these different plant epoxide hydrolases, further work will be needed to achieve the cloning of the soybean and apple enzymes and to better characterize the catalytic and biochemical properties of the potato and *Arabidopsis* hydrolases.

Soluble plant epoxide hydrolases exhibit regioselectivity toward the isomers of LA monoepoxides; for instance, the enzyme purified from soybean and the hydrolase from germinating *Euphorbia lagascae* showed a preference for the 12,13-monoepoxide and was hydrolyzed six to ten times faster than 9,10-epoxystearate (39,54). It is interesting to recall that for the epoxidation of LA the peroxygenase presented a regioselectivity in favor of the 9,10-unsaturation (15). Soybean epoxide hydrolase also hydrates diepoxystearate into 9,10-epoxy-12,13-dihydroxy- and 12,13-epoxy-9,10-dihydroxystearates and the corresponding tetradiols (Blée, unpublished results). Similarly, mammalian soluble epoxide hydrolase was found to convert diepoxystearate into the tetradiol derivatives. Under limiting enzyme concentrations, the epoxydiols intermediate was found to be converted in tetrahydrofurandiols (55). The formation of such products during diepoxystearate hydrolysis by soybean epoxide hydrolase was not detected. In conclusion, it appears that the selectivity of the soybean hydrolase seems to be in good agreement with the observation that this enzyme is primarily used in the biosynthesis of endogenous compounds, as opposed to the mammalian epoxide hydrolases that are mainly involved in xenobiotic detoxication.

Enantioselectivity and Stereospecificity of the Epoxide Hydrolase

As mentioned previously, soybean hydrolase catalyzes a remarkably efficient trans-hydration of *cis*-9,10-epoxystearic acid into *threo*-9,10-dihydroxystearic acid. It implies, as found with most epoxide hydrolases, that the reactive carbon of the oxirane ring undergoes a nucleophilic attack by a water molecule or by a carboxyl group of the active site *(vide infra)* resulting in its complete stereochemical inversion (56). In addition, the hydration reaction is characterized by an important enantioselectivity, that was shown by studying the kinetics of hydrolysis of racemic fatty acid epoxides and confirmed by determining the absolute configuration of the unreacted epoxides by chiral-phase HPLC.

We have demonstrated that soybean enzyme preferentially hydrates 9*R*,10*S* enantiomers of the epoxides derived from oleic acid and LA and the 12*R*,13*S*-enantiomer derived from the LA monoepoxide regioisomer (11,39). It should be remembered that these enantiomers are precisely the major metabolites formed by epoxidation of oleic acid and LA by the peroxygenase, again suggesting a physiological role for these enzymes in soybean cells. Free carboxylate at C(1) of the substrates seems to be a major factor for this selectivity, for example enantioselection in favor of 9*R*,10*S*-epoxystearic acid is much reduced when using methyl 9,10-epoxystearate as a substrate.

In contrast, such enantioselectivity in favor of the *9R,10S*-enantiomer was still observed for 18-hydroxy-9,10-epoxy-stearic acid. However, the enantioselectivity of plant enzymes seems to differ greatly depending on the species (Table 8.1). While soluble epoxide hydrolases from maize, rice, or wheat (monocotyledons from the family of poacae, ex-graminae) hydrate at the same rate, those from potato or banana present a very strong enantioselectivity in favor of *9R,10S*-epoxystearic acid. Thus, these plants are particularly appropriate for kinetic resolution of racemic epoxides derived from fatty acids that yield highly enriched *S,R* epoxides. Similarly, a bacterial (*Pseudomonas*) hydrating enzyme has been shown to attack one enantiomer of racemic *cis* and *trans*-epoxystearate specifically (57).

Trans-addition of water to the epoxide could lead to the formation of products with high optical purity if the enzyme were associated with a high degree of stereoselectivity. For example, enzymatic hydration of racemic benzo(a)pyrene 4,5-oxide by a microsomal mammalian enzyme yields (-)benzo(a)pyrene 4,5-dihydrodiol almost exclusively (58). A structure-activity relationship investigation of the hydration of a series of substituted stilbene oxides revealed a high selectivity for the attack at the carbon atom with (*S*)-stereochemistry (59). However, it seems that the regiospecificity of mammalian epoxide hydrolase is largely governed by steric factors, and hydration occurs at the sterically least hindered epoxide carbon atom (60). It can be inferred that such a regiospecific discrimination should be abolished when an almost complete symmetry exists around the oxirane ring as in 9,10-epoxystearate, a quasi-*meso* molecule. Indeed, this epoxide is hydrated by rabbit liver (microsomal) epoxide hydrolase with no apparent enantioselectivity and with a low regioselectivity, that is *threo*-*9R,10R*-dihydroxystearic acid is formed with a slight (1.5-fold) preference (61).

TABLE 8.1 Enantioselectivity of 9,10 Epoxystearate Hydrolysis Catalyzed by Plant Epoxide Hydrolases

Source of plant epoxide hydrolase	Enantioselectivity Ratio[a] (*9R,10S*) / (*9S,10R*)
Maize	0.89
Rice	3.9
Wheat	0.99
Arabidopsis	9.6
Jerusalem artichoke tuber	14.4
Soybean	20
Celery	44.7
Banana	100
Potato	>300

[a]Expressed as V/K_m ratio for the respective 9,10 epoxystearate enantiomers. *Source*: Blée and Schuber (11).

For soybean epoxide hydrolase, experiments performed with ^{18}O-labeled epoxides indicate that both enantiomers of *cis*-9,10-epoxystearic acid attack the oxirane ring exclusively at the carbon 9 or 10 that has the (*S*)-chirality (11). This result was confirmed by chiral HPLC resolution of the dihydrodiols formed during the reaction. Thus, the remarkable result was obtained that complete hydrolysis of both racemic and pure enantiomers of 9,10-epoxystearate leads to the exclusive formation of *threo*-9*R*,10*R*-dihydroxystearate.

No steric or electronic differences at the two potential epoxide reaction centers can explain the observed enantioselectivity. But since enantio- and regioselectivity of the hydration of epoxyeicosatrienoic acid by soluble mammalian enzyme was recently reported to differ from one enantiomer to the other (62), we have extended our studies to both enantiomers of the LA monoepoxides that possess a (*Z*)-double bond adjacent to the oxirane ring. Likewise, complete hydrolysis of either racemic or soybean peroxygenase formed (i.e., enriched with one enantiomer) regioisomers of LA monoepoxides leads to the formation of highly enantiomerically enriched (>90% e.e.) dihydrodiols, that is 9*R*,10*R*-dihydroxy-12(*Z*)-octadecenoic and 12*R*,13*R*-dihydroxy-9(*Z*)-octadecenoic acids (Blée and Schuber, unpublished results). Thus, the stereospecificity of soybean enzyme is not influenced by the introduction of a (Z)-double bond in 9,10-epoxystearic acid, and its active site can handle both 9,10- and 12,13-regioisomers of the LA monoepoxides with the same stereochemical outcome.

Our results are in agreement with a stereocontrolled hydrolysis of the epoxide by soybean epoxide hydrolase. In such a mechanism, and compared with a regioselective-controlled epoxide hydrolysis, the hydration of the two pure enantiomeric forms of the epoxide leads to the formation of the same diol product; the e.e. of this product remains insensitive to the reaction progress. It also leads to the remarkable result that a racemic epoxide can be converted to a diol of very high enantiomeric excess at 100% yield. Such a stereocontrolled hydrolysis imposes a precise topology on the active site that provides a unique reaction pathway for the nucleophilic attack and the oxirane ring opening. One can envision that the prevailing interactions between the substrates and the active site are centered around the oxirane ring which, regardless the nature of its substituents, is docked in a unique position that permits the configuration inversion reaction only at the carbon that has the (*S*)-configuration (Fig. 8.3).

Two mechanisms for transforming the oxirane ring into the diol can be invoked depending on the nature of the nucleophile that attacks the (*S*)-carbon. According to the classical mechanism, the enzyme catalyzes a nucleophilic attack directly with a water molecule, possibly activated by a general-base mechanism and assisted by a general-acid activation of the oxirane ring. Alternatively, as recently shown for the mammalian hydrolase (63), a catalytic carboxylic group of the enzyme active site could open the oxirane ring, yielding an intermediate ester that is hydrolyzed to release the diol. Regardless of the precise mechanism of ring opening, our hypothetical model imposes a closed topology to the enzyme active site, in which the catalytic residues are at the bottom of a pocket, and is also in agreement with our previous observation that, when compared to epoxides derived from fatty acids with *cis*-double

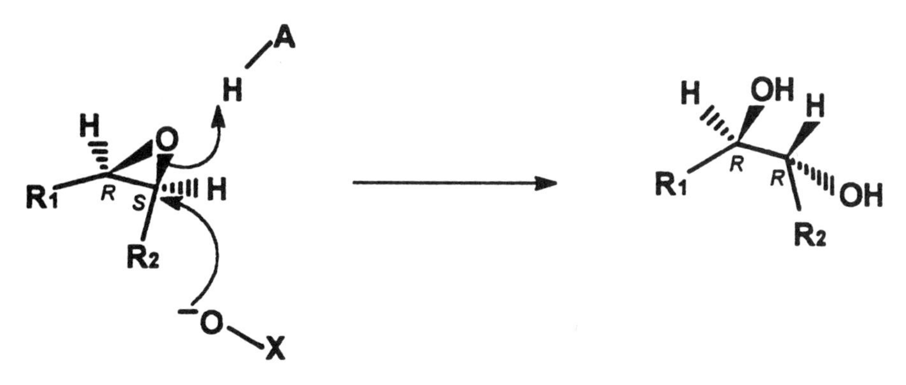

Figure 8.3. Stereochemical course and hypothetical mechanism of soybean epoxide hydrolase–catalyzed hydration of fatty acid epoxides.

bonds that can adopt hairpin-type configurations, the epoxides of *trans* fatty acids are particularly poor substrates for the soybean hydrolase.

In comparison, it should be noted that mammalian epoxide hydrolase active sites that can easily accommodate *trans*-epoxides do not fit such a model. An insect epoxide hydrolase also has been shown to catalyze stereocontrolled epoxide hydration of pheromones, such as disparule (64). However, further work with a single probe, such as the ubiquitous substrate 9,10-epoxystearic acid, will be needed before concluding the steric and electronic requirements of the catalytic binding sites of epoxide hydrolases from different origins, since the reaction stereochemistry catalyzed by some epoxide hydrolases can vary considerably depending on the structures of the substrate.

Possible Physiological Roles of the Peroxygenase Pathway

In mammals, oxygenation of arachidonic acid implies different pathways initiated by cyclooxygenase, lipoxygenases, and cytochrome P450. This latter enzyme, associated with an epoxide hydrolase, leads to the formation of epoxy and dihydroxy-derivatives of eicosanoids with potent physiological properties. In plants, the major oxidative pathway of unsaturated fatty acids is the lipoxygenase pathway. As described previously (Fig. 8.1), this enzyme produces fatty acid hydroperoxides that are the precursors for two metabolic routes that involve two different enzymes, a hydroperoxide lyase and a hydroperoxide dehydrase (2).

It was suggested that octadecanoids could have a similar regulatory role in plants as eicosanoids do in animals (65). Indeed, jasmonic acid, the ultimate metabolite of the hydroperoxide dehydrase pathway, was reported to be implicated in the reaction of higher plants to a number of external stimuli, such as wounding (66), fungal elicitation (67), mechanical forces (68), and osmotic stresses (69). Furthermore, volatile C6 and C9 aldehydes emitted via the hydroperoxide lyase pathway as a plant response to wounding possess bactericidal properties and may be a part of the plant disease resistance mechanism (70). In this context, the peroxygenase/epoxide hydrolase cascade that composes the peroxygenase pathway appears to constitute a new branch of the so-called "lipoxygenase pathway" (Fig. 8.1). Likewise, this new pathway may participate in plant defense responses, since it is likely to be involved in the biosynthesis of precursors of cutin monomers, in the production of antifungal compounds, and in some detoxication mechanisms.

Biosynthesis of Cutin Monomers

The cuticle that covers all the aerial parts of a plant constitutes the first barrier against invasion by pathogens; it also limits the penetration of compounds, such as pesticides, and controls water fluctuations. Its chief structural component, a biopolymer named cutin, is formed by a reticulated network of oxygenated fatty acids generally of 16 and 18 carbon atoms, cross-linked by ester and ether bonds. Oleic acid, 18-hydroxyoleic acid, 9,10-epoxy-18-hydroxystearic acid and 9,10,18-trihydroxystearic acid are the major constituents of the C18 family for most cutins of plants, together with some of their 12,13-unsaturated analogs (71).

Pioneering work from the group of Kolattukudy showed that these oxylipins are derived from unsaturated fatty acids (oleic acid or LA) by hydroxylation and epoxidation of a double bond followed by the hydrolysis of the corresponding epoxide (Fig. 8.4; pathway d,e,f [71]). The epoxidation step was suggested to be catalyzed by a cytochrome-P450 enzyme, while the epoxide was found to be hydrolyzed by a particulate fraction from apple skin (49). The observation that [1-^{14}C]oleic acid could be incorporated into C18 diols, triols, and tetraols by apple skin slices, led certain authors to discard earlier reports that revealed that cutin monomers could be derived from lipoxygenase and unsaturated fatty acid reaction products (72–74).

The biosynthetic scheme we have proposed (Fig. 8.4) involves the peroxygenase pathway, which reconciles most of the results related to that subject in the literature (33). In addition to this pathway, hydroxylation of oleic acid, and its epoxide and diol derivatives were found to be catalyzed by a cytochrome P450 mono-oxygenase (75). The order of the different enzymatic steps was determined by studying the selectivity of the enzymes (expressed by the V/K_m ratio for two competing substrates) and the stereochemistry of the compounds formed by the different enzymes associated with soybean seedling microsomes (33).

Despite the fact that the composition of cutin changes with the age of the cuticles (76), the major monomer remains 9,10-epoxy-18-hydroxystearic acid. We found that this oxylipin is predominantly formed according to pathway a,g in Figure 8.4 and yields a chiral epoxide. The minor route (pathway d,e) leads to racemic epoxide. The formation of 9,10,18-trihydroxystearic acid can result from either step f or pathway b,c, depending on the respective activities of epoxide hydrolase and cytochrome-P450 dependent ω-hydroxylase in the plant compartment responsible for the cutin monomer biosynthesis.

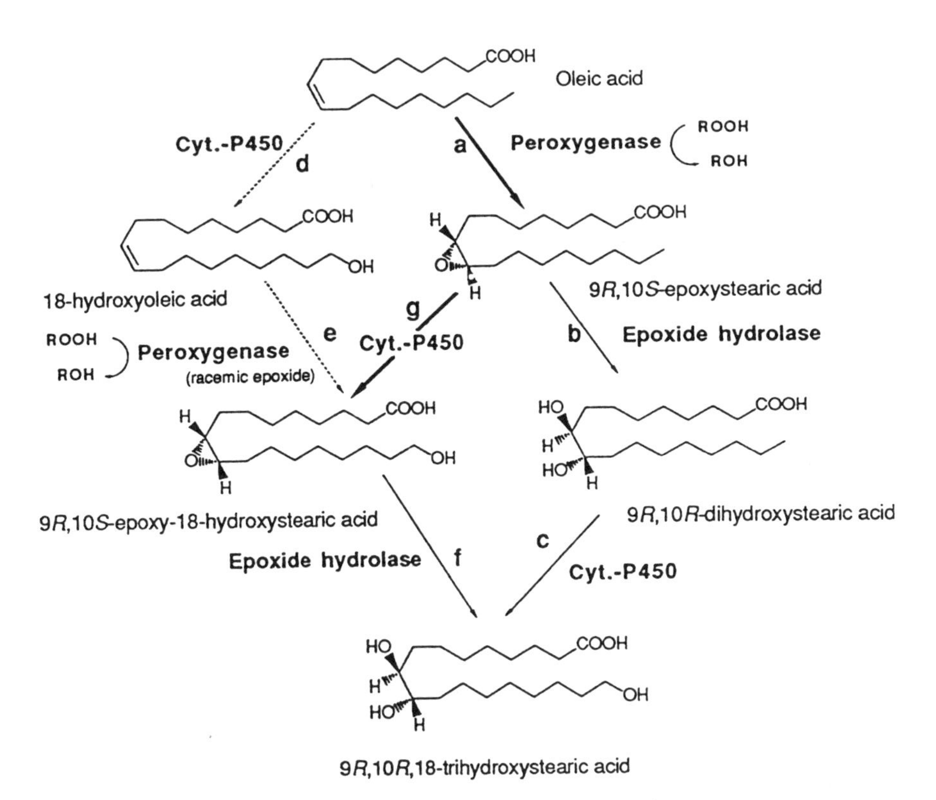

Figure 8.4. Biosynthetic scheme proposed for the formation of C18 cutin monomers.

It is important to stress here that the biosynthesis and composition of cutin will depend not only on the presence of particulate substrates in cells but also on the relative expression, activities, and compartmentation of the biosynthetic enzymes at a given time of development. For example, 9,10,18-trihydroxystearate was found in older leaves rather than in young ones (76); this observation is in agreement with the detection of gene expression for soluble epoxide hydrolase (*Arabidopsis thaliana*) in aged leaves and stems (51). Likewise, a particularly low amount of this trihydroxy derivative in spinach could be correlated with a very low level of fatty acid epoxide hydrolase in this plant (33,77). We have also predicted that the stereochemistry of the cutin monomers should overwhelmingly be determined by the stereoselectivity of the peroxygenase and the epoxide hydrolase (33). Thus, a single stereoisomer, for example 9*R*,10*R*,18-trihydroxystearic acid, should be formed. This stereochemistry corresponds to the one reported for a compound already described in *Chamaepeuce* seed oil (78), but unfortunately the absolute stereochemistry of cutin monomers is presently unknown.

The scheme given in Figure 8.4 represents a dynamic combination of possible pathways that are modulated, for example by the levels of the hydroxylating enzyme and the peroxygenase and lipoxygenase activities. It is interesting to note that incorporation of labeled fatty acid precursors was observed only with rapidly growing tissues (79), such as young fruits or immature leaves, precisely under the conditions where high levels of lipoxygenase activity are classically found (80). It should also be noted that Heinen and Van den Brand had already reported the stimulation of lipoxygenases during the synthesis of cutin in wounded leaves (73).

The composition of cutin was generally studied after hydrolysis of its ester bonds, but very little is known about the ether-linked core portion of this polymer that can represent up to 45% of its mass (81). Only LA and LNA, and not oleic acid, are incorporated into these structures, and it was suggested that a *cis*-1,4-pentadiene arrangement was a requirement for the fatty acid precursors (79). We have hypothesized that the hydroperoxides produced by a lipoxygenase from LA or LNA are converted by the peroxygenase into the precursors of the ether-linked structures (33), which could be formed by reaction of the epoxy groups with hydroxyl-substituents of adjacent monomers, as suggested by Schmidt and Schönherr (82).

In conclusion, several lines of evidence strongly suggest the involvement of the peroxygenase pathway in the biosynthesis of cutin monomers. However, the determination of the stereochemistry of these components isolated from plant cuticles and the chemical nature of the monomers engaged in nonester cutin are necessary to validate our hypothesis.

Biosynthesis of Antifungal Compounds

Plants respond to stress, such as wounding and attacks by pathogens, with the activation of a battery of defense reactions including the synthesis of low molecular weight, antimicrobial compounds called phytoalexins (83). In rice, the phytoalexins, known to be produced and accumulated in response to infection by pathogenic fungi, include oryzalexins and momilactones (84,85). But the oxylipins with the structures represented in Figure 8.5 and derived from LA should also be considered to be rice phytoalexins. Kato et al. isolated such compounds from sensitive rice infected with *Magnaporthe grisea,* the cause of pyriculariose, one of the most destructive rice diseases (86). These oxygenated derivatives were also found in rice resistant to the fungus and were suggested to be self-defense substances (87). Indeed, they were shown to inhibit spore germination and tube germ growth of rice blast fungus (88,89). One of these oxylipins, 9,12,13-trihydroxy-10 (*E*)-octadecenoic acid, has been found in tubers of taro inoculated with a black rot fungus; it also possesses antifungal properties (90).

In spite of the fact that the structures of these different compounds were determined many years ago, the enzymatic systems responsible for their biosynthesis remained unknown. However, the LA cascade involving the peroxygenase pathway described in this chapter explains the biosynthesis of the epoxy-, hydroxy-, dihydroxy-, and trihydroxy-derivatives of LA endowed with antifungal properties (Fig. 8.6).

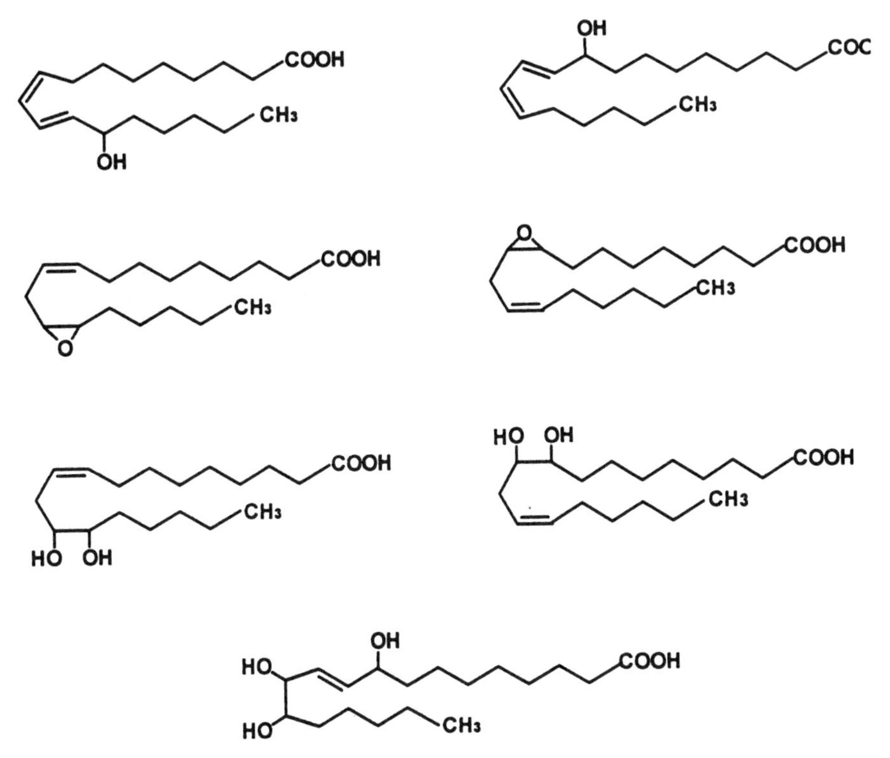

Figure 8.5. Structures of phytooxylipins possessing antifungal properties isolated in rice plants. *Source:* Kato et al. (87), Namai et al. (8).

Moreover, it seems likely that this cascade is also responsible for the uncharacterized "lipid-hydroperoxide decomposing activity" that leads to the formation of monohydroxy and trihydroxy fatty acids in rice infected by blast fungus (91).

Two of the fungitoxic oxylipins, an epoxy- and a trihydroxy-derivative, were isolated in sufficient quantities to allow determination of their fine structure. The epoxide was characterized as 12*R*,13*S*-epoxy-9(*Z*)-octadecenoic acid (86), which is precisely the enantiomer that is predominantly formed by peroxygenase from LA and preferentially hydrated by the epoxide hydrolase (34,39,54).

The product of the combined action of peroxygenase and epoxide hydrolase does not usually accumulate in the plant cell as we would expect for a signal (physiologically active) molecule. In contrast, the opposite 12*S*,13*R*-enantiomer, vernolic acid, has been found as a seed storage compound in *Euphorbiacae* and as a seed oil constituent in numerous species of plant families, such as *Valerianaceae* (92). The trihydroxy compound assigned as the 9*S*,12*S*,13*S*-trihydroxyoctadeca-10(*E*)-enoic acid is probably derived from 12,13-epoxy-9(*S*)-hydroxyoctadeca-10(*E*)-enoic acid (87). This hydrolysis could be spontaneous, since allylic epoxy alcohols have been

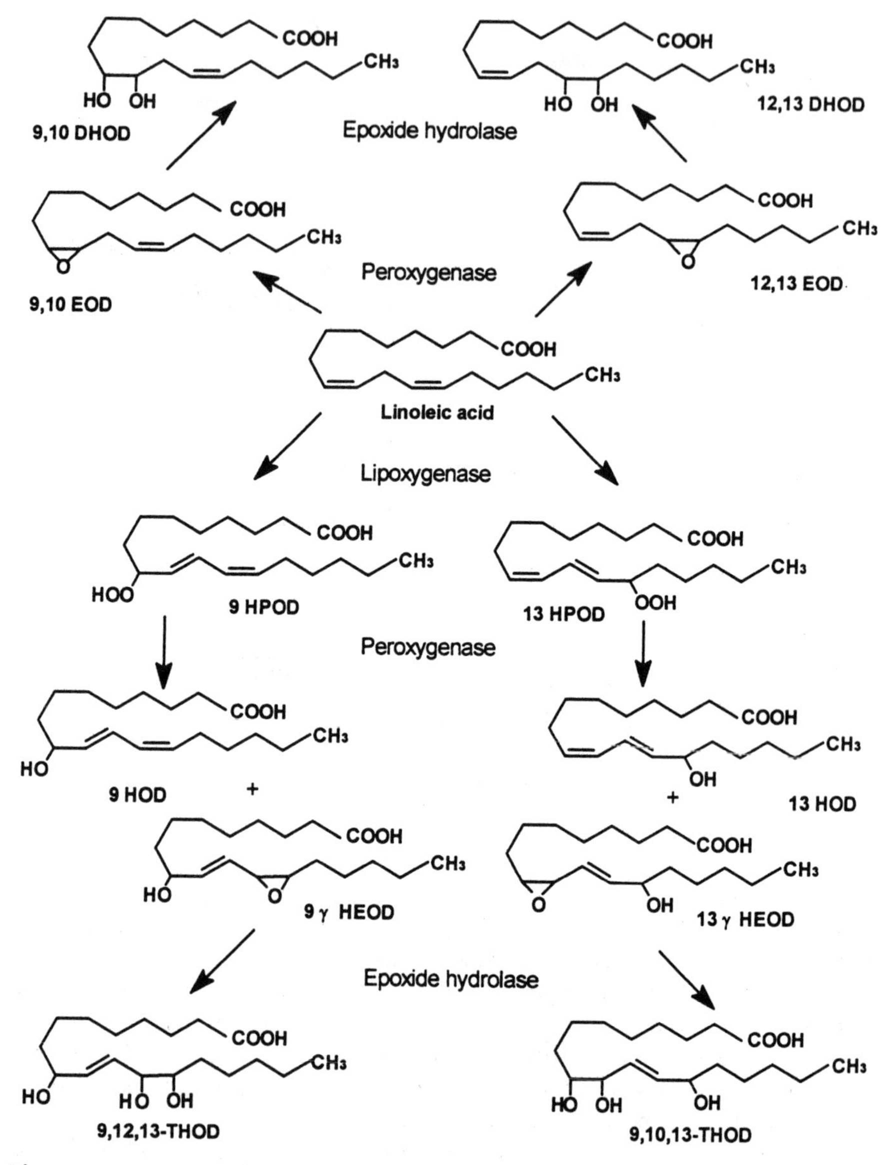

Figure 8.6. Reactions of the peroxygenase pathway with LA as the substrate.

described as quite unstable compounds or catalyzed by an epoxide hydrolase (93). We have found that the epoxy alcohols formed by the peroxygenase from 13(*S*)- and 9(*S*)-hydroperoxides of LA could be substrates for the epoxide hydrolase, but the stereochemistry of the resulting trihydroxyoctadecenoic acids has not yet been determined.

Detoxication Mechanisms

One of the primary characteristics in host-parasite interactions is the formation of peroxides and oxygenated free radicals that can result from lipoxygenase action on polyunsaturated lipids released from membranes during wounding (94). Protection of the cells against the cytotoxicity of such active oxygen species is thought to involve superoxide dismutase (EC 1.15.1.1.), catalase (EC 1.11.1.6.), and peroxidase (EC 1.11.1.7.). But peroxygenase, by reducing hydroperoxides into their corresponding alcohols may also play a protective role in vivo against the deleterious effect of fatty acid hydroperoxides. Interestingly, such oxygenated lipids have been reported to trigger phytoalexin accumulation in rice (95). Importantly, it should be noted that H_2O_2, formed during the initial and rapid oxidative burst accompanying plant/pathogen interaction, may also be used as a co-oxidant by peroxygenase instead of fatty acid hydroperoxides (13).

On the other hand, the peroxygenase pathway may also contribute to the more general detoxication mechanism of xenobiotics. In mammals such a role is carried out by a family of specialized enzymes that includes cytochrome P450s and epoxide hydrolases (56). In plants, cytochrome-P450 dependent enzymes have been shown to detoxify some pesticides via hydroxylation reactions (96). Conversely, lipoxygenase was reported to catalyze a spectrum of oxidative reactions (17), such as sulfoxidation of thiobenzamide (a hepatotoxic chemical), epoxidation of aldrin (organochloride insecticide) or cyclophosphamide (teratogen), desulfuration and dearylation of parathion (an organophosphate insecticide and acaricide), *N*-oxidation of 2-aminofluorene (a potent carcinogenic compound).

These co-oxidation reactions are generally mediated by free radicals generated during enzymatic catalysis and produce no chiral products. For example, we have found that soybean membrane-bound lipoxygenase co-oxidation of methyl *p*-tolyl sulfide yields the corresponding racemic sulfoxide (Blée, unpublished data). In sharp contrast, peroxygenase catalyzes the formation of the sulfoxide with high optical purity. Biological oxidation of the thioether pesticide group has been assumed to lead to optically active products. It has been shown that rice plants using an uncharacterized enzymatic system metabolized propaphos, an insecticide, into a chiral sulfoxide that is more effective against some insects than its optical isomer (97,98). However, few studies have been dedicated to the impact of the fine structure of pesticides on their toxicity.

Peroxygenase appears to be a good candidate to catalyze biological oxidation of sulfides in plants since it stereoselectively transfers oxygen to a sulfur atom. In fact, we have already shown that thiobenzamide, the herbicide ethyl propyl thiocarbamate, and the insecticides methiocarb and parathion are substrates for peroxygenase, which converts them to their sulfoxides (25,31,99). No further oxidation seems to occur, since the corresponding sulfones were not detected. The ferryl-oxo intermediate, generated during the catalytic process of peroxygenase, is a powerful oxidant and is also able to mediate oxidation of other functions, such as *N*-oxidations. However, no attempts have been made to screen the xenobiotics susceptible to co-oxidation by this enzyme, nor is the full spectrum of the reactions that can be catalyzed by the peroxygenase known.

Conclusions

During these last few years, oxylipins have been proposed to play a major role in the reactions of higher plants to pathogen attack or wounding. For example, it was shown that volatile C6 aldehydes (70), jasmonate (100), epoxy- and hydroxy-octadecanoids (86), derived from hydroperoxide lyase, dehydrase, and peroxygenase pathways, respectively, possess antifungal properties. In addition, recent evidence indicates that metabolites produced by the peroxygenase pathway play a primary role in plant disease resistance. First, it was reported that two cutin monomers that lack fungitoxic activity, *cis*-9,10-epoxy-18-hydroxy-stearic acid and 9,10,18-trihydroxy-stearic acid, act as endogenous inducers of acquired resistance in cereals, possibly through a mechanism involving host transcription (101). Second, epoxides derived from LA and LNA were shown to induce resistance of rice to infection by *Magnaporthe grisea*. At low concentrations, these epoxy fatty acids exhibit potent inhibitory activity toward germination and germ tube elongation of the spores of the fungi, induction of resistance was shown to be clearly due to the uptake of these epoxides by rice plants (8). However, the underlying biochemical mechanisms leading to the activation of plant defenses by these molecules have not yet been defined.

An understanding of the role of the phytooxylipins and their impact on the plant/microorganism interaction is developing. An increasing number of studies show that jasmonic acid acts as a mediator in signal transduction. The key precursor of this mediator is 13(*S*)-hydroperoxy-9(*Z*),11(*E*),15(*Z*)-octadecatrienoic acid (Fig. 8.1) which is generated from LNA either enzymatically by lipoxygenases or nonenzymatically, under stress induced by fungal pathogen aggression. It should be noted that since hydroperoxide lyase and peroxygenase also seem to be constitutively present in plant cells, they may compete with the hydroperoxide dehydrase for such a common substrate. Therefore, these three enzymes should be under strict control.

Presently, the understanding of the regulation of the different branches of the lipoxygenase pathway is still in its infancy. What could be the signals that favor one enzymatic system in response to mechanical injury and other stress conditions? Are different isoforms of lipoxygenases more related to one pathway and preferentially involved in wounding or pathogen-attack responses? Could oxylipins derived from lyase or peroxygenase pathways, other than jasmonic acid, play similar roles as signal molecules? One important issue is the compartmentation of the different enzymes in plant cells. For example, we have recently shown that the plast envelope from spinach contains hydroperoxide lyase and dehydrase activities but not the peroxygenase. Nevertheless, most of the answers are lacking today but their determination in the future will advance our understanding of the role of the newly described peroxygenase pathway in plant responses to biotic and abiotic stresses.

References

1. Hamberg, M. (1993) *J. Lipid Med. 6,* 375–384.
2. Vick, B.A. (1993) in *Lipid Metabolism in Plants,* Moore, T.S., CRC Press, Boca Raton, pp. 167–191.
3. Zimmerman, D.C., and Coudron, C.A. (1979) *Plant Physiol. 63,* 536–541.
4. Song, W.C., and Brash, A.R. (1991) *Science 253,* 781–784.
5. Hamberg, M., and Fahlstadius, P. (1989) *Arch. Biochem. Biophys. 276,* 518–526.
6. Holloway, P.J. (1982) in *The Plant Cuticle,* Cutler, D.F., Alvin, K.L., and Price, C.E., Academic Press, London, pp. 45–85.
7. Kolattukudy, P.E. (1981) *Ann. Rev. Plant Physiol. 32,* 539–567.
8. Namai, T., Kato, T., Yamaguchi, Y., and Hirukawa, T. (1993) *Biosci. Biotech. Biochem. 57,* 611–613.
9. Ohta, H., Shida, K., Peng, Y.L., Furusawa, J., Aibara, S., and Morita, Y. (1990) *Plant Cell Physiol. 31,* 1117–1122.
10. Blée, E., Wilcox, A.L., Marnett, L.J., and Schuber, F. (1993) *J. Biol. Chem. 268,* 1708–1715.
11. Blée, E., and Schuber, F. (1992) *J. Biol. Chem. 267,* 11881–11887.
12. Ishimaru, A., and Yamazaki, I. (1977) *J. Biol. Chem. 252,* 199–204.
13. Ishimaru, A., and Yamazaki, I. (1977) *J. Biol. Chem. 252,* 6118–6124.
14. Blée, E., Casida, J.E., and Durst, F. (1985) *Biochem. Pharmacol. 34,* 389–390.
15. Blée, E., and Schuber, F. (1990) *J. Biol. Chem. 265,* 12887–12894.
16. Hamberg, M., and Hamberg, G. (1990) *Arch. Biochem. Biophys. 283,* 409–416.
17. Akhilender, K., Abhinender, K., and Kulkarni, A.P. (1994) *Prostaglandins Leukotrienes Essential Fatty Acids 50,* 155–159.
18. Belvedere, G., Tursi, F., Elovaara, E., and Vainio, H. (1983) *Toxicol. Lett. 18,* 39–44.
19. Hollenberg, P.F. (1992) *Fed. Am. Soc. Exp. Biol. J. 6,* 8510–8517.
20. Nordblom, G.D., White, R.E., and Coon, M.J. (1976) *Arch. Biochem. Biophys. 175,* 524–533.
21. Gardner, H.W. (1989) *Free Radicals Biol. Med. 7,* 65–86.
22. Blée, E., and Schuber, F. (1989) *Biochemistry 28,* 4962–4967.
23. Lemberg, R., and Barrett, J. (1973) in *Cytochromes,* Academic Press, New York, pp. 8–16.
24. Ishimaru, A. (1979) *J. Biol. Chem. 254,* 8427–8433.
25. Blée, E., and Durst, F. (1987) *Arch. Biochem. Biophys. 254,* 43–52.
26. White, R.E., Sligar, S.G., and Coon, M.J. (1980) *J. Biol. Chem. 255,* 11108–11111.
27. Blake, R.C., and Coon, M.J. (1981) *J. Biol. Chem. 256,* 12127–12133.
28. Thompson, J.A., and Wand, M.D. (1985) *J. Biol. Chem. 260,* 10637–10644.
29. Thompson, J.A., and Yumibe, N.P. (1989) *Drug Metab. Rev. 20,* 365–378.
30. Dix, T.A., and Marnett, L.J. (1985) *J. Biol. Chem. 260,* 5351–5357.
31. Blée, E. (1991) *Z. Naturforsch. 46c,* 920–925.
32. Hamberg, M., and Fahlstadius, P. (1991) *Plant Physiol. 99,* 987–995.
33. Blée, E., and Schuber, F. (1993) *Plant J. 4,* 113–123.
34. Blée, E., and Schuber, F. (1990) *Biochem. Biophys. Res. Commun. 173,* 1354–1360.
35. Waxman, D.J., Light, D.R., and Walsh, C. (1982) *Biochemistry 21,* 2499–2501.
36. Kobayashi, S., Nakano, M., Kimura, T., and Schapp, A.P. (1987) *Biochemistry 26,* 5019–5022.

37. Ortiz de Montellano, P.R. (1987) *Acc. Chem. Res. 20,* 289–294.
38. Blée, E., and Schuber, F. (1992) *Biochem. J. 282,* 711–714.
39. Blée, E., and Schuber, F. (1992) *Biochem. Biophys. Res. Commun. 187,* 171–177.
40. Lau, S.M.C., Harder, P.A., and O'Keefe D.P. (1993) *Biochemistry 32,* 1945–1950.
41. Wixtrom, R.N., and Hammock, B.D. (1981) *Biochem. Pharmacol. Toxicol. 1,* 1–93.
42. Meijer, J., and DePierre, J.W. (1988) *Chem. Biol. Interact. 64,* 207–249.
43. Seidegard, J., and De Pierre, J.W. (1983) *Biochim. Biophys. Acta 695,* 251–270.
44. Watabe, T., Kanai, M., Isobe, M., and Osawa, N. (1981) *J. Biol. Chem. 256,* 2900–2907.
45. McGee, J., and Fitzpatrick, F. (1985) *J. Biol. Chem. 260,* 12832–12837.
46. Pace-Asciak, C.R., and Lee, W.S. (1989) *J. Biol. Chem. 264,* 9310–9313.
47. Scott, W.E., Krewson, C.F., and Riemenshneider, R.W. (1962) *Chem. Ind.* 2038–2039.
48. Ross, M.S.F., Lines, D.S., Stevens, R.G., and Brain, K.R. (1978) *Phytochemistry 17,* 45–48.
49. Croteau, R., and Kolattukudy, P.E. (1975) *Arch. Biochem. Biophys. 170,* 61–72.
50. Stapleton, A., Beetham, J.K., Pinot, F., Garbarino, J.E., Rockhold, D.R., Friedman, M., Hammock, B.D. and Belknap, W.R. (1994) *Plant J. 6,* 251–258.
51. Kiyosue, T., Beetham, J.K., Pinot, F., Hammock, B.D., Yamaguchi-Shinozaki, K., and Shinozaki, K. (1994) *Plant J. 6,* 259–269.
52. Guenthner, T.M., and Oesch, F. (1983) *J. Biol. Chem. 258,* 15054–15061.
53. Gill, S.S., and Hammock, B.D. (1979) *Biochem. Biophys. Res. Commun. 89,* 965–971.
54. Blée, E., Stahl, U., Schuber, F., and Stymne, S. (1993) *Biochem. Biophys. Res. Commun. 197,* 778–784.
55. Zadeh, J.N., Uematsu, T., Borhan, B., Kurth, M., and Hammock, B.D. (1992) *Arch. Biochem. Biophys. 294,* 675–685.
56. Armstrong, R.N. (1987) *CRC Critical Reviews in Biochemistry 22,* 39–88.
57. Niehaus, W.G., Kisic, A., Jr., Torkelson, A., Bednarczyk, D.J., and Schroepfer, J. (1970) *J. Biol. Chem. 245,* 3802–3809.
58. Lu, A.Y.H., and Miwa, G.T. (1980) *Ann. Rev. Pharmacol. Toxicol. 20,* 513–531.
59. Dansette, P.M., Makedonska, V.B., and Jerina, D.M. (1978) *Arch. Biochem. Biophys. 187,* 290–298.
60. Hanzlik, R.P., Edelman, M., Michaely, W.J., and Scott, G. (1976) *J. Am. Chem. Soc. 98,* 1952–1955.
61. Watabe, T., and Ajamatsu, K. (1972) *Biochim. Biophys. Acta 279,* 297–305.
62. Zeldin, D.C., Kobayashi, J., Falk, J.R., Winder, B.S., Hammock, B.D., Snapper, J.R., and Capdevila, J.H. (1993) *J. Biol. Chem. 268,* 6402–6407.
63. Lacourcière, G.M., and Armstrong, R.N. (1993) *J. Am. Chem. Soc. 115,* 10466–10467.
64. Prestwich, G.D., Graham, S. McG. and König, W.A. (1989) *J. Chem. Soc., Chem. Commun.* 575–577.
65. Vick, B.A., and Zimmerman, D.C. (1984) *Plant Physiol. 75,* 458–461.
66. Farmer, E.E., and Ryan, C.A. (1990) *Proc. Natl. Acad. Sci. USA 87,* 7713–7716.
67. Gundlach, H., Müller, M.J., Kutchan, T.M., and Zenk, M.H. (1992) *Proc. Natl. Acad. Sci. USA 89,* 2389–2393.
68. Falkenstein, E., Groth, B., Mithöfer, A., and Weiler, E.W. (1991) *Planta 185,* 316–322.
69. Sembdner, G., and Parthier, B. (1993) *Ann. Rev. Physiol. Plant Mol. Biol. 44,* 569–589.
70. Croft, K.P.C., Jüttner, F., and Slusarenko, A.J. (1993) *Plant Physiol. 101,* 13–24.
71. Kolattukudy, P.E. (1981) *Ann. Rev. Plant Physiol. 32,* 539–567.

72. Holloway, P.J., and Deas, A.H.B. (1973) *Phytochemistry 12,* 1721–1735.
73. Heinen, W., and Van den Brand, I. (1963) *Z. Naturforsch. 18b,* 67–79.
74. Bredemeijer, G., and Heinen, W. (1968) *Acta Bot. Neerl. 17,* 15–25.
75. Pinot, F., Salaün, J.P., Bosch, H., Lesot, A., Mioskowski, C., and Durst, F. (1992) *Biochem. Biophys. Res. Commun. 184,* 183–193.
76. Riederer, M., and Schönherr, J. (1988) *Planta 174,* 127–138.
77. Holloway, P.J. (1982) in *The Plant Cuticle,* Cutler, D.F., Alvin, K.L., and Price, C.E., London, Academic Press, pp. 45–85.
78. Morris, L.J., and Crouchman, M.L. (1972) *Lipids 7,* 372–379.
79. Kolattukudy, P.E., Walton, T.J., and Kushwaha, R.P.S. (1973) *Biochemistry 12,* 4488–4498.
80. Siedow, J.N. (1991) *Ann. Rev. Plant Physiol. Plant Mole. Biol. 42,* 145–188.
81. Kolattukudy, P.E., and Walton, T.J. (1972) *Biochemistry 11,* 1897–1907.
82. Schmidt, H.W., and Schönherr, J. (1982) *Planta 156,* 380–384.
83. Paxton, J.D. (1981) *Phytopathol. Z. 101,* 106–109.
84. Sekido, H., Endo, T., Suga, R., Kodama, O., Akatsuda, T., Kono, Y., and Takeushi, S. (1986) *J. Pesticide Sci. 11,* 369–372.
85. Cartwright, D., Langcake, P., Pryce, R.J., Leworthy, D.P., and Ride, J.P. (1977) *Nature 267,* 511–513.
86. Kato, T., Yamaguchi, Y., Namai, T., and Hirukawa, T. (1993) *Biosci. Biotech. Biochem. 57,* 283–287.
87. Kato, T., Yamaguchi, Y., Uyehara, T., Yokoyama, T, Namai, T., and Yamanaka, S. (1983) *Tetrahedron Lett. 24,* 4715–4718.
88. Kato, T., Yamaguchi, Y., Uyehara, T., Yokoyama, T., Namai, T., and Yamanaka, S. (1983) *Naturwiss. 70,* 200–201.
89. Kato, T., Yamaguchi, Y., Hirano, T., Yokoyama, T., Uyehara, T., Namai, T., Yamanaka, S., and Harada, N. (1984) *Chem. Lett.* 409–412.
90. Masui, H., Kondo, T., and Kojima, M. (1989) *Phytochemistry 28,* 2613–2615.
91. Ohta, H., Shida, K., Peng, Y.L., Furusawa, I., Shishiyama, J., Aibara, S., and Morita, Y. (1990) *Plant Cell Physiol. 31,* 1117–1122.
92. Wolff, I.A. (1966) *Science 154,* 1140–1149.
93. Hamberg, M. (1991) *Lipids 26,* 407–415.
94. Mehdy, M.C. (1994) *Plant Physiol. 105,* 467–472.
95. Li, W.X., Kodama, O., and Akatsuka, T. (1991) *Agric. Biol. Chem. 55,* 1041–1047.
96. Bolwell, P.G., Bozak, K., and Zimmerlin, A. (1994) *Phytochemistry 37,* 1491–1506.
97. Miyazaki, A., Nakamura, T., and Marumo, S. (1985) *J. Pestic. Sci. 10,* 727–728.
98. Miyazaki, A., Nakamura, T., and Marumo, S. (1989) *Pestic. Biochem. Physiol. 33,* 11–15.
99. Blée, E., and Schuber, F. (1992) *Biochem. Soc. Trans. 20,* 223.
100. Neto, G.C., Kono, Y., Hyakutake, H., Watanabe, M., Suzuki, Y., and Sakurai, A. (1991) *Agric. Biol. Chem. 12,* 3097–3098.
101. Schweizer, P., Jeanguenat, A., Mösinger, E., and Métraux, J.P. (1994) *Adv. Mol. Gen. Plant-Microbe Inter. 3,* 371–374.

Chapter 9

Oxylipin Pathway in Soybeans and Its Physiological Significance

Harold W. Gardner[a], Hitoshi Takamura[b,1], David F. Hildebrand[c], Kevan P.C. Croft[c], Thomas D. Simpson[a], and Yangkyo P. Salch[a]

[a]Phytoproducts Research, National Center for Agricultural Utilization Research, USDA, ARS, Peoria, IL 61604; [b]Department of Food Science and Nutrition, Nara Women's University, Nara 630, Japan; and [c]Department of Agronomy, University of Kentucky, Lexington, KY 40546-0091, USA.

Introduction

This review is concerned with the oxylipin pathway of soybeans, specifically regarding research completed in our laboratory. We apologize for overlooking much of the excellent international work completed in this area. Pertinent references are given, although they are not discussed, when it specifically impacts the research topic. A number of recent reviews outline various aspects of the oxylipin pathway in plants (1–5).

Soybean Seed Lipoxygenase

Axelrod's laboratory was the first to show the existence of three lipoxygenase (LOX) isoenzymes in soybean seed called SBL-1, SBL-2, and SBL-3 (6,7). The SBL-1 isoenzyme is synonymous with the "lipoxidase" first crystallized by Theorell's laboratory (8). Subsequently, the oxidation specificity of SBL-1 was defined by Hamberg and Samuelsson as antarafacial hydrogen-removal at the C-11 methylene of linoleic acid (LA) followed by oxygen-placement at the ω-6 carbon (9). With either LA or linolenic acid (LNA) substrates, the result was 13(*S*)-hydroperoxides.

A number of workers noted that LA was oxidized to a minor product, 9(*S*)-hydroperoxy-10(*E*),12(*Z*)-octadecadienoic acid (9*S*-HPOD), especially at low pH values. According to Funk et al. (10), this dual behavior of ω-6- and ω-10–oxidation of SBL-1 was due to a head-first or tail-first orientation of the substrate to the enzyme based on the formation of 9(*R*)- and 13(*R*)-hydroperoxides from the unusual substrates, 9(*E*),12(*Z*)- and 9(*Z*),12(*E*)-octadecadienoic acids, respectively. In our laboratory (11), it was found that the extent of 9(*S*)-oxidation from LA by SBL-1 was dependent on low pH. The extent of 9*S*-HPOD formation paralleled the titration curve of the carboxylate anion of LA (Fig. 9.1). Essentially at high pHs with all the substrate existing as carboxylate anion only the hydrocarbon tail-first orientation can occur,

[1]Research completed while on sabbatical leave at the National Center for Agricultural Utilization Research, USDA, ARS, Peoria, IL 61604, USA.

affording only 13(*S*)-hydroperoxides. At acidic pHs where only the carboxylic acid existed, the carboxylic acid head-first orientation led to 9(*S*)-hydroperoxide formation to a maximum extent of 25%. The oxidation specificity of SBL-2 has been reported by van Os and co-workers (13), but we are unaware of any previous report of SBL-3 specificity.

Recently, we examined the oxidation specificity of highly purified SBL-2 and SBL-3, and our results (Croft, K.P.C., Wang, C., Hildebrand, D.F., Simpson, T.D., and Gardner, H.W.; unpublished results) indicate that the products of SBL-2 obtained by van Os et al. (13) may have been due to the inadvertent use of a mixture of SBL-2 and SBL-3 (Table 9.1). Our findings showed that SBL-2 functioned more like an "acidic pH SBL-1"; that is, at its pH optimum, 13(*S*)-hydroperoxy-9(*Z*),11(*E*)-octadecadienoic acid (13*S*-HPOD) was the predominant product with 9*S*-HPOD being the next most abundant component. On the other hand, the products of SBL-3 were nearly racemic.

The soybean plant has multiple LOX isoenzymes that are distinct from those residing in the seed (1). Although the positional specificity of at least one of these has been determined, their stereospecificity has not yet been assessed.

Chain Cleavage Catalyzed by LOX

More than two decades ago it was discovered that SBL-1 could catalyze the cleavage of its product of oxidation, 13*S*-HPOD, provided its normal substrate, LA, and anaerobic conditions were present (14). It was theorized (15,16) that LA donated an electron to the ferric form of LOX reducing it to a ferrous state, which in turn oxidized LA to its pentadienyl radical. Subsequently, the ferrous form of the enzyme reduced 13*S*-HPOD to its alkoxyl radical, and completed the cycle. By this theory the alkoxyl radical derived from 13*S*-HPOD then would undergo chemical cleavage by a well-known radical process called β-scission into 13-oxo-9(*Z*),11(*E*)-tridecadienoic acid (13-OTA) and pentane, that were observed as products (14). Recently, we noted a similar cleavage of 13(*S*)-hydroperoxy-9(*Z*),11(*E*),15(*Z*)-octadecatrienoic acid (13*S*-HPOT) in soybean extracts (17).

Although the cleavage was stimulated by anaerobic conditions, there were important differences compared to the LA/13*S*-HPOD reaction. First, there was no requirement for added LA or LNA. When 13*S*-HPOT was added to the extract in the absence of polyunsaturated fatty acids, 13-OTA was produced in amounts five times greater than that generated by addition of 13*S*-HPOD. The products of 13*S*-HPOT were found to be 13-OTA, 2(*Z*)-penten-1-ol, and 1-penten-3-ol. A report by Vaz et al. (18), that an animal microsomal cytochrome P450 capable of cleaving 13*S*-HPOD into 13-OTA and pentane in the presence of NADPH, prompted us to examine the effects of NADPH and cytochrome P450 inhibitors, but there was no significant effect by any of these reagents.

Retention of the activity in the 150,000 × *g* supernatant also argued against a microsomal cytochrome P450. The cleavage enzyme was successfully purified by

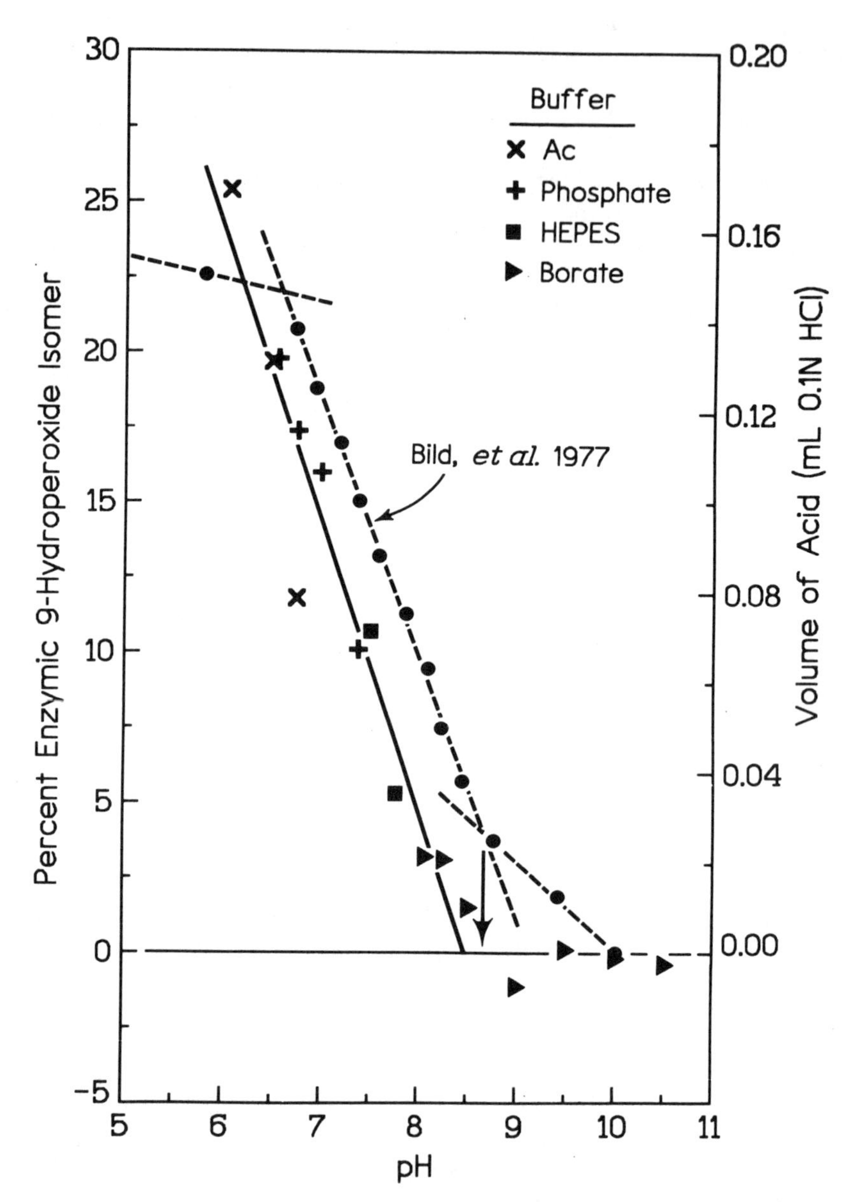

Figure 9.1. Percent "enzymic" 9-hydroperoxide produced by soybean lipoxygenase SBL-1 oxidation of LA as a function of pH (bottom curve). The percent "enzymic" 9-hydroperoxide was determined by subtracting the contribution of *R*-stereoisomeric hydroperoxides arising from autoxidation. The top curve is the acid titration of sodium linoleate dispersed by Tween 20 as described by Bild et al. *Source:* Gardner (11), Bild et al. (12).

TABLE 9.1 Oxidation Specificity of Soybean Seed Lipoxygenases at their pH Optima[a]

	Percent composition							
	13-Hydroperoxides				9-Hydroperoxides			
Isoenzymes	*S,Z,E*[b]	*R,Z,E*	*S,E,E*	*R,E,E*	*S,Z,E*	*R,Z,E*	*S,E,E*	R,E,E
SBL-1[c]	94.1	0.9	0.5	0.4	1.4	1.6	0.6	0.5
SBL-2[d]	76.6	2.4	2.5	1.4	11.5	2.3	1.8	1.6
SBL-3[d]	12.5	5.5	11.0	10.1	23.5	18.5	9.3	9.6
SBL-2[e]	12.5	12.5	n.d.	n.d.	33.7	41.3	n.d.	n.d.

[a]Enzymic oxidation was under 1 atm O_2 at pH 10.0 for SBL-1 and at pH 6.6 for SBL-2 and SBL-3.
[b]Stereoisomers *R* or *S* according to Cahn–Prelog–Ingold convention; double bond configuration *Z,E* (*cis,trans*) or *E,E* (*trans,trans*).
[c]*Source:* Gardner (11).
[d]*Source:* Croft, K.P.C., Wang, C., Hildebrand, D.F., Simpson, T.D., and Gardner, H.W.; unpublished data.
[e]*Source:* van Os et al. (13).
Abbreviation: n.d., not determined.

$(NH_4)_2SO_4$ fractionation (30–60%) followed by gel filtration (Sephacryl S-200), but activity was lost after subsequent purification on either DEAE-Sepharose or DEAE-Sephacel. Activity in the eluant of DEAE-type columns was restored by either including dithiothreitol in the eluting solvent or assaying in the presence of dithiothreitol. Under these conditions cleavage activity exactly corresponded to LOX activity. As seen in Figure 9.2, two cleavage activity peaks were obtained corresponding to SBL-1 activity, and especially type-2 LOX (SBL-2 and/or SBL-3). The LOX inhibitors, nordihydroguaiaretic acid, *n*-propylgallate, salicylhydroxamate, and H_2O_2, were nearly equally effective in inhibiting both LOX and cleavage activities.

Since the cleavage of 13*S*-HPOT was catalyzed by LOX, it was felt that the differential response of 13*S*-HPOD and 13*S*-HPOT should be explained. Thus, the two hydroperoxides as their methyl esters were subjected to alkoxyl-radical-generating conditions of elevated temperature (260°C) in the injector of a gas chromatograph (GC). Results were similar to those obtained in the enzymic reaction; that is, 13*S*-HPOD gave much reduced yields of 13-OTA compared to 13*S*-HPOT. Like the enzymic reaction, heat-decomposed 13*S*-HPOT afforded 13-OTA, 2(*Z*)-penten-1-ol, and 1-penten-3-ol. Among the heat-generated products, we discovered isomeric pentene dimers, and a reexamination of the enzymic products also revealed the presence of pentene dimers. Thus, it seemed plausible that the reaction in question was a classical example of an anaerobic generation of alkoxyl radicals from 13*S*-HPOT, and the alkoxyl radicals then reacted by a purely chemical route.

The compound(s) supplying reducing equivalents to the ferric form of LOX is the remaining mystery, and evidently they are retained on DEAE-type columns. Linoleic acid can provide the necessary reducing equivalents to cycle the ferric form (15,16), and it was determined that lipids, including fatty acids, were being eluted with the active fractions by a gel filtration column prior to DEAE chromatography. It was shown that dithiothreitol can also supply these reducing equivalents (17), even though

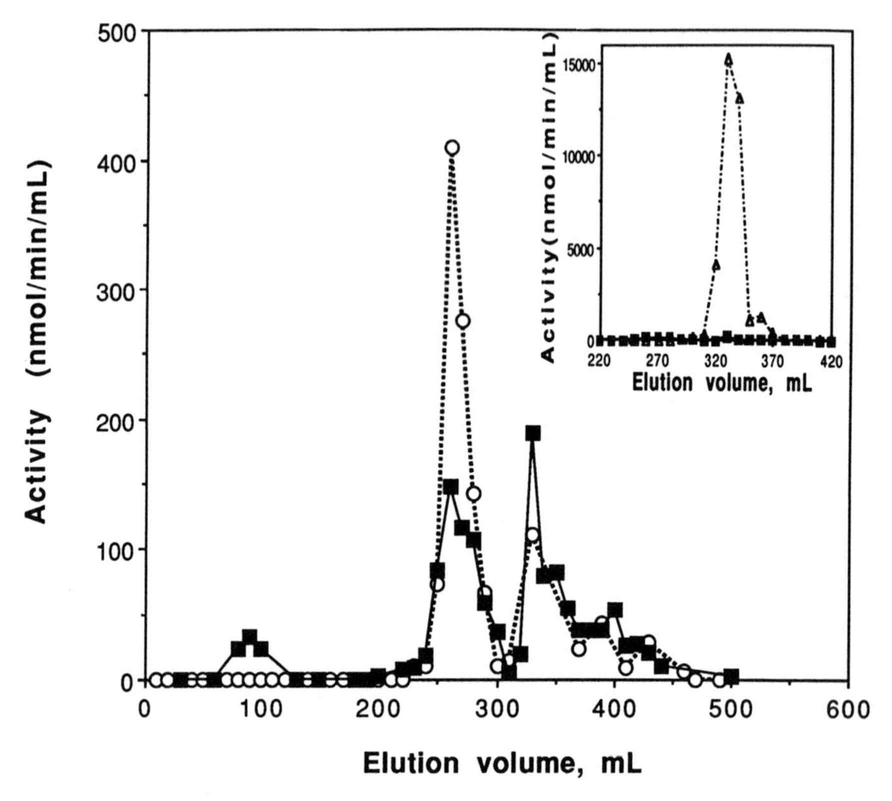

Figure 9.2. Coincidence of cleavage activity (50 mM Hepes, pH 7.5, 10 mM dithiothreitol, 1 mM 13*S*-HPOT) and lipoxygenase activity (25 mM K Pipes, pH 6.45, 1 mM LA , 10 mM dithiothreitol) in fractions separated by DEAE-Sepharose chromatography. Inset shows relative lipoxygenase activity measured at pH 9.8 (25 mM K borate, 1 mM LA, 10 mM dithiothreitol). Lipoxygenase activity (O_2 uptake) at pH 6.45 (■–■); cleavage activity by A_{287} absorbance (O-O); lipoxygenase activity at pH 9.8 (Δ-Δ). Elution of the column was with a linear gradient of NaCl (0–0.4 M) in 600 mL 50 mM Hepes, pH 7.5, and 10 mM dithiothreitol. *Source:* Salch, et al. (17).

sulfhydryl reagents are inhibitory to LOX activity via the generation of H_2O_2 (19). A unified scheme of the proposed reaction mechanism is shown in Figure 9.3.

Since the chain-cleavage reaction of LOX furnishes 2(*Z*)-penten-1-ol and 1-penten-3-ol as products, it is possible that soybean alcohol dehydrogenase and NAD^+ may be transforming these two alcohols into the important odor volatiles, ethylvinylketone (EVK) and 2(*Z* or *E*)-pentenal. It is known from the literature that EVK contributes to the "raw bean" flavor of soybeans (20). The odor of EVK straight from the reagent bottle is extremely unpleasant. Preliminary research (Salch, Y.P. and Gardner, H.W., unpublished research) revealed the presence of an alcohol dehydrogenase activity, that in the presence of NAD^+ and 1-penten-3-ol furnished EVK as a

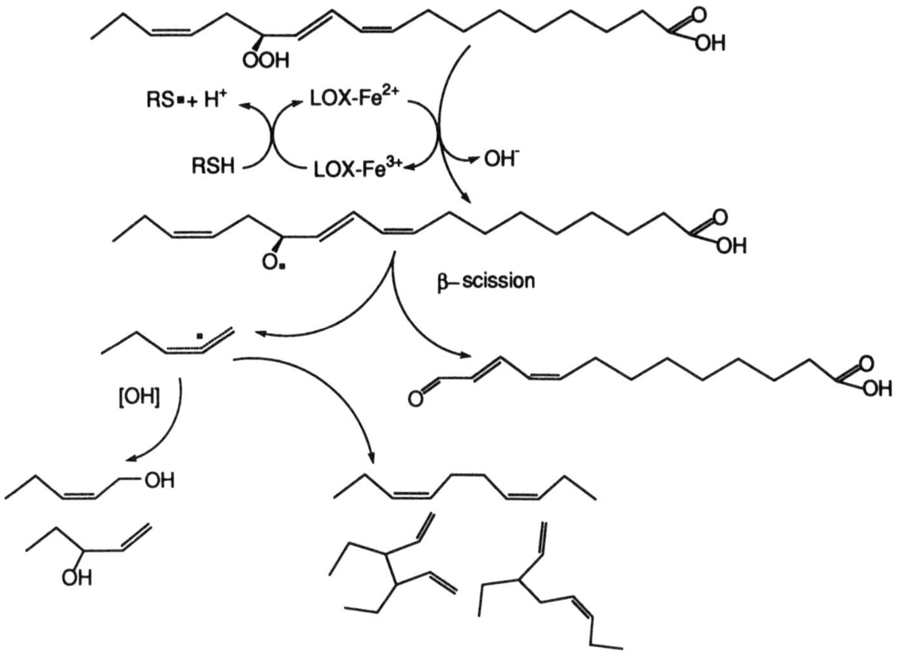

Figure 9.3. Proposed reaction mechanism leading to cleavage products as catalyzed by dithiothreitol-reduced iron in the active-site of lipoxygenase. *Source:* Salch, et al. (17).

product. Ethylvinylketone was positively identified by gas chromatography/mass spectrometry (GC/MS) as the *syn* and *anti* isomers of its benzyloxime derivative when compared with an authentic standard. However, the activity of alcohol dehydrogenase in the presence of 1-penten-3-ol was much lower than comparable reactions with ethanol and 2(*E*)-hexen-1-ol. Work is now focused on isolation of alcohol dehydrogenase isoenzymes in order to determine which of these show specificity for 1-penten-3-ol.

Allene Oxide Synthase

The action of allene oxide synthase on 13*S*-HPOT leads to the formation of an unstable allene oxide, 12,13-epoxy-11,9(*Z*),15(*Z*)-octadecatrienoic acid (21). This intermediate mainly hydrolyzes into ketols, 13-hydroxy-12-oxo-9(*Z*),15(*Z*)-octadecadienoic acid and 9-hydroxy-12-oxo-10(*E*),15(*Z*)-octadecadienoic acid. However, a small percentage of the allene oxide cyclizes into racemic 9(*S*),13(*S*) and 9(*R*),13(*R*) 12-oxo-phytodienoic acids. In the presence of allene oxide cyclase, the predominant product is 9(*S*),13(*S*)-isomer of 12-oxo-phytodienoic acid (22). The two enzymes, allene oxide synthase and allene oxide cyclase, are two important inaugural steps in the biosynthesis of the phytohormones, 7-iso-jasmonic and jasmonic acids (Fig. 9.4). Jasmonic acid has been identified in soybean plants (23), and this phytohormone is known to regulate the expression of vegetative storage protein in soybean leaves (24).

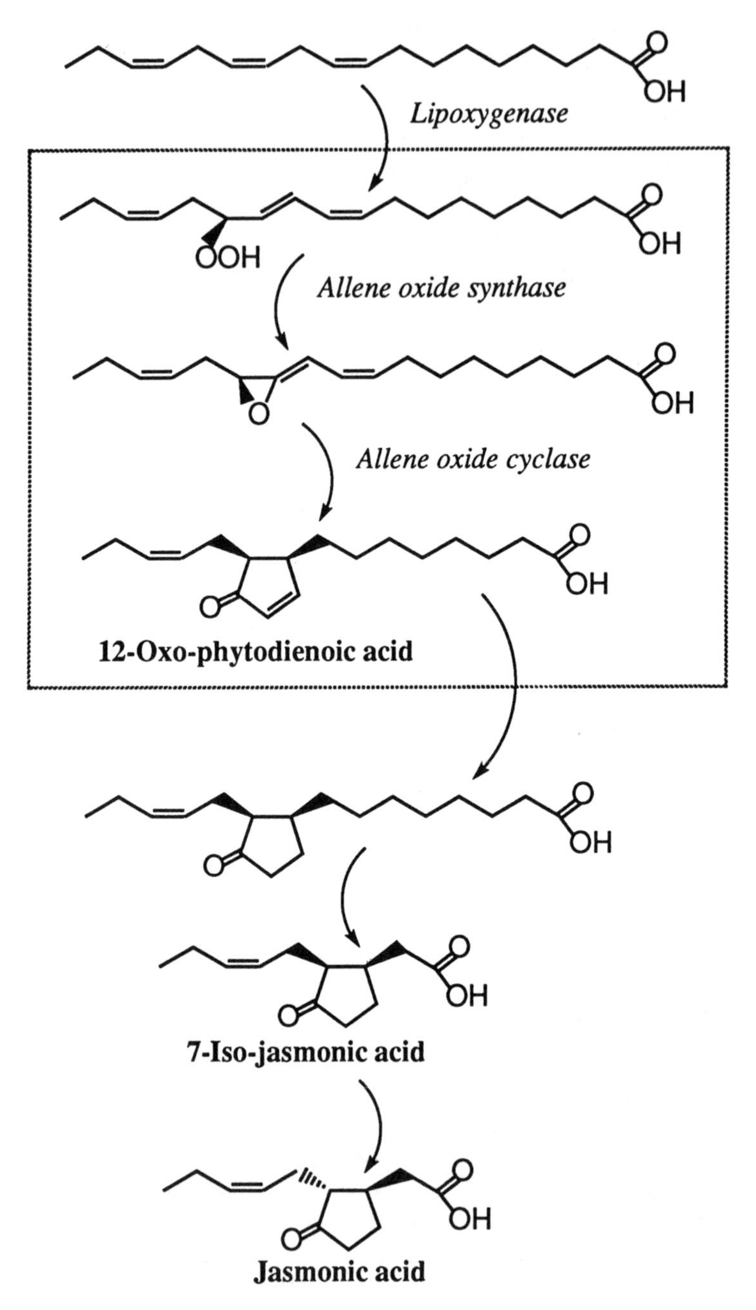

Figure 9.4. Pathway to jasmonic acid. Portions within the boxed area were the enzymic reactions examined in soybean tissues. *Source:* Simpson and Gardner (25).

Although jasmonic acid has been identified in soybeans, there has been no definitive proof that allene oxide synthase and allene oxide cyclase are present in soybeans. We were unable to detect allene oxide synthase in mature soybean seeds in our laboratory (25), although a trace of activity may exist in the embryo (hypocotyl-radicle axis). Instead, the greatest activity found was in the immature fruit in the following order: seed coat > pericarp >> seed hypocotyl-radicle axis > seed cotyledon. Because the formation of ketols was very low in these immature fruit tissues, it was implied that allene oxide cyclase activity was very high. After the pericarp matured to the brown stage, the α-ketol became the prevalent product indicating the loss of allene oxide cyclase, but not allene oxide synthase. The activity in leaves from the fruiting plant was only one-seventh that of the seed coats on a per mg protein basis. The data for the activity in various tissues of the fruiting plant is summarized in Table 9.2. Furthermore, Lopez et al. found that the prevalence of jasmonic acid in soybean tissues followed the same general pattern as 12-oxo-phytodienoic acid biosynthesis (26).

As stated previously, it is known that jasmonic acid induces expression of vegetative storage protein in leaves (24). It is believed that the vegetative storage proteins are then mobilized from the leaves to be assimilated into seeds during seed-set. Interestingly, we found that 12-oxo-phytodienoic acid biosynthetic activity in leaves prior to flowering was about three times higher than leaves from the fruiting plant, which tends to confirm the connection between jasmonic acid and leaf vegetative storage protein. The question remains: Why is 12-oxo-phytodienoic acid biosynthesis so high in the vascular tissue that delivers assimilate to seed, namely the seed coat and pericarp? Wilen et al. showed that jasmonic acid induced the expression of seed storage proteins in *Brassica napus* embryo cultures (27). If this example is universal, perhaps high 12-oxo-phytodienoic acid biosynthesis could lead to expression of storage protein in soybean seeds, or induce *N*-assimilation into the seed.

TABLE 9.2 Net 12-Oxo-Phytodienoic Acid (12-OPDA) Biosynthesis (minus endogenous activity) of Selected Tissues from the Fruiting Soybean Plant (per mg protein)

Tissue[a]	Net 12-OPDA µg/mg protein	Relative 12-OPDA biosynthesis
Seed coat	21.3	100
Membranous endocarp	16.0	75
Pericarp minus membranous endocarp	12.3	58
Leaves	3.0	14
Radicle axis	0.5	2.3
Cotyledon	0.3	1.4

[a]All plant material was from greenhouse-grown plants 38 days after flowering (fruiting plants), except the leaves were from field-grown plants, about 38 days after flowering; the radicle axis was the hypocotyl-radicle axis plus plumule. *Source:* Simpson and Gardner (25).

Hydroperoxide Lyase Cascade

The membrane-bound hydroperoxide lyase causes a cleavage reaction distinct from the cleavage obtained in the anaerobic reaction of LOX. This enzyme cleaves 13*S*-HPOD and 13*S*-HPOT into hexanal and 3(*Z*)-hexenal, respectively, as well as 12-oxo-9(*Z*)-dodecenoic acid. 3(*Z*)-Hexenal is particularly notable as it has an intense "grassy" odor with a particularly low threshold of detection. This type of hydroperoxide lyase has been detected in soybean seeds, seedlings (28-30), and leaves (30,31). Hydroperoxide lyase activity in soybean leaves was found to be associated with the chloroplasts (30).

Plant hydroperoxide lyase is also known to cleave 9*S*-HPOD into 3(*Z*)-nonenal and 9-oxononanoic acid. According to some workers (28,29), this 9*S*-HPOD-specific hydroperoxide lyase is not present in soybean seed and seedlings. However, we found that 9*S*-HPOD was an equally effective substrate as 13*S*-HPOD (30). We readily found the product, 9-oxononanoic acid, but 3(*Z*)-nonenal was absent. It was then found that soybeans effectively converted 3(*Z*)-nonenal into both 2(*E*)-nonenal and 4-hydroxy-2(*E*)-nonenal (30, Takamura, H., and Gardner, H.W., unpublished results). Exposing a seedling preparation to a racemic mixture of LA hydroperoxides (*rac*-HPOD), isolated from autoxidation of LA, showed that there was a preferential utilization of 9*S*-HPOD as well as 13*S*-HPOD from the mixture of eight isomers (Fig. 9.5*a*). On the other hand, activity from chloroplasts of the soybean leaves were devoid of 9*S*-HPOD-specific hydroperoxide lyase activity (30). After exposing the chloroplast enzyme to *rac*-HPOD, it was shown that only 13*S*-HPOD was utilized (Fig. 9.5*b*). These data implied that seed/seedlings possessed different isoenzyme(s) compared to leaf tissue. To emphasize this point, the chloroplast hydroperoxide lyase V_{max} for 13*S*-HPOT was sixfold higher than for 13*S*-HPOD, but with hydroperoxide lyase from soybean seedlings the same ratio, V_{max} 13*S*-HPOT/V_{max} 13*S*-HPOD, was only 0.4 (30).

As mentioned previously, 3(*Z*)-nonenal was further metabolized to 2(*E*)-nonenal and 4-hydroxy-2(*E*)-nonenal. Since there is significant literature regarding the cytotoxic effects of 4-hydroxy-2(*E*)-nonenal in mammalian systems (32), we concentrated on its biosynthesis in plants. The biosynthetic pathway in mammalian systems, a microsomal-dependent reaction, has not yet been resolved, but it is thought that 4-hydroxy-2(*E*)-nonenal originates from ω-6 polyunsaturated fatty acids (32). Inasmuch as there are striking similarities in the requirements for the plant and mammalian systems, the pathways may be similar. The biosynthesis of 4-hydroxy-2(*E*)-nonenal from 3(*Z*)-nonenal has been characterized with a microsomal preparation from broad bean seed (*Vicia faba* [33]). As shown in Figure 9.6, the pathway involved a dual reaction scheme that required the participation of an 3(*Z*)-alkenal dioxygenase and a hydroperoxide-dependent epoxygenase. An understanding of the biosynthetic pathway led to the development of a "biomimetic" strategy for the chemical synthesis of 4-hydroxy-2(*E*)-nonenal (34).

Further work demonstrated that the 4-hydroxy-2(E)-nonenal biosynthetic system was also localized in microsomes of soybean seed (Takamura, H., and Gardner, H.W.,

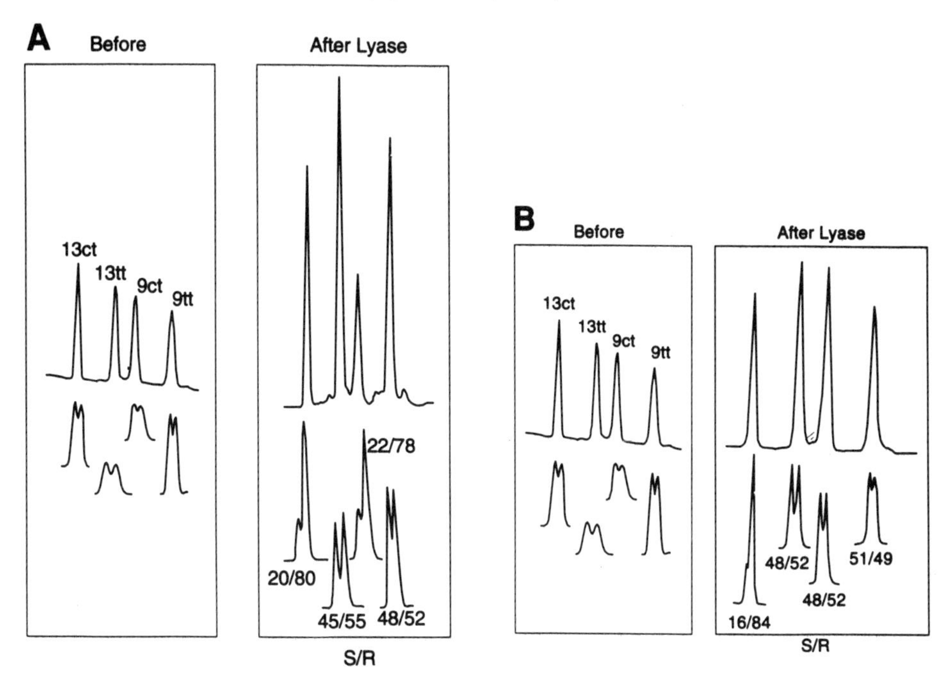

Figure 9.5. A racemic mixture of eight LA hydroperoxides (9- and 13-hydroperoxides; *cis,trans-* and *trans,trans*-dienes; and *R-* and *S*-stereoisomers) was exposed to soybean preparations to determine substrate specificity. Analyses were done by straight-phase silica (top) and chiral-phase (bottom) HPLC. *a*) Seed enzyme; *b*) Leaf chloroplast enzyme. *Source:* Gardner et al. (30).

unpublished data). Additionally, this preparation could transform 3(*Z*)-hexenal into 4-hydroxy-2(*E*)-hexenal, but at reduced rates compared to the conversion of 3(*Z*)-nonenal. Also, it was found that many properties of the soybean system were different from the one in the broad bean.

While the 4-hydroxy-2(*E*)-alkenal biosynthetic system was localized in microsomes, the conversion of 3(*Z*)-alkenals to their corresponding 2(*E*)-alkenals was primarily due to a soluble enzyme (Takamura, H., and Gardner, H.W., unpublished data). This type of double bond isomerization has been known for some time, and it can be caused by either nonenzymic catalysis (acid, heat) or by a 3(*Z*):2(*E*)-enal isomerase (35).

It is also known that aldehydes can be reduced to alcohols in the presence of alcohol dehydrogenase and NADH. Alcohol dehydrogenase has been studied in soybeans (36,37), and we are currently examining this enzyme in our laboratory (Salch, Y.P., and Gardner, H.W., unpublished results).

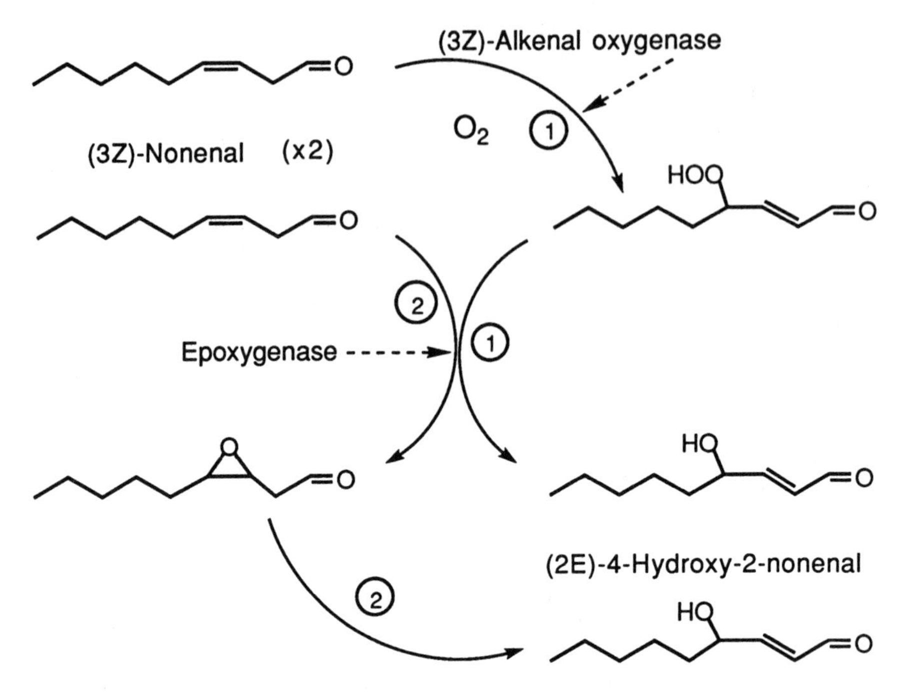

Figure 9.6. Proposed mechanism for the biosynthesis of 4-hydroxy-2(*E*)-nonenal from 3(*Z*)-nonenal in broad bean microsomes. *Source:* Gardner and Hamberg (33).

Physiological Significance

The best known physiological function for LOX is its role in formation of jasmonic acid and related phytohormones via the oxylipin "cascade." As discussed previously, jasmonic acid in soybeans leads to the expression of vegetative storage proteins. If the literature regarding the effect of jasmonic acid is instructive (21), then it is likely that other physiological roles will be found in soybean.

The function of aldehyde products of the oxylipin pathway is less well studied; however, the literature predominantly points to the inhibition of the growth of pathogenic organisms. In our work with soybeans, aldehydes were shown to inhibit seed germination and seedling growth, and it was found that α,β-unsaturated aldehydes, 2(*E*)-hexenal and 2(*E*)-nonenal were notably more inhibitory than hexanal (38). Compared to the controls germinated without aldehydes, the aldehyde-treated seeds were remarkably free of fungal growth on their seed coats at doses well below those lethal to soybeans (< 1.9 μg hexanal/mL air).

The question arose as to whether soybeans might actually employ this system in defense against fungal attack. It was known from previous research that glyceride fatty acids are ineffective precursors to aldehyde formation while the free fatty acids, LA and LNA, are precursors to aldehyde production in soybean extracts (30,39). Apparently, lipase activity is the rate-limiting step in aldehyde production. It is well known that seed storage fungi produce copious lipase in order to parasitize their oilseed hosts. *Aspergillus flavus* is one of those organisms capable of growing on oilseed hosts, such as the scutellum of corn seeds, but for some reason soybeans are resistant to *A. flavus*.

In our laboratory an experiment was designed to discover if soybeans resisted *A. flavus* infestation due to LOX/hydroperoxide-lyase reactions (40). Essentially, an agar-block was inoculated with *A. flavus* spores, and this block was situated in the headspace above a freshly prepared homogenate of soybean cotyledons. The cotyledon preparation alone was insufficient to inhibit the germination of *A. flavus* spores, but inclusion of increasing quantities of a commercial lipase preparation resulted in increasing inhibition of spore germination. It was also demonstrated that hexanal production by the cotyledon preparation was dependent on the lipase added. Addition of either LA or LNA could serve as a replacement for lipase addition, but fatty acids incapable of being oxidized by LOX had no effect (oleic acid, palmitic acid, and stearic acid). Finally, it was shown that addition of the LOX inhibitor, nordihydroguiairetic acid, partly reversed the effect of adding lipase. In conclusion, the data supported our hypothesis that lipase produced by oilseed-feeding fungi trigger a cascade resulting in fungal suicide.

Inasmuch as *A. flavus* is not normally a pathogen of soybeans, the question arose as to what effect LOX-derived aldehydes would have on soybean pathogens. Thus, the effects of 2(*E*)-hexenal, 2(*E*)-nonenal, 3(*Z*)-nonenal, and 4-hydroxy-2(*E*)-nonenal on the growth of *Colletotrichum truncatum, Rhizoctonia solani,* and *Sclerotium rolfsii* were tested (41). *C. truncatum* is a foliar pathogen of soybeans; whereas, *R. solani* and *S. rolfsii* usually invade the soybean plant at or below the soil level. All four aldehydes were very inhibitory (usually 100%) at < 2 μmol/mL agar medium. Although 2(*E*)-hexenal, 2(*E*)-nonenal, and 3(*Z*)-nonenal were inhibitory as headspace volatiles, 4-hydroxy-2(*E*)-nonenal was not, as it does not have an appreciable vapor pressure.

As the interest in LOX-mediated reactions in plants increases, it seems certain that many startling discoveries of its biological significance will be forthcoming.

References

1. Shibata, D., and Axelrod, B. (1995) *J. Lipid Mediators Cell Signaling 12,* 213–228.
2. Vick, B.A. (1993) in *Lipid Metabolism in Plants,* Moore, T.S., Jr., CRC Press, Boca Raton, FL, pp. 167–191.
3. Gardner, H.W. (1991) *Biochim. Biophys. Acta 1084,* 221–239.
4. Siedow, J.N. (1991) *Ann. Rev. Plant Physiol. Plant Mol. Biol. 42,* 145–188.
5. Hildebrand, D.F. (1989) *Physiol. Plant. 76,* 249–253.

6. Christopher, J.P., Pistorius, E.K., and Axelrod, B. (1970) *Biochim. Biophys. Acta 198,* 12–19.
7. Christopher, J.P., Pistorius, E.K., and Axelrod, B. (1972) *Biochim. Biophys. Acta 284,* 54–62.
8. Theorell, H., Holman, R.T., and Åkeson, Å. (1947) *Acta Chem. Scand. 1,* 571–576.
9. Hamberg, M., and Samuelsson, B. (1967) *J. Biol. Chem. 242,* 5329–5335.
10. Funk, M.O., Jr., Andre, J.C., and Otsuki, T. (1987) *Biochemistry 26,* 6880–6884.
11. Gardner, H.W. (1989) *Biochim. Biophys. Acta 1001,* 274–281.
12. Bild, G.S., Ramadoss, C.S., and Axelrod, B. (1977) *Lipids 12,* 732–735.
13. van Os, C.P.A., Rijke-Schilder, G.P.M., and Vliegenthart, J.F.G. (1979) *Biochim. Biophys. Acta 575,* 479–484.
14. Garssen, G.J., Vliegenthart, J.F.G., and Boldingh, J. (1971) *Biochem. J. 122,* 327–332.
15. Gardner, H.W. (1975) *J. Agric. Food Chem. 23,* 129–136.
16. de Groot, J.J.M.C., Veldink, G.A., Vliegenthart, J.F.G., Boldingh, J., Wever, R., van Gelder, B.F. (1975) *Biochim. Biophys. Acta 377,* 71–79.
17. Salch, Y.P., Grove, M.J., Takamura, H., and Gardner, H.W. (1995) *Plant Physiol. 108,* 1211–1218.
18. Vaz, D.N.V., Roberts, E.S., and Coon, M.J. (1990) *Proc. Natl. Acad. Sci. USA 87,* 5477–5503.
19. Mitsuda, H., Yasumoto, K., and Yamamoto, A. (1967) *Agric. Biol. Chem. 31,* 853–860.
20. Mattick, L.R., and Hand, D.B. (1969) *J. Agric. Food Chem. 17,* 15–17.
21. Hamberg, M., and Gardner, H.W. (1992) *Biochim. Biophys. Acta 1165,* 1–18.
22. Hamberg, M., and Fahlstadius, P. (1990) *Arch. Biochem. Biophys. 276,* 518–526.
23. Meyer, A., Miersch, O., Büttner, C., Dathe, W., and Sembdner, G. (1987) *J. Plant Growth Regul. 3,* 1–8.
24. Staswick, P.E. (1990) *Plant Cell 2,* 1–6.
25. Simpson, T.D., and Gardner, H.W. (1995) *Plant Physiol. 108,* 199–202.
26. Lopez, R., Brückner, C., Miersch, O., and Sembdner, G. (1987) *Biochem. Physiol. Pflanz. 179,* 317–325.
27. Wilen, R.W., van Rooijen, G.J.H., Pearce, D.W., Pharis, R.P., Holbrook, L.A., and Moloney, M.M. (1991) *Plant Physiol. 95,* 399–405.
28. Matoba, T., Hidaka, H., Kitamura, K., Kaizuma, N., and Kito, M. (1985) *J. Agric. Food Chem. 33,* 856–858.
29. Olias, J.M., Rios, J.J., Valle, M., Zamoraa, R., Sanz, L.C., and Axelrod, B. (1990) *J. Agric. Food Chem. 38,* 624–630.
30. Gardner, H.W., Weisleder, D., and Plattner, R.D. (1991) *Plant Physiol. 97,* 1059–1072.
31. Sekiya, J., Kajiwara, T., Munechika, K., and Hatanaka, A. (1983) *Phytochemistry 22,* 1867–1869.
32. Esterbauer, H., Schauer, R.J., and Zollner, H. (1991) *Free Radical Biol. Med. 11,* 81–128.
33. Gardner, H.W., and Hamberg, M. (1993) *J. Biol. Chem. 268,* 6971–6977.
34. Gardner, H.W., Bartelt, R.J., and Weisleder, D. (1992) *Lipids 27,* 686–689.
35. Phillips, D.R., Matthew, J.A., Reynolds, J., and Fenwick, G.R. (1979) *Phytochemistry 18,* 401–404.
36. Tihanyi, K., Talbot, B., Brzezinski, R., and Thirion, J.-P. (1989) *Phytochemistry 28,* 1335–1338.

37. Matoba, T., Sakurai, A., Taninoki, N., Saitoh, T., Kariya, F., Kuwahata, M., Yukawa, N., Fujino, S., and Hasegawa, K. (1989) *J. Food Sci. 54,* 1607–1610.
38. Gardner, H.W., Dornbos, D.L., Jr., and Desjardins, A.E. (1990) *J. Agric. Food Chem. 38,* 1316–1320.
39. Zhuang, H., Hildebrand, D.F., Andersen, R.A., and Hamilton-Kemp, T.R. (1991) *J. Agric. Food Chem. 39,* 1357–1364.
40. Doehlert, D.C., Wicklow, D.T., and Gardner, H.W. (1993) *Phytopathology 83,* 1473–1477.
41. Vaughn, S.F., and Gardner, H.W. (1993) *J. Chem. Ecol. 19,* 2337–2345.

Chapter 10

The Role of Lipoxygenase in Plant Resistance to Infection

Alan J. Slusarenko

Institut für Biologie III (Pflanzenphysiologie), RWTH, Worringerweg, D-52076 Aachen, Germany.

Introduction

The successful colonization of a plant by a pathogen leads to disease; the plant is said to be "susceptible" and the interaction is described as "compatible." Effective resistance of the plant is expressed in the "incompatible" interaction and disease fails to develop.

When a plant is confronted by a pathogen, it has at its disposal a number of active and passive defense strategies. Passive defenses include a thick cuticle, constitutive antimicrobial chemicals such as alkaloids or easily hydrolyzable and oxidizable phenolic glycosides; these factors must be overcome before colonization and disease can ensue. The active, second line of plant defense depends upon successful recognition of the pathogen by the plant. On recognition, a plethora of defense reactions is activated, encompassing a switch-over from the normal housekeeping pattern of gene expression to a kind of "pathogen shock" state in many ways reminiscent of the well-known "heat shock" response. Thus, the level of some mRNA species is down regulated, the level of others is increased, and some new mRNAs accumulate. This results in a realignment of the plant's metabolism so that novel enzymes are synthesized and metabolites are channeled into newly activated biosynthetic pathways culminating in the accumulation of antimicrobial substances called phytoalexins. Some of the newly synthesized plant enzymes, for example chitinase (CHT) and β-1,3-glucanase, can degrade pathogen cell walls and are considered antimicrobial in their own right. Plant cells at the site of pathogen ingress can undergo a form of programmed cell death called hypersensitive response (HR) that is associated with stopping pathogen spread. Linking or bridging the recognition of the pathogen to the activation of defense gene expression are signal transduction pathways that have been poorly characterized up to now.

The topic of this review is to ask the question "What are the *possible, probable*, and *actual* roles of lipoxygenase in resistance" against the background of plant defense metabolism. In order to ask, and hopefully to answer these questions satisfactorily, it is necessary to consider what happens to lipoxygenase (LOX) during infection and to describe some plant defense responses in a little more detail and consider how the enzyme's properties and direct reaction products, as well as the cascade of metabolites produced in the LOX pathway, can contribute to defense.

Lipoxygenase

Lipoxygenase (lineolate:oxygen oxidoreductase, E.C. 1.13.11.12) is a nonheme iron-containing dioxygenase that catalyzes the hydroperoxidation of unsaturated fatty acids containing a (Z,Z)-1,4-pentadiene moiety. Thus, linoleic [(LA); 18:2(9,12(all Z))] and α-linolenic [(LNA); 18:3(9,12,15(all *Z*))] acids, that are common in plant membrane phospholipids, are substrates. Arachidonic acid [(AA); 20:4(5,8,11,14(all *Z*))] is also a substrate that, while not found in plants, occurs in the cell walls of oomycete plant pathogenic fungi, such as *Phytophthora infestans*. The carbon atom where the hydroperoxy group is preferentially added by a particular enzyme species is denoted by the number that prefixes the enzyme name. Thus, a LOX from potato (*Solanum tuberosum*) has 9-LOX activity with LA but 5-LOX activity with AA (1). Lipoxygenase isoenzymes fall into two classes, those with optimum activity at relatively high pH 8–9 (type-1) and those most active at near neutral pH (type-2 [2]).

The fatty acid hydroperoxides formed by LOX activity feed via hydroperoxide lyase (HPL) activity into aliphatic acids, alcohols, and aldehydes, or via allene oxide synthase into ketol acids or after cyclization into prostaglandin-like oxophytodienoic acids and the related jasmonates (3,4; Chapter 9). This forms a major branch point in the lipoxygenase pathways in plants.

Jasmonic acid (JA) and methyl jasmonate (MeJ), ultimately derived from the 13(*S*)-hydroperoxide of LNA (5,6), are potent plant growth regulators with many pronounced physiological effects (7). Several other substances on the linolenate cascade initiated by LOX activity, such as the wound hormone traumatin (12-*oxo*-10(*E*)-dodecanoic acid [8]) and some volatile aldehydes (9), are also physiologically active in plants. Since some LOX pathway products are structurally similar to important mammalian signal substances, for example leukotrienes and prostaglandins (10), it is tempting to speculate that they may play analogous roles in signal transduction in plant disease. Some volatiles that are derived from LOX activity, such as (*E*)-2-hexenal, are potent antimicrobials and may be directly involved in plant defense (11).

Further, there is good evidence that the plant growth regulatory substance abscisic acid (ABA) may be synthesized from the carotenoid violaxanthin via a co-oxidation reaction with linoleate and LOX to produce the plant-growth inhibitor xanthoxin (12), which Taylor and Burden (13) showed could be converted by the Dwarf bean to ABA. Altered levels of ABA in diseased plants have been correlated with various effects on disease development and resistance (14,15).

In general, it seems that the highest levels of LOX are found in young, rapidly growing tissues (2). Localization studies showed that LOX protein is often prevalent in the outer regions of organs such as hypocotyls, sepals, and the pericarp, as in the French bean (16), where it might best be suited for serving a protective function against pathogen colonization or grazing insects (17,18). A further, intriguing speculation on the role of LOX in host-pathogen interactions comes from the unpublished observations of Loers and Grambow (personal communication) which show that fatty acids can replace H_2O_2 as a co-substrate in some peroxidase-catalyzed reactions.

What Happens to LOX Activity on Infection?

An increase in LOX activity, protein, or mRNA levels has been reported in several host-pathogen combinations (Table 10.1) and after elicitor treatment (19,20). An early report of LOX involvement in plant disease is that for tobacco infected with *Erysiphe cichoracearum*, the cause of powdery mildew disease (21). The authors detected an increase in LOX activity over uninfected controls from 2 to 3 days after inoculation peaking at 11 days. Data for changes in LOX activity during a resistance response was not given. They concluded that the increase in LOX might "cause a distortion" of host membranes that would allow a transfer of solutes from host mesophyll cells into fungal haustoria in epidermal cells, and bring about enhanced senescence and necrotization of the colonized host tissue. The former would enable the pathogen to obtain a better supply of nutrients and the latter could contribute to the symptom development

TABLE 10.1 Reported Instances of LOX Involvement in Plant-Pathogen Interactions[a]

Pathogen	Host	Induction of:			Reference
		mRNA	protein	activity	
Viruses					
TMV	tobacco			x	(137)
TNV	cucumber			x	(100)
Bacteria					
P. s. pv. pisi	cucumber			x	(71)
P. s. pv. phaseolicola	French bean	x	x	x	(28,37)
P. s. pv. syringae	tomato	x	x	x	(23)
P. s. pv. tomato	tomato	x	x	x	(23)
P. s. pv. maculicola	Arabidopsis	x			(40)
P.s. pv. tomato	Arabidopsis	x			(40)
Fungi					
Phoma exigua	potato			x	(17)
Phytophthora infestans	potato			x	(50)
Erysiphe chicoracearum	tobacco			x	(21)
Phytophthora parasitica	tobacco			x	(58)
Colletotrichum lagenarium	cucumber			x	(100)
Leveilla taurica	tomato			x	(24)
Ceratocystis fimbriata	taro			x	(18)
Puccinia coronata avenae	oats			x	(73)
Puccinia coronata avenae	wheat			x	(72)
Puccinia graminis triticae	wheat			x	(72)
Pyricularia oryzae	rice			x	(44)
Pyricularia oryzae	rice	x	x	x	(39)
Erysiphe graminis f.sp. hordei	barley			x	(138)

[a]Where compatible and incompatible interactions were compared, the magnitude or speed of the response was usually greater in the incompatible interaction.

seen in the later stages of infection. Interestingly, the same phenomenon of membrane damage caused by LOX activity is invoked to explain resistance-related effects, such as HR, in an incompatible interaction. This illustrates one of the difficulties of research into the effects of LOX; that is, several plausible hypotheses can be made to fit a given set of results, and critical testing of each hypothesis has often remained elusive.

When compatible and incompatible interactions have been compared, it was generally observed that LOX activity increases more quickly and to a higher level in resistant than susceptible plants (22). A summary of the salient features for some of the pathogen and elicitor studies in host plants for which more complete information is available follows. It should be noted that a role for LOX in resistance does not preclude a contribution to symptom development at later stages in a compatible combination as suggested by Lupu et al. for powdery mildew of tobacco (21).

Tomato

Tomato seedlings inoculated with *Pseudomonas syringae* pv. *syringae* (incompatible interaction associated with HR) or *P. syringae* pv. *tomato* (compatible interaction) were monitored for changes in LOX activity, LOX protein, and mRNA in the first 48 hours after infection (23). Lipoxygenase activity and mRNA levels were very low in healthy, control tomato leaves, but there was a clear accumulation of LOX mRNA by 3 hours after inoculation with a maximum around 12 hours in the incompatible combination. Increases in LOX activity followed mRNA transcript increases by a few hours. Hypersensitive cell collapse was observed around 12 hours after inoculation when the induced LOX activity had reached approximately 2 nkat mg^{-1} protein.

In the compatible interaction, LOX transcripts accumulated more slowly and reached a maximum around 24 hours with increases in enzyme activity again following mRNA accumulation by a few hours. Necrosis was first observed around 48 hours after infection, rather early for this host/pathogen combination, but a consequence of the high inoculum levels used (1×10^8 colony-forming units mL^{-1}) in order to be comparable with the incompatible combination where sufficient bacteria were used to give macroscopically visible HR lesions. Interestingly, necrosis in the compatible combination occurred when LOX activity was measured to be about twice that for HR cell collapse in the incompatible interaction (Fig. 10.1). These results show that the LOX response to plant pathogenic bacteria in tomato is more pronounced in the incompatible than the compatible interaction, that enhanced LOX activity correlates with necrosis, and are consistent with the hypotheses that the LOX response is regulated at least in part at the level of transcription.

Kato et al. reported a rapid increase in LOX activity in dwarf tomato leaves inoculated with the powdery mildew pathogen *Leveilla taurica* (24). Activity increased approximately fivefold within an hour of inoculation and the enzyme exclusively produced 9-hydroperoxylinolenic acid with LNA. However, although the authors state that the oxidation products had antimicrobial activity towards several plant pathogenic fungi they do not present data for 9-hydroperoxylinolenic acid but only for 13-

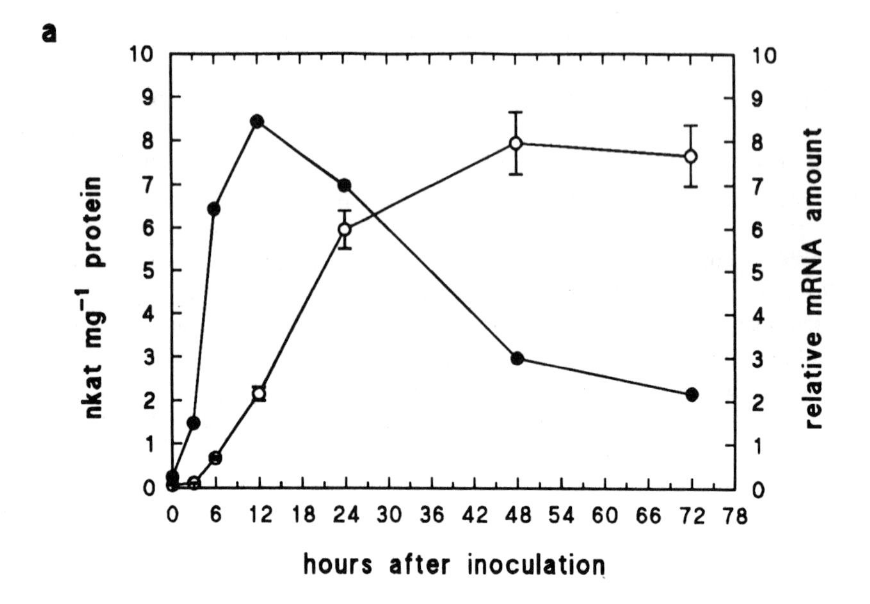

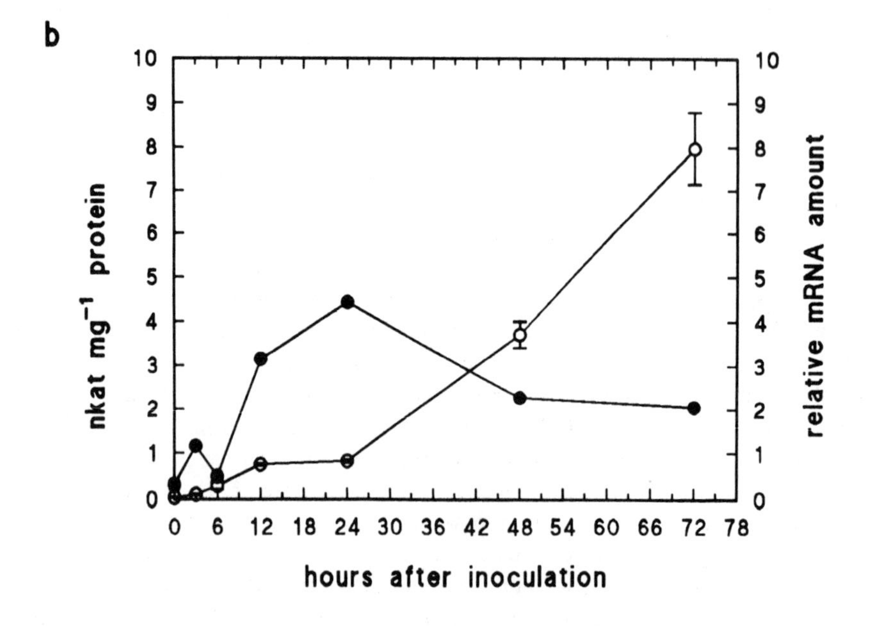

Figure 10.1. Time course of changes in LOX enzyme activity (○) and relative amount of LOX mRNA (●) in leaves inoculated with *a) P. syringae* pv. *syringae* (HR), and *b) P. syringae* pv. *tomato* (susceptible reaction). Enzyme activity is expressed as nkat mg^{-1} protein. Used with permission. *Source:* Koch et al. (23).

hydroperoxylinolenic acid and some of its derivatives. Thus, it is difficult to judge whether the induced LOX activity is involved in the production of defense compounds in tomato or has some other function.

Pathogen-derived molecules that elicit responses in the plant similar to those that occur in response to the pathogen itself can be used to simplify the investigation of host-pathogen interactions. In this way, Peever and Higgins showed that a race-specific elicitor (SE) from the plant pathogenic fungus *Cladosporium fulvum* induced LOX activity and lipid peroxidation in resistant but not in susceptible tomato genotypes (25). However, necrosis induction by SE was dependent upon light, while LOX activity and lipid peroxidation were also induced in the dark. Further, a nonspecific elicitor (NSE) that induced LOX and lipid peroxidation in all tomato genotypes tested did not require light. The LOX inhibitor piroxicam reduced electrolyte leakage caused by NSE but not SE. When considered together, all these observations suggest that the mechanisms by which SE and NSE cause cell death are different and that another factor, in addition to LOX, is necessary in the case of SE.

Since tomato is genetically transformable and the LOX response is relatively uncomplicated (23), it would be a suitable host in which to investigate the regulation of LOX genes at the molecular level and use molecular methods to test some of the hypotheses concerning the role of LOX in host-pathogen interactions. A first stage in this would have to be the cloning and characterization of the pathogen-induced LOX from tomato.

French Bean and Other Legumes

Legumes are singled out in the plant world because they not only have large quantities of LOX in seeds and vegetative parts but also possess a complex array of different LOX isoforms encoded in large gene families that appear to be expressed in a developmental- and tissue-specific manner (16,26). Some LOX isoforms seem to function as vegetative storage proteins or VSPs (27), while the roles of other isoforms remain to be determined with certainty.

Phaseolus vulgaris inoculated with virulent or avirulent races of *Pseudomonas syringae* pv. *phaseolicola* showed an increase in LOX activity in the incompatible combination but not in the compatible combination (28). The increase in LOX activity was inhibited by treatment of the leaves with the protein synthesis inhibitor cycloheximide, which also inhibits HR cell collapse (29). Lipid peroxidation during HR was demonstrated by measuring the production of several volatiles from the LOX pathway (28,30); some products, such as (*E*)-2-hexenal, were shown to be produced in vivo in amounts that show clear inhibition of pathogen growth with in vitro assays.

Rogers et al. demonstrated that lipid peroxidation in bean cell suspension cultures was a consequence of treatment with a crude elicitor preparation from the bean pathogen *Colletotrichum lindemuthianum* (31). They suggested that, since phytoalexins were also induced to accumulate by cell treatment with xanthine/xanthine oxidase and Fe^{+2}/ethylenediamine di-*o*-hydroxyphenyl acetic acid, active oxygen species

(AOS) might be involved in lipid peroxidation triggered by elicitors. However, Kondo et al. showed that the transient accumulation of phosphatidylcholine hydroperoxide (PC–OOH) observed in elicitor-treated soybean cotyledons was probably initiated by oxygenation of membrane lipids in situ by the direct action of LOX rather than AOS (32,33). These authors also showed a transient increase in a LOX isoform with direct activity on membranes peaking 6 hours after elicitor treatment of cotyledons.

In addition to contributing to membrane damage in HR, LOX activity per se might be important to produce fatty acid hydroperoxides and other metabolites on the LOX pathway. Some of these substances have been shown to act as elicitors of plant defense responses in their own right. Thus, LA hydroperoxide has been shown to elicit the accumulation of the soybean phytoalexin glyceollin (34). In addition, Longland et al. showed that some PUFA caused necrosis and elicited the accumulation of the phytoalexins phaseolin and coumestrol in bean leaves (35). The authors speculated that the effect was mediated by the fatty acid hydroperoxides but did not support this claim with any data.

In order to study the regulation of the LOX response in bean to *P.s.* pv. *phaseolicola* more closely, a bean LOX cDNA was isolated using a probe from pea (36) and used to probe RNA blots to measure transcript accumulation. The bean cDNA was also expressed in *E. coli* and the LOX fusion protein was used for raising anti-LOX antibodies to monitor LOX protein levels in infected leaves. Bean leaves were inoculated with bacteria using a hypodermic syringe and leaf tissue was sampled in three zones: the directly inoculated zone that undergoes HR collapse, the surrounding 5–7 mm, and the remainder of the leaf (37).

Young primary bean leaves have a background level of LOX transcripts that declines as the leaves age (16). In HR in zone 1, LOX transcript levels were rapidly down-regulated in the whole leaf and were essentially undetectable by 6 hours after inoculation. However, approximately 14 hours after inoculation a pronounced accumulation of LOX transcripts in zones 2 and 3 was observed. In the compatible combination a strong accumulation of LOX transcripts was also observed in zones 2 and 3 at 14 hours and, in marked contrast to the HR, also in zone 1 (38).

One might have expected LOX transcript accumulation to have been up-regulated in zone 1 in the incompatible combination if it played an important role in the HR. It was concluded that the bean LOX cDNA probe was probably detecting a vegetative storage form of LOX that was induced to accumulate in the leaves by the stress of inoculation. Presumably, the VSP form of LOX did not accumulate in zone 1, that is the HR tissue itself, because the cells were switched over to defense metabolism as evidenced by the accumulation of known defense-related transcripts, such as those coding for phenylalanine ammonia lyase (PAL), chalcone synthase (CHS), and CHT (37,38). In this regard it should be noted that the LOX isoform reported to be induced in the resistance response in rice only has low homology to other LOX isoforms characterized so far (39). If the same is true in bean, then our probe would not necessarily detect the pathogen-induced LOX transcripts. All in all, it seems that the legume system with its multiplicity of LOX genes and recalcitrance to genetic transformation is not ideally suited to study the regulation of the LOX response to pathogens at the molecular level. Pathosystems for *Arabidopsis* or tomato are potentially much better suited to this purpose.

Arabidopsis

This diminutive crucifer's increasingly important role in plant biology research holds real promise that the physiological roles of LOX might be fully defined in this plant first. To this end, a full-length LOX cDNA and its corresponding gene (*LOX1*) were cloned from *Arabidopsis* (ecotype Columbia [40]). During development *LOX1* was expressed in roots and young seedlings and in mature plant leaves in response to ABA and MeJ. No differences in LOX activity were found in our work with *Arabidopsis* downy mildew caused by *Peronospora parasitica*. No change in transcript levels between resistant and susceptible interactions was observed using the probe supplied by Melan et al. (40,41). However, a sixfold accumulation of *LOX1* transcripts were induced in leaves by inoculation with virulent or avirulent *Pseudomonas syringae* isolates, although the response was faster with an avirulent *P.s.* pv. *tomato* isolate than with a virulent *P.s.* pv. *maculicola* strain (40).

In the incompatible interaction the maximum level of transcript accumulation was around 12 hours after inoculation whereas in the compatible combination the maximum was after 48 h. It will be interesting to see data for changes in LOX activity and lipid peroxidation in this pathosystem, and in situ hybridization studies coupled with promoter-reporter gene analysis to localize expression of the *LOX1* gene in relation to infection. Such experiments will provide further data of a correlative nature; in addition, the *Arabidopsis* model system has the potential for critically testing the action of LOX in the plant by making phenocopy mutants. In other words, the advantages of the simple genome organization and ease of genetic transformability of *Arabidopsis* can be exploited to prepare transgenic plants containing overexpression and antisense constructs for LOX under the control of constitutively active or inducible promoters.

The transgenic plants, that behave as LOX-overexpressed, LOX-reduced, or even LOX-null mutants, can then be studied to see if predicted phenotypes are observed. Thus, one might expect the antisense plants to have less LOX, be reduced in their HR expression, and perhaps show a shift in the direction of increased susceptibility to pathogens. The LOX-overexpressed plants might be expected to show the reverse. Of course all this depends on the plant's ability to survive the condition where its usual regulation of LOX genes, that may have essential roles in development and normal physiological functioning, is overridden.

Rice

Hydroxy and epoxy unsaturated fatty acids have been reported to act as antifungal phytoalexins in rice (42,43). Ohta et al. showed that these substances were derived from linoleate hydroperoxide via a lipid hydroperoxide-decomposing activity (LHDA

[44]). Three LOX isozymes were purified from *Magnaporthe grisea*-infected and healthy rice leaves and the LOX-3 isoform was the most strongly induced during attempted infection by an avirulent isolate of the rice blast pathogen (45). The authors showed that the LOX-3 isoform produced the 13-hydroperoxide with LA as substrate.

Peng et al. cloned a pathogen-induced LOX (*Lox2:Os:1*) from rice infected with *M. grisea* (39). This LOX gene has relatively low homology with other known LOX, and in addition is the first reported to have a leader sequence for chloroplast targeting. The enzyme produces exclusively 13-hydroperoxides of LA and LNA, and its mRNA and protein were detectable by 15 hours after inoculation in the incompatible combination; much lower levels were observed in the compatible combination (39). It is possible, but not yet proven, that the protein encoded by *Lox2:Os:1* is identical with the LOX-3 isozyme reported by Ohta et al. (45). The increase in LOX accumulation coincides with penetration of the rice leaf by the pathogen, and the authors speculated that the enzyme might play an important role in the plant's defense response, possibly by providing fatty acid hydroperoxides for the synthesis of rice phytoalexins.

Potato

It has long been known that AA and eicosapentaenoic [(EPA); 20:5(5,8,11,14,17(all Z))] acid, that are present in the cell walls of *Phytophthora infestans* and are substrates for potato LOX, elicit HR and phytoalexin accumulation in potato (46). Several lines of evidence indicate that LOX activity is crucial to the action of these fatty acids as elicitors in altering secondary metabolism in potato tissues. Thus, structure-activity studies with several fatty acids related to AA and EPA correlated their elicitor activity with their suitability as LOX substrates and LOX-null potato lines were reduced in their response to AA (14,47,48).

Lipoxygenase activity was also shown to be stimulated in *Phytophthora*-infected and AA-treated leaves and tuber discs (49,50); LOX inhibitors suppressed phytoalexin accumulation and HR in potato tubers (51,52). Ricker and Bostock demonstrated that ^{14}C-AA and its metabolites were released from sporangia of *P. infestans* within 9 hours after inoculation of potato leaves (53), however, only 5% of the radioactive label was found in the peroxidized fatty acid fraction. As little as 5 µg AA applied to potato tuber discs elicits the accumulation of phytoalexins, and the maximum response can be achieved with 50 µg AA (46).

More recently Ricker and Bostock showed that the production of eicosanoid derivatives of AA was inversely proportional to the amount of AA applied and that the 5-LOX–derived metabolites themselves had no elicitor activity (54). Further apparent contradictions to the hypothesis that AA elicitor activity depends on LOX is that preparations of β-glucans from *P. infestans* that enhance AA elicitor activity inhibited LOX and the formation of eicosanoids, whereas ABA which suppresses phytoalexin accumulation and induces susceptibility in potato had no effect on the spectrum or quantities of eicosanoid production (55). Essentially, these results suggest that the pool of eicosanoids that forms rapidly after AA treatment may not be associated with the elicitor activity of AA. It is difficult to reconcile these results with the earlier observations that correlate LOX activity with that of AA as an elicitor. Nevertheless, the earlier data remain to be explained and the authors showed that several of the eicosanoids produced by LOX activity on AA were themselves inhibitory to *P. infestans* cystospore germination (55), thus indicating a potential further role for LOX in resistance in this pathosystem.

A recent interpretation of the effect of LOX-inhibiting nonsteroidal anti-inflammatory drugs (NSAID) in suppressing phytoalexin accumulation in potato tuber slices was given by Ellis et al. (56). These authors showed that the suppression of phytoalexin accumulation was probably due to the superoxide scavenging properties of the NSAID. Other active oxygen scavengers were also effective at inhibiting phytoalexin accumulation induced by AA or even the nonlipid elicitor chitosan. These results suggest that both fatty acid and nonlipid elicitors have a similar mode of action that probably involves the production of superoxide anions or other AOS.

Tobacco

Changes in LOX activity have been investigated in tobacco cell suspension cultures after treatment with crude elicitor from *Phytophthora parasitica* var. *nicotianae*, and in tobacco plants infected with the same pathogen (57,58). An increase in LOX activity was found in cell suspension cultures in response to elicitor, and the increase was suppressed by cycloheximide but not by actinomycin D. It was suggested that the LOX response in tobacco depended upon host protein synthesis but not on transcription (57). However, elicitor-stimulated ethylene production, which is known to be regulated by the transcriptional activation of the genes for 1-aminocyclopropane-1-carboxylic acid synthase (ACC synthase) and the ethylene-forming enzyme (EFE) (59,60), was also not suppressed by actinomycin D under the conditions used.

It seems likely that transcription was not effectively inhibited in the experiments, and the conclusion that the LOX response to elicitor in tobacco is not transcriptionally regulated might be premature. The elicitor-induced LOX activity was purified to homogeneity and characterized biochemically (58). The major products obtained from C18 and C20 fatty acids were 5- and 9-hydroperoxides, but small amounts of 13-hydroperoxides (7% for LNA) were also formed from C18 fatty acids (58). The authors suggest that the enzyme might be involved in signaling phenomena via JA and other fatty acid–derived products leading to defense. However, this remains to be shown, and it should be noted that LNA 13-hydroperoxide, the JA precursor, was only a minor product of the reaction. Thus it is possible that LOX plays another, as yet undefined, role in resistance in tobacco.

The Hypersensitive Reaction and Associated Plant Defense Responses

In an incompatible interaction (i.e., a resistant host challenged by an avirulent pathogen) there is often a rapid localized plant cell necrosis at the site of pathogen ingress. This reaction is associated with the halting of pathogen spread; since the plant behaves as though it were "more than usually sensitive" to the presence of the pathogen, it is described as "hypersensitive," and one talks of the "hypersensitive response" (61). Hypersensitive response is an active defense response, dependent upon host protein synthesis and can be viewed as a form of programmed cell death or apoptosis (29,62–68). In the case of biotrophic pathogens, HR of the host cells prevents the establishment of the finely tuned nutritional relationship with the host; thus it is clear how HR serves as a powerful resistance mechanism. In contrast, the role of HR in resistance to pathogens that are not obligate biotrophs is less obvious, and it is thought that in these cases the final necrosis per se may not be as important as other HR-associated changes in host metabolism (22,64).

The first physiological changes that have been detected in cells that will undergo hypersensitive cell collapse, such as electrolyte leakage and the failure to plasmolyze, indicate irreversible membrane damage or "IMD" (22,69). Keppler and Novacky suggested that superoxide anion ($O_2^{\cdot -}$) mediated lipid peroxidation might account for IMD in HR (70,71). Certainly, a large amount of lipid peroxidation accompanies HR cell collapse (30), and it is clear that both enzymic and nonenzymic mechanisms could contribute to the process (22). One of the earliest reports linking LOX activity with HR was in the resistance response of wheat to black stem rust caused by the basidiomycete fungus *Puccinia graminis tritici* (72), and in oats to crown rust caused by *P. coronata avenae* (73). Croft et al. measured an increase in LOX activity preceding HR cell collapse in *Phaseolus vulgaris* after inoculation with an avirulent isolate of a bacterial bean pathogen and suggested that the already high background level of lipolytic acyl hydrolase (LAH) activity in bean might be working in conjunction with LOX to degrade membranes during HR (28).

It is generally believed that free fatty acids are the best substrates for LOX. However, a growing number of examples show that some LOX isozymes can work directly on membrane lipids or esterified fatty acids. Thus, Eskola and Laakso and Brash et al. showed that soybean LOX-1 could catalyze the oxygenation of PUFA esterified in phosphatidyl choline (74,75). Indeed, Maccarrone et al. recently showed that soybean LOX-2 can oxygenate esterified, unsaturated fatty acid moieties in preparations of soybean plasmalemma, chloroplast, and golgi membranes in vitro (76). Also, rabbit 15-LOX is highly reactive towards phospholipids both in liposomes and biological membranes and is responsible for the degradation of intracellular organelle membranes in reticulocytes during their maturation; it is the second most abundant protein after hemoglobin in erythrocytes (77–79). In a very significant paper, Kondo et al. showed that the LOX activity induced in soybean cotyledons in response to elicitor treatment was more active with fatty acids esterified in phosphatidylcholine lipid than with free fatty acids (33), suggesting that the pathogen-induced LOX activity might be specifically targeted at membranes.

Lipoxygenase has often been reported to be associated with membranes of various types, and it is possible that this is an artifact of the purification procedure. However, there are well-documented reports of specialized membrane-bound isoforms of LOX that differ in biochemical properties from the soluble forms. For example in soybean (80), and in green, unripe tomato fruits membrane-associated forms of LOX were shown in plasmalemma and tonoplast fractions (81,82), and particulate and soluble LOX isoforms have been reported in cucumber (83). Lipoxygenase activity associated with the plasmalemma of sunflower protoplasts has also been described recently (84).

The picture now emerging is that membrane-associated LOX may play an important role in the "remodeling" and degradation of plant membranes, particularly in senescence and in stress responses of the plant to wounding and pathogens (80), strengthening the case for LOX to have a role in causing membrane damage in plant HR to pathogens.

In causing membrane damage in HR, LOX probably works in conjunction with other lipolytic enzymes. Thus, in addition to the high constitutive level of LAH in bean, Croft et al. also detected a slight, but relatively slow, increase in LAH activity in the incompatible combination of bean with *Pseudomonas syringae* pv. *phaseolicola* (28). However, given the high constitutive background LAH activity in bean, the increase is probably insignificant for HR. There is also the possibility that LOX works in conjunction with a more specific phospholipase to degrade cell membranes during HR, but evidence for this has not yet been presented. Once formed, the peroxidized fatty acids can further contribute to membrane damage by decreasing membrane fluidity and increasing the gel phase, thus contributing to increased ion leakage (85).

An alternative view of the events leading to membrane damage in HR was proposed by Keppler and Novacky, who suggested that fatty acid hydroperoxidation was initiated by $O_2^{\cdot-}$ and that LOX might metabolize and detoxify the potentially harmful free fatty acids released by membrane destruction (71). Although not an integral part of the LOX mechanism, both $O_2^{\cdot-}$ and singlet oxygen ($^1O_2$) are produced as by-products in a peripheral manner (86,87). However, it seems unlikely that LOX activity is a major source of AOS in host-pathogen interactions. Further, H_2O_2 and $O_2^{\cdot-}$ are generally considered not to be sufficiently reactive to abstract the hydrogen from the methylene group in the unsaturated fatty acid to initiate the hydroperoxidation process (88). Thus, many authors suggest either that Fenton chemistry, in which the hydroxyl radical ($OH\cdot$) can be produced from H_2O_2 and $O_2^{\cdot-}$ in an iron-catalyzed Haber–Weiss process, may be involved; or that singlet oxygen ($^1O_2$) might play a role in membrane lipid peroxidation (89,90). Fenton chemistry demands the availability of iron or other transition metal, but an alternative might be the protonation of $O_2^{\cdot-}$ in acidic environments (e.g., at membrane surfaces) to form the perhydroxyl radical ($HO_2\cdot$) that, when compared to O_2, is sufficiently reactive to initiate in situ lipid peroxidation in membranes (22,88,91,92). On the whole, the evidence that LOX activity is an essential component of cell collapse in HR seems good, but this view is not universally held, and the contribution of active oxygen species towards membrane damage cannot be ignored.

LOX and Systemic Acquired Resistance

Plants that are susceptible to a particular pathogen can often be made resistant to it by a prior infection with a pathogen that causes necrosis. The resistance can work systemically throughout the whole plant after a predisposing infection limited to a single leaf. This phenomenon of "systemic acquired resistance" (SAR) has been known for many years and can also be induced by treating the plant with certain chemicals that are not antimicrobial themselves (93–95). Once a plant has been immunized, the acquired resistance state can last for several weeks and functions against a broad spectrum of pathogenic viruses, bacteria, and fungi, however, it is not universally effective against all pathogens.

Immunization increases the levels of some enzymes systemically, such as CHT and β-1,3-glucanases, so that these defense-related hydrolases are already present when the challenge with the second pathogen occurs. Some defense responses that occur locally in a resistance response, such as lignification and phytoalexin accumulation, do not occur systemically in immunized plants. However, on challenge with the second pathogen these phenomena manifest very rapidly, whereas under normal circumstances this would not occur. Thus, immunized plants can be described as being "sensitized" to contact with the challenge pathogen and transcripts of defense-associated genes accumulate rapidly in response to the challenge pathogen. What would normally be a compatible interaction becomes an incompatible interaction in which the expression of resistance can be very extreme, up to and including a phenocopy of HR (96,97). Exactly how this "sensitization" is controlled in the plant is still not known for certain, but it appears to be related to the level of salicylic acid (SA) which increases in immunized plants and has been shown to be essential for resistance (98,99).

Interestingly, LOX is one of the enzymes shown to increase systemically in immunized cucumber (100), rice, and tobacco plants and has been described as a "molecular marker for the immunized state" in rice (95). It was also reported that treatment of potato tubers with inhibitors of LOX activity suppressed the induction of SAR (101). Although the precise role of LOX in the systemic resistance response is unclear, there is the potential for it to act in the three spheres mentioned previously, in the production of antimicrobial metabolites, signal substances, or in contributing to membrane damage in the phenocopy HR induced in response to the challenge pathogen. Thus, it could be argued that enhanced levels of LOX present systemically in the leaves of immunized plants could contribute to the plant's state of sensitized anticipation that helps speed its response to the challenge pathogen.

LOX-Pathway Metabolites and Disease

Metabolites Affecting the Plant

Jasmonates. In a landmark publication, Farmer and Ryan (102) demonstrated the potential for MeJ, a metabolite of 13-hydroperoxylinolenic acid, to act as a volatile signal between plants, and for MeJ and JA to act within the plant as signal molecules in the regulation of proteinase inhibitor (PI) gene expression associated with resistance of tomato plants to insect grazing (103,104, Pub: IPR). Peña-Cortez et al. (105) showed very elegantly that wound-induced gene expression in tomato leaves was prevented by blocking JA synthesis with salicylhydroxamic acid (SHAM) and aspirin, substances that inhibit LOX and allene oxide synthase (hydroperoxide dehydrase), respectively.

The proposal that MeJ/JA might act in plant resistance as signal substances to pathogens received some support from the observation that elicitor-induced accumulation of antimicrobial secondary metabolites in cell suspension cultures of *Rauwolfia* and other dicotyledons, monocots, and gymnosperms was mediated by MeJ (103,106,107). The authors tested cell suspension cultures from 36 plant species and reported that MeJ elicited the accumulation of secondary plant products. Methyl jasmonate was shown to cause PAL transcripts to accumulate and was associated with an increase of PAL activity in cell suspension cultures of *Glycine max.* (107). Mitchell and Walters (108) reported that treatment of the first leaves of barley seedlings with MeJ or MeJ vapor reduced powdery mildew infection on the upper second leaves. In addition, PAL, POX and LOX activities were stimulated in second leaves. In a field trial, MeJ applied at 412 g ha-1 controlled powdery mildew infection (ibid).

In other situations, data implicating a role for MeJ/JA in resistance to pathogens are more equivocal. Thus, although Schweizer et al. protected barley from *Erysiphe* infection with MeJ (109), they showed that this was probably due to a direct antifungal effect of the MeJ-containing preparations and not due to any MeJ-induced effects in the plant. Kogel et al., also working with the barley/powdery mildew pathosystem, showed that MeJ and JA did not induce any detectable host defense response (110). Interestingly, the latter authors could not detect any direct toxic effect of MeJ or JA on the pathogen. In addition, the authors measured endogenous levels of MeJ and JA by radioimmunoassay, and although an increase was observed in osmotically stressed plants, no changes were observed either in an incompatible interaction with *E. graminis* or in plants with chemically induced resistant states (110). The authors concluded that resistance in barley to powdery mildew was not associated with enhanced levels of endogenous jasmonates, and hence that in this case there was no evidence to support the model of a lipid-based signaling pathway via jasmonates in plant resistance to infection (103,107).

We investigated the role of MeJ/JA in the *Phaseolus vulgaris/Pseudomonas syringae* pv. *phaseolicola* pathosystem and found that neither substance induced the accumulation of isoflavonoid phytoalexins or transcripts for the PAL or CHS enzymes involved in their synthesis. In contrast, transcripts of LOX and CHT were induced to accumulate and served as an internal control to show that MeJ/JA were effective in mediating at least some responses in the system at the concentrations used (Eiben, Croft, and Slusarenko, unpublished).

The most conclusive evidence that MeJ, although involved in wound-induced signaling, is not a component of pathogen-induced signaling was provided by Choi et al. (111). Gene-specific probes were used to follow the accumulation of transcripts for the *hmg1* and *hmg2* (3-hydroxy-3-methylglutaryl-coenzyme A reductase) genes in AA-treated or MeJ-treated potato tuber discs. The isozymes encoded by these two genes provide substrates alternatively for two branches of the terpenoid pathway, that is, to steroids that accumulate on wounding or in response to pathogens or elicitors, or to the terpenoid phytoalexins, such as lubimin. Methyl jasmonate treatment led to a doubling of the wound-induced accumulation of *hmg1* transcripts and caused steroid-glycoalkaloid (SGA) accumulation, whereas *hmg2* transcript abundance was reduced and phytoalexins did not accumulate. Evidence that endogenous jasmonates might be involved in signaling in wound response was obtained by showing that LOX inhibitors reduced the wound-induced accumulation of *hmg1* transcripts and suppressed SGA levels; in both cases the effect was counteracted by applying MeJ. In contrast, AA in a concentration-dependent manner strongly induced *hmg2* transcripts to accumulate whereas *hmg1* was down-regulated; SGA levels were suppressed, and terpenoid phytoalexin accumulation was induced. These results demonstrate clearly that two distinct signaling pathways are operating—one for wound responses (via jasmonate) and another for pathogen responses (not via jasmonate).

Another perspective on the potential role of jasmonates in host-pathogen interactions was given by Kauss et al. (112). Working with cell suspension cultures of parsley, it was shown that while MeJ does not elicit hydroxycoumarin phytoalexins or other defense-related phenomena, such as the incorporation of esterified phenolics or "ligninlike" polymers into the plant cell wall, it does condition the parsley cells so that they give an increased response to a subsequent treatment with an elicitor preparation from *Phytophthora megasperma* f.sp. *glycinea*. A similar enhancement was noted for the elicitation of AOS (113). The analogy to the "sensitization" seen in plants induced to the SAR state is striking, but unfortunately there is no information yet available as to whether MeJ can induce SAR in parsley plants.

It is interesting to note that when effects have been ascribed to MeJ or JA, it has usually been with cell suspension cultures. Whereas, when information is available for whole plants there is often no apparent effect of jasmonates relevant to defense responses against pathogens.

The work of Peña-Cortez et al. can also be considered here (105). These authors showed not only that aspirin (acetyl salicylic acid), but also SA itself, inhibits wound-induced gene expression via the suppression of JA synthesis. Since both aspirin and SA are strong inducers of SAR in plants (95,114), this suggests that endogenous control of SAR is unlikely to be regulated via jasmonates.

The question remains open as to whether effects due to exogenously applied MeJ or JA have biological relevance; endogenous jasmonates must be demonstrated to regulate the effect in vivo before it can be concluded that they are important in SAR or other resistance phenomena.

Abscisic Acid. Another important metabolite that influences plant-pathogen interactions is the C15 molecule ABA. Two biosynthetic pathways were proposed for ABA, the "direct pathway," as for other sesquiterpenoids, from mevalonic acid via farnesyl pyrophosphate; and the "indirect pathway" involving a LOX-mediated co-oxidation of (C40) xanthophylls with xanthoxin as an intermediate. Evidence from the characterization of several allelic *aba* mutants in *Arabidopsis* has shown conclusively that the indirect pathway is the operative one, at least for this plant (115,116). In addition, Creelman et al. showed that LOX inhibitors were able to inhibit stress-induced accumulation of ABA in soybean cell suspension cultures and seedlings, whereas CO, a potent inhibitor of heme-containing monooxygenases was not inhibitory (117). The authors concluded that the in vivo oxidative cleavage reaction involved in ABA synthesis is catalyzed by a non-heme oxygenase having LOX-like properties.

In host-pathogen interactions, raised ABA levels have sometimes been correlated with resistance (118,119), although there are exceptions. Thus, treatment of tobacco with ABA caused susceptibility to *Peronospora tabacina* (120), although the authors reported that the effect was inconsistent, and that in some experiments resistance was enhanced in so far as lesion area and sporulation was reduced. In potato and soybean, aside from increasing susceptibility to *Phytophthora infestans* and *P. megasperma* f.sp. *glycinea* respectively, it was reported that phytoalexin accumulation in both hosts was suppressed by ABA treatment (14,121). In contrast, Dixon and Fuller reported that ABA enhanced phytoalexin accumulation in *Phaseolus vulgaris* cell suspension cultures, although cell growth was inhibited (122). It is obvious that there is no simple consensus regarding the effects of ABA on host-pathogen interactions.

Whenham and Fraser showed that in the local necrosis response of *Nicotiana tabacum* cv. White Burley to the *flavum* strain of TMV, free ABA levels increased approximately 18-fold in or very close to the lesions and also increased systemically in uninfected leaves (15). Similarly, increased ABA content has also been shown to be associated with physiologically induced resistance of *P. vulgaris* to the pathogenic fungus *Colletotrichum lindemuthianum* (119). Thus, since LOX increases in SAR, and ABA synthesis may depend on LOX, this might explain the importance of LOX in acquired resistance.

Both bacterial infection and ABA treatment cause transcripts of the *Arabidopsis LOX1* gene to accumulate (40), and since ABA deficient mutants of *Arabidopsis* are available, it will be interesting to see if these mutants are altered in their responses to pathogens, particularly in a well-characterized pathosystem, such as downy mildew caused by *P. parasitica* (123). In addition, SAR is a well-established phenomenon in *Arabidopsis* (96,97,124), and it will be revealing to see whether ABA-deficient mutants are compromised in their ability to be immunized to the SAR state.

Metabolites Affecting Pathogens

A major biocidal product formed on the LOX pathway is (*E*)-2-hexenal. This C6 α-β unsaturated aldehyde is chemically quite reactive and has pronounced effects on biological systems (125). The α-β unsaturated bond is important for activity because the saturated aldehydes are less effective biocides (125,126). The earliest report known to this author of a suggested role for (*E*)-2-hexenal in plant resistance to infection was from Major et al. (127). The authors isolated (*E*)-2-hexenal from leaves of the maidenhair tree (*Ginkgo biloba*) and showed that it could inhibit growth of the plant pathogenic fungus *Monilinia fructicola* at 300 ppm. (*E*)-2-hexenal was shown to be antiprotozoal by Von Schildknecht and Rauch (128).

Zeringue and McCormick demonstrated that wounded cotton plants produced (*E*)-2-hexenal that had inhibitory activity against the saprophytic fungus *Aspergillus flavus* (129), and it was later shown that C6–C10 alkenals were able to elicit phytoalexin accumulation in wounded cotton bolls (130). Evidence that hexanal produced via the LOX pathway helps make soybeans resistant to spoilage in storage due to *Aspergillus flavus* was given by Doehlert et al. (131). Vaughn and Gardner investigated the effects of LOX-derived aldehydes on the growth of several plant pathogenic fungi and showed that (*Z*)-3- and (*E*)-2-nonenal were more toxic than (*E*)-2-hexenal but were less effective because of their lower volatilities (132). Croft et al. demonstrated that (*E*)-2-hexenal and (*E*)-3-hexen-1-ol were both antibacterial, with (*E*)-2-hexenal being about 20 times more effective than (*Z*)-3-hexen-1-ol, and that both were produced in significant amounts during the cell collapse phase of HR (30). In contrast, in a compatible interaction, in which HR does not occur, only low amounts of these volatiles were produced by the plant. The authors concluded that the volatiles might play a role in reducing pathogen growth at a time before other defense responses, such as the accumulation of isoflavonoid phytoalexins, were operative.

Localization studies in *Phaseolus vulgaris* and tomato have shown that LOX protein is often concentrated in cell layers towards the periphery of plant organs, such as hypocotyls, sepals and fruits (16,133); recently Shibata et al. (134), using tissue-print immunoblots, demonstrated that fatty acid HPL was most abundant in the outer parenchymal cells of the pericarp of bell pepper fruits. Thus, the fact that these two enzymes are located in tissues that will be among the first to be damaged by pathogen or pest attack, is perhaps a good, albeit circumstantial, indication that an important function of LOX is to provide protective volatiles after mechanical or pathogen-caused damage.

LOX-derived volatiles do not always have destructive effects on pathogens. For example, it was reported that some LOX-derived C6 alcohols, for example (*Z*)-3-hexen-1-ol and (*E*)-2-hexen-1-ol, in the presence of an as yet unidentified epicuticular leaf factor, induce the differentiation of infection structures in the wheat stem rust fungus *Puccinia graminis* f.sp. *tritici*. Differentiation-inducing activity was observed at 0.1 mM whereas higher concentrations were inhibitory to pathogen growth (135).

Conclusions

There is now an almost overwhelming number of reports correlating increased LOX activity, increased accumulation of LOX transcripts, and occurrence of lipid peroxidation with resistance rather than susceptibility of plants to pathogens. Encouraged by a wealth of analogous data from animal systems (77,136), a number of feasible roles for

LOX in plant resistance have been proposed. However, as with investigations of LOX function in the healthy plant, the data used as a basis for formulating hypotheses are mostly correlative and the hypotheses await critical testing. In addition, the overall picture has often been clouded by contradictory results.

Perhaps the most likely and best-supported role for LOX in resistance to pathogens is to produce lipid-derived antimicrobial substances as shown in rice, bean, and several other species. A role for LOX in contributing directly to IMD in HR is gaining more support from recent work showing the susceptibility of esterified fatty acids in membranes to oxygenation via LOX, but definitive proof is lacking. Although the idea that LOX might provide signal molecules to coordinate plant defense responses is attractive, it awaits further clarification. Certainly, the picture that seems to be emerging is that McJ, although important as a signal substance in wound responses, is perhaps not physiologically relevant in host pathogen interactions. The next generation of experiments, using molecular techniques in transgenic plants to produce phenocopy mutants to assess LOX function in host-pathogen interactions, and the characterization of mutants altered in LOX-dependent fatty acid metabolism, might bring the first critical tests for some of the hypotheses of LOX action in host-pathogen interactions outlined in this review.

Acknowledgments

John Friend (University of Hull, U.K.), Hans-Jürgen Grambow (RWTH, Aachen, BRD), Felix Mauch (University of Berne, CH), and Ian Whitehead (Firmenich SA, Geneva, CH) are all thanked for constructive criticism of the manuscript.

References

1. Shimizu, T., Honda, Z.-I., Miki, I., Seyama, Y., Izumi, T., Rådmark, O., and Samuelsson, B. (1990) *Methods Enzymol. 187,* 296–306.
2. Siedow, J.N. (1991) *Ann. Rev. Plant Physiol. Plant Mol. Biol. 42,* 145–188.
3. Song, W.-C., Funk, C., and Brash, A.R. (1993) *Proc. Natl. Acad. Sci. USA 90,* 8519–8523.
4. Hamberg, M., and Gardner, H.W. (1992) *Biochim. Biophys. Acta 1165,* 1–18.
5. Vick, B.A., and Zimmerman, D.C. (1984) *Plant Physiol. 75,* 458–461.
6. Vick, B.A., and Zimmerman, D.C. (1987) in *The Metabolism Structure and Function of Lipids,* Stumpf, P.K., Mudd, J.B., and Nes, W.D., Plenum Press, New York, pp. 383–390.
7. Parthier, B. (1991) *Bot. Acta 104,* 446–454.
8. Zimmerman, D.C., and Coudron, C.A. (1979) *Plant Physiol. 63,* 536–541.
9. Gardner, H.W., Dornbos, D.L., and Desjardins, A.E. (1990) *J. Agric. Food. Chem. 38,* 1316–1320.
10. Anderson, J.M. (1989) in *Second Messengers in Plant Growth and Development,* Boss, W.F., and Morré, D.J., Alan R. Liss, New York, pp. 181–212.
11. Lyr, H., and Banasiak, L. (1983) *Acta Phytopathol. Acad. Sci. Hung. 18,* 3–12.
12. Firn, R.D., and Friend, J. (1972) *Planta 103,* 263–266.

13. Taylor, H.F., and Burden, R.S. (1973) *J. Exp. Bot. 24,* 873–880.
14. Bostock, R.M., and Stermer, B.A. (1989) *Annu. Rev. Phytopathol. 27,* 343–371.
15. Whenham, R.J., and Fraser, R.S.S. (1981) *Physiol. Plant Pathol. 18,* 267–278.
16. Eiben, H.G., and Slusarenko A.J. (1994) *Plant J. 5,* 123–135.
17. Galliard, T. (1978) in *Biochemistry of Wounded Plant Tissues,* Kahl, G., Walter de Gruyter, Berlin, pp. 155–199.
18. Masui, H., and Kojima, M. (1990) *Agric. Biol. Chem. 54,* 1689–1695.
19. Esquerré-Tugayé, M., Fournier, J., Pouenat, M.L., Veronesi, C., Rickauer, M., and Bottin, A. (1993) in *Mechanisms of Plant Defence Responses,* Fritig, B., and Legrand, M., Kluwer, Dordrecht, pp. 202–210.
20. Slusarenko, A.J., Meier, B.M., Croft, K.P.C., and Eiben, H.G. (1993) in *Mechanisms of Plant Defence Responses,* Fritig, B., and Legrand, M., Kluwer, Dordrecht, pp. 211–220.
21. Lupu, R., Grossman, S., and Cohen, Y. (1980) *Physiol. Plant Pathol. 16,* 241–248.
22. Slusarenko, A.J., Croft, K.P.C., and Voisey, C.R. (1991) in *Biochemistry and Molecular Biology of Host-Pathogen Interactions,* Smith, C.J., Clarendon Press, Oxford, pp. 126–143.
23. Koch, E., Meier, B.M., Eiben, H.G., and Slusarenko, A.J. (1992) *Plant Physiol. 99,* 571–576.
24. Kato, T., Maeda, Y., Hirukawa, T., Namal, T., and Yoshoioka, N. (1992) *Biosci. Biotech. Biochem. 56,* 373–375.
25. Peever, T.L., and Higgins, V.J. (1989) *Plant Physiol. 90,* 867–875.
26. Altschuler, M., Grayburn, W.S., Colins, G.B., and Hildebrand, D.F. (1989) *Plant Science 63,* 151–158.
27. Tranbarger, T.J., Franchesci, V.R., Hildebrand, D.F., and Grimes, H.D. (1991) *Plant Cell 3,* 973–987.
28. Croft, K.P.C., Voisey, C.R., and Slusarenko, A.J. (1990) *Physiol. Mol. Plant Pathol. 36,* 49–62.
29. Lyon, F.M., and Wood, R.K.S. (1977) *Ann. Bot. 41,* 479–491.
30. Croft, K.P.C., Jüttner, F., and Slusarenko, A.J. (1993) *Plant Physiol. 101,* 13–24.
31. Rogers, K.R., Albert, F., and Anderson, A.J. (1988) *Plant Physiol. 86,* 547–553.
32. Kondo, Y., Miyazawa, T., and Mizutani, J. (1992) *Biochim. Biophys. Acta 1127,* 227–232.
33. Kondo, Y., Kawai, Y., Hayashi, T., Ohnishi, M., Miyazawa, T., Itoh, S., and Mizutani, J. (1993) *Biochim. Biophys. Acta 1170,* 301–306.
34. Montillet, J.-L., and Degousée, N. (1991) *Plant Physiol. Biochem. 29,* 689–694.
35. Longland, A., Slusarenko, A.J., and Friend, J. (1987) *J. Phytopathol. 120,* 289–297.
36. Casey, R., Domoney, C., and Nielsen, N.C. (1985) *Biochem. J.* 232, 79–85.
37. Meier, B.M., Shaw, N., and Slusarenko, A.J. (1993) *Mol. Plant-Microb. Interact. 6,* 453–466.
38. Meier, B.M. (1992) Ph.D thesis, University of Zurich, Switzerland.
39. Peng, Y.-L., Shirano, Y., Ohta, H., Hibino, T., Tanaka, K., and Shibata, D. (1994) *J. Biol. Chem. 269,* 3755–3761.
40. Melan, M.A., Dong, X., Endara, M.E., Davis, K.R, Ausubel, F.M., and Peterman, T.K. (1993) *Plant Physiol. 101,* 441–450.
41. Mauch-Mani, B., Croft, K.P.C., and Slusarenko, A.J. (1993) in *Arabidopsis thaliana as a Model for Plant-Pathogen Interactions* Davis, K.R., and Hammerschmidt, R., APS Press, St. Paul, Minnesota, pp. 5–20.

42. Kato, T., Yamaguchi, Y., Uyehara, T., Nakai, T., and Yamanaka, S. (1983) *Tetrahedron Lett. 24,* 4715–4718.
43. Kato, T., Yamaguchi, Y., Abe, N., Uyehara, T., Nakai, T., Kodama, M., and Shiobara, Y. (1985) *Tetrahedron Lett. 26,* 2357–2360.
44. Ohta, H., Shida, K., Peng, Y.-L., Furusawa, I., Shishiyama, J., Aibara, S., and Morita, Y. (1990) *Plant Cell Physiol. 31,* 1117–1122.
45. Ohta, H., Shida, K., Peng, Y.-L., Furusawa, I., Shishiyama, J., Aibara, S., and Morita, Y. (1991) *Plant Physiol. 97,* 94–98.
46. Bostock, R.M., Kuc, J.M., and Laine, R.A. (1981) *Science 212,* 67–69.
47. Preisig, C.L., and Kuc, J.A. (1985) *Arch. Biochem. Biophys. 236,* 379–389.
48. Vaughn, S.F., and Lulai, E.C. (1992) *Plant Science 84,* 91–98.
49. Bostock, R.M., Yamamoto, H., Choi, D., Ricker, K.E., and Ward, B.L. (1992) *Plant Physiol. 100,* 1448–1456.
50. Fidantsef, A.L., Bostock, R.M., and Choi, D. (1993) *Plant Physiol. Suppl. 102,* 113 (Abstract)
51. Stelzig, D.A., Allen, R.D., and Bhatia, S.K. (1983) *Plant Physiol. 72,* 746–749.
52. Preisig, C.L., and Kuc, J.A. (1987) *Plant Physiol. 84,* 891–894.
53. Ricker, K.E., and Bostock, R.M. (1992) *Physiol. Mol. Plant Pathol. 41,* 61–72.
54. Ricker, K.E., and Bostock, R.M. (1994) *Physiol. Mol. Plant Pathol. 44,* 65–80.
55. Henfling, J.W.D.M., Bostock, R.M., and Kuc, J.A. (1980) *Phytopathol. 70,* 1074–1078.
56. Ellis, J.S., Keenan, P.J., Rathmell, W.G., and Friend, J. (1993) *Phytochem. 34,* 649–645.
57. Rickauer, M., Fournier, J., Pouénat, M.-L., Berthalon, E., Bottin, A., and Esquerré-Tugayé, M.-T. (1990) *Plant Physiol. Biochem. 28,* 647–653.
58. Fournier, J., Pouenat, M.-L., Rickauer, M., Rabinovitch-Chable, H., Rigaud, M., and Esquerré-Tugayé, M.-T. (1993) *Plant J. 3,* 63–70.
59. Sato, T., and Theologis, A. (1989) *Proc. Natl. Acad, Sci. USA 86,* 6621–6625.
60. Spanu, P., Reinhardt, D., and Boller, T. (1991) *EMBO J. 10,* 2007–2013.
61. Stakman, E.C. (1915) *J. Agric. Res. IV,* 193–201.
62. Keen, N.T., Ersek, T., Long, M., Bruegger, R., and Holliday, M. (1981) *Physiol. Plant Pathol. 18,* 325–337.
63. Slusarenko, A.J., and Longland, A. (1986) *Physiol. Mol. Plant Pathol. 29,* 79–94.
64. Collinge, D.B., and Slusarenko, A.J. (1987) *Plant Mol. Biol. 9,* 389–410.
65. Greenburg, J.T., and Ausubel, F.M. (1993) *Plant J. 4,* 327–341.
66. Dietrich, R.A., Delaney, T.P., Ukness, S.J., Ward, E.R., Ryals, J.A., and Dangl, J.L. (1994) *Cell 77,* 565–577.
67. Greenburg, J.T., Guo, A., Klessig, D.F., and Ausubel, F.M. (1994) *Cell 77,* 551–563.
68. Levine, A., Tenhaken, R., Dixon, R.A., and Lamb, C.J. (1994) *Cell 79,* 583–593.
69. Woods, A.M., Fagg, J., and Mansfield, J. (1989) *Physiol. Mol. Plant Pathol. 32,* 483–497.
70. Keppler, L.D., and Novacky, A. (1986) *Phytopathol. 76,* 104–108.
71. Keppler, L.D., and Novacky, A. (1987) *Physiol. Mol. Plant Pathol. 30,* 233–245.
72. Ocampo, C.A., Moersbacher, B., and Grambow, H.J. (1986) *Z. Naturforsch. 41c,* 559–563.
73. Yamamoto, H., and Tani, T. (1986) *J. Phytopathol. 116,* 329–337.
74. Eskola, J., and Laakso, S. (1983) *Biochim. Biophys. Acta 751,* 305–311.
75. Brash, A.R., Ingram, C.D., and Harris, T.M. (1987) *Biochem. 26,* 5465–5471.

76. Maccarrone, M., van Aarle, P.G.M., Veldink, G.A., and Vliegenthart, J.F.G. (1994) *Biochim. Biophys. Acta 1190,* 164–169.
77. Schewe, T., Wiesner, R., and Rappoport, S.M. (1981) *Methods Enzymol. 71,* 430–451.
78. Hunt, T. (1990) *Trends Biochem. Sci. 14,* 393–394.
79. Kuhn, H., Belkner, J., Wiesner, R., and Brash, A.R. (1990) *J. Biol. Chem. 265,* 18351–18361.
80. Macrì, F., Braidot, E., Petrussa, E., and Vianello, A. (1994) *Biochim. Biophys. Acta 1215,* 109–114.
81. Todd, J.F., Paliyath, G., and Thompson, J.E. (1990) *Plant Physiol. 94,* 1225–1232.
82. Droillard, M.-J., Rouet-Mayer, M.-A., Bureau, J.-M., and Lauière, C. (1993) *Plant Physiol. 103,* 1211–1219.
83. Feussner, I., and Kindl, H. (1994) *Planta 194,* 22–28.
84. Vianello, A., Braidot, E., Bassi, G., and Macrì F. (1995) *Biochim. Biophys. Acta 1255,* 57–62.
85. Paliyath, G., and Droillard, M.J. (1992) *Plant Physiol. Biochem. 30,* 789–812.
86. Lynch, D.V., and Thompson, J.E. (1984) *FEBS Lett. 173,* 251–254.
87. Thompson, J.E., Legge, R.L., and Barber, R.F. (1987) *New Phytol. 105,* 317–344.
88. Bielski, B.H.J., Arudi, R.L., and Sutherland, M.W. (1983) *J. Biol. Chem. 258,* 4759–4761.
89. Epperlein, M.M., Noronha-Dutra, A.A., and Strange, R.N. (1986) *Physiol. Mol. Plant. Pathol. 28,* 67–77.
90. Halliwell, B. *Trends Biochem. Sci. 7,* 270–272.
91. Fridovich, I. (1988) in *Oxygen Radicals and Tissue Injury,* Halliwell, B., Proceedings of a Brook Lodge Symposium, Upjohn, pp. 1–5.
92. Sutherland, M.W. (1991) *Physiol. Mol. Plant Pathol. 39,* 79–93.
93. Chester, K.S. (1933) *Q. Rev. Biol. 8,* 275–324.
94. Kuc, J. (1982) *BioSci. 32,* 854–860.
95. Kessmann, H., Staub, T., Hofmann, C., Maetzke, T., and Herzog, J. (1994) *Annu. Rev. Phytopathol. 32,* 439–459.
96. Ukness, S., Mauch-Mani, B., Moyer, M., Potter, S., Williams, S., Dincher, S., Chandler, D., Slusarenko, A., Ward, E., and Ryals, J. (1992) *Plant Cell 4,* 645–656.
97. Mauch-Mani, B., and Slusarenko, A.J. (1994) *Mol. Plant-Microbe Interact. 7,* 378–383.
98. Vernooij, B., Friedrich, L., Morse, A., Reist, R., Kolditz-Jawhar, R., Ward, E., Ukness, S., Kessmann, H., and Ryals, J. (1994) *Plant Cell 6,* 959–965.
99. Delaney, T.P., Ukness, S., Vernooij, B., Friedrich, L. Weymann, K., Negrotto, D., Gaffney, T., Gut-Rella, M., Kessmann, H., Ward, E., and Ryals, J. (1994) *Science 266,* 1247–1250.
100. Avdiushko, S.A., Ye, X.S., Hildebrand, D.F., and Kuc, J. (1993) *Physiol. Mol. Plant Pathol. 42,* 83–95.
101. Avdiushko, S.A., Chalova, L.I., Ozeretskovskay, O.L., Chalenko, G.I., Imbs, A.B., Latyshev, N.A., Bezuglov, V.V., and Bergelson, L.D. (1987) *Dokl. Akad. Nauk. SSSR 296,* 1012–1014.
102. Farmer, E.E., and Ryan, C.A. (1990) *Proc. Natl. Acad. Sci. USA 87,* 7713–7716.
103. Farmer, E.E., and Ryan, C.A. (1992) *Plant Cell 4,* 129–134.
104. Farmer, E.E., and Ryan, C.A. (1992) *Trends Cell Biol. 2,* 236–241.
105. Peña-Cortez, H., Albrecht, T., Prat, S., Weiler, E., and Willmitzer, L. (1993) *Planta, 191,* 123–128.

106. Gundlach, H., Müller, M.J., Kutchan, T.M., and Zenk, M.H. (1992) *Proc. Nat. Acad. Sci. USA 89,* 2389–2393.
107. Mueller, M.J., Brodschelm, W., Spannagl, E., and Zenk, M.H. (1993) *Proc. Natl. Acad. Sci. USA 90,* 7490–7494.
108. Mitchell, A.F., and Walters, D.A. (1995) *Aspects of App. Biol. 42*, 323–326.
109. Schweizer, P., Gees, R., and Mösinger, E. (1993) *Plant Physiol. 102,* 503–511.
110. Kogel, K.-H., Ortel, B., Jarosch, B., Atzorn, R., Schiffer, R., and Wasternack, C. (1995) *Eur. J. Plant Pathol. 101,* 319–332.
111. Choi, D., Bostock, R.M., Avdiushko, S., and Hildebrand, D.F. (1994) *Proc. Natl. Acad. Sci. USA 91,* 2329–2333.
112. Kauss, H., Krause, K., and Jeblick, W. (1992) *Biochem. Biophys. Res. Commun. 189,* 304–308.
113. Kauss, H., Jeblick, W., Ziegler, J., and Krabler, W. (1994) *Plant Physiol. 105,* 89–94.
114. White, R.F. (1979) *Virology 99,* 410–412.
115. Rock, C.D., and Zeevaart, J.A.D. (1991) *Proc. Natl. Acad. Sci. USA 88,* 7496–7499.
116. Finkelstein, R.R., and Zeevaart, J.A.D. (1994) in *Arabidopsis,* Meyerowitz, E.M., and Somerville, C.R., Cold Spring Harbour Laboratory Press, New York, pp. 523–553.
117. Creelman, R.A., Bell, A., and Mullet, J.E. (1992) *Plant Physiol. 99,* 1258–1260.
118. Fraser, R.S.S., and Whenham, R.J. (1982) *Plant Growth Regulation 1*, 37–59.
119. Dunn, R.M., Hedden, P., and Bailey, J.A. (1990) *Physiol. Mol. Plant Pathol. 36,* 339–349.
120. Salt, S.D., Tuzun, S., and Kuc, J. (1986) *Physiol. Mol. Plant Pathol. 28,* 287–297.
121. Ward, E.W.B., Chaill, D.M., and Bhattacharyya, M.K. (1989) *Plant Physiol. 91,* 23–27.
122. Dixon, R.A., and Fuller, K.W. (1978) *Physiol. Plant Pathol. 12,* 279–288.
123. Koch, E., and Slusarenko, A.J. (1990) *Plant Cell 2,* 437–445.
124. Cameron, R.K., Dixon, R.A., and Lamb, C.J. (1994) *Plant J. 5,* 715–725.
125. Schauenstein, E., Ersterbauer, H., and Zollner, H. (1977) in *Aldehydes in Biological Systems—Their Natural Occurrence and Biological Activity,* translated by Gore, P.H., Pion Limited, London, pp. 32–35, 40–41.
126. Andersen, R.A., Hamilton-Kemp, T.R., Hildebrand, D.F., McCracken, C.T., Collins, R.W., and Fleming, P.D. (1994) *J. Agric. Food. Chem. 42,* 1563–1568.
127. Major, R.T., Marchini, P., and Sproston, T. (1960) *J. Biol. Chem. 235,* 3298–3299.
128. Von Schildknecht, H., and Rauch, G. (1961) *Z. Naturforsch. 16b,* 422–429.
129. Zeringue, H.J., Jr., and McCormick, S.P. (1989) *J. Am. Oil Chem. Soc. 66,* 581–585.
130. Zeringue, H.J., Jr. (1992) *Phytochem. 31,* 2305–2308.
131. Doehlert, D.C., Wicklow, D.T., and Gardner, H.W. (1993) *Phytopathol. 83,* 1473–1477.
132. Vaughn, S.F., and Gardner, H.W. (1993) *J. Chem. Ecol. 19,* 2337–2345.
133. Hatanaka, A., Kajiwara, T., Matsui, K., and Kitamura, A. (1992) *Z. Naturforsch. 47c,* 369–374.
134. Shibata, Y., Matsui, K., Kajiwara, T., and Hatanaka, A. (1995) *Plant Cell Physiol. 36,* 147–156.
135. Grambow, H.J. (1977) *Z. Pflanzenphysiol. 85*, 361–372.
136. Samuelsson, B., Dahlén, S.-E., Lindgren, J.A., Rouzer, C.A., and Serhan, C.A. (1987) *Science 237,* 1171–1176.
137. Ruzicska, P., Gombos, Z., and Farkas, G.L. (1983) *Virology 128*, 60–64.
138. Frič, F. (1993) *Acta Phytopath. Entomol. Hung. 28*, 173–183.

Chapter 11

Immobilization of Soybean Lipoxygenase and Promotion of Fatty Acid and Ester Oxidation

George J. Piazza

Eastern Regional Research Center, Agricultural Research Service, U.S. Department of Agriculture, Philadelphia, PA 19118, USA.

Introduction

While other chapters in this monograph have discussed the structural and mechanistic features of lipoxygenase and the physiological impact of its metabolites, this chapter consists of a discussion of work on the immobilization of lipoxygenase from soybeans and its use in oxidizing fatty acid esters. Others have demonstrated that hydroperoxides are useful chiral synthons for further chemical modification (1–5). The study of enzymes that use fatty acid hydroperoxides as their substrates is still in a developing stage. As our knowledge of these enzymes grows, ways of using them for practical syntheses will be devised, and the richness of synthetically useful chemical structures derived from hydroperoxides will expand manyfold. As indicated in part by the large number of free and immobilized lipases that are currently available from commercial sources, the age of chemical synthesis with enzymes has arrived. Other chapters in this monograph have alluded to diseases that are linked to the action of lipoxygenase, and certainly the food industry views lipoxygenase activity as a potential problem in product quality (6). However, this chapter emphasizes the potential utility of a much-maligned enzyme in chemical synthesis.

Practical Problems Associated with the Use of Free Lipoxygenase

Free soybean lipoxygenase is unstable in solution, losing most of its activity within a day. This instability problem is exacerbated when high oxygen levels are achieved by bubbling oxygen into an aqueous buffer containing a fatty acid salt. Fatty acid salts have detergent properties, and much foam is formed. The resulting surface tension in the medium surrounding each bubble has a propensity to promote enzyme denaturation. Existing laboratory procedures for promoting high conversion to hydroperoxide require the repeated addition of small amounts of lipoxygenase to compensate for rapid enzyme inactivation. To circumvent this problem, syntheses have been con-

---

Mention of brand or firm names does not constitute an endorsement by the U.S. Department of Agriculture over others of a similar nature not mentioned.

ducted in high-pressure bombs (7), in the presence of antifoam (3), and in media containing high levels of organic solvents since oxygen is much more soluble in nonpolar organic solvents than in aqueous buffers (8–10). Also it has been reported that the presence of an organic solvent is beneficial because it reduces substrate aggregation (9). If the source of lipoxygenase is inexpensive enough to be compensated for by the increased value of the resulting product, then the use of free lipoxygenase is acceptable. However, it has been recognized that enzyme recycling by binding it to a matrix could result in added economy (11). In addition, immobilized enzymes are sometimes more stable than their free counterparts.

Immobilization of Lipoxygenase

Published literature on lipoxygenase immobilization is very limited. Lipoxygenase was immobilized on cyanogen bromide–activated agarose (12). The activity of the immobilized preparation was nearly identical to that of free lipoxygenase. It retained activity at room temperature for at least 1 month. A functional bioreactor was prepared using lipoxygenase immobilized on cyanogen bromide–activated Sepharose (13). The flow rate and the concentration of the substrate, linoleic acid (LA), were fixed, and the amount of immobilized material was adjusted to determine the amount needed to most efficiently convert LA to its corresponding hydroperoxide. The conversion efficiency, in terms of product formed per unit time, could be further increased by increasing the flow rate up to a point, but if the flow rate was too high, the yield of oxygenated product was reduced.

Experiments were performed using lipoxygenase adsorbed through hydrophobic interactions, covalently bound by cyanogen bromide and physically adsorbed, then cross-linked by glutaraldehyde (14). Lipoxygenase adsorbed onto wetted porous glass or on porous silica showed low catalytic activity in *n*-decane, whereas lipoxygenase adsorbed onto the amphiphilic gels octyl- and phenyl Sepharose exhibited high activity in *n*-decane. All of the immobilized lipoxygenase preparations, adsorbed and covalently bound, had half-lives of several hours. Soybean lipoxygenase-1 (SBL-1) was covalently immobilized on oxirane acrylic beads (15). Very high yields of the LA hydroperoxide were obtained when the beads were used in an oxygenated, aqueous buffer. The beads could be reused, but information on the stability of this lipoxygenase preparation was not presented. Very recent work has investigated the adsorption of plant lipoxygenases onto talc (16,17). Lipoxygenases isolated from *Solanum tuberosum* tuber and *Lupinus albus, Cicer arietinum*, and *Pisum sativum* seeds were utilized. When stored at 4°C the adsorbed lipoxygenases from all sources showed greatly enhanced stability compared to free controls. The preparation derived from *S. tuberosum* was tested for reusability. Its activity dropped by only 34% after 10 uses.

Our laboratory has investigated the immobilization of soybean lipoxygenase (Sigma, Lipoxidase, Type 1-B) on a commercially available carbonyldi-imidazole-activated support (Pierce, ReactiGel® 6X [18]). The urethane linkage formed when a protein is bound to this support is about 20 times more stable than the *N*-substituted isourea linkage formed during protein immobilization onto cyanogen bromide–activated matrices (19). The immobilized preparation of lipoxygenase was extremely stable at 5°C, losing only 5% of its activity in 6 months. At 15°C, the immobilized lipoxygenase had reduced stability, but as shown in Figure 11.1, its stability was still much greater than that of free lipoxygenase. The half-life of free lipoxygenase was approximately 7 hours. The half-life of immobilized lipoxygenase was estimated to be 75 hours by replotting the activity data on a logarithmic scale and using a linear least-squares regression fit (20).

Catalytic Properties of Immobilized Lipoxygenase

The enzymatic activity of lipoxygenase immobilized on ReactiGel was examined at 15°C in mixtures containing organic solvent and aqueous buffer in which air was the sole source of oxygen (21). The amount of hydroperoxide, hydroperoxyoctadecadienoic acid (HPOD), formed in water-saturated *n*-octane from LA in 3 hours as the amount of aqueous buffer increased is shown in Figure 11.2. Although some HPOD formed even when no aqueous buffer was added, having 35% v/v aqueous buffer resulted in a threefold increase in the amount of HPOD.

The pH of the aqueous buffer also had a strong influence on the amount of HPOD formed (Fig. 11.3). As in all aqueous media, the amount of HPOD formed was maximized at pH 9–9.5. Hydroperoxyoctadecadienoic acid formation rapidly decreased as the pH was lowered. At pH values higher than 9.5, a more gradual decrease in HPOD formation occurred.

Figure 11.4 shows the influence of a variety of organic solvents on HPOD formation. The most HPOD was formed in 1,1,2-trichlorotrifluoroethane. Approximately equal amounts of HPOD were formed when the solvents hexane, heptane, octane, and 2,2,4-trimethylpentane were used. Still lower levels of HPOD were formed in toluene and cyclohexane; the lowest detectable levels of HPOD were found in reaction mixtures that contained diethyl ether and di-isopropyl ether. No HPOD formation was detected with the solvents 2-butanone or 2-octanone. These results are consistent with other work showing that polar organic solvents generally are detrimental to enzymic activity (22).

With some enzymes, the solvent can have a profound effect upon regioselectivity (23). As has been discussed previously, a fatty acid substrate in media containing both aqueous and organic components would have a propensity to be oriented in such a way that its polar head group is found in or is oriented toward the aqueous fraction where lipoxygenase resides. This might result in a reversal of the regiospecificity of lipoxygenase, such that the Δ-9 carbon receives the hydroperoxide functionality, rather than the Δ-13 carbon as is observed in all aqueous media. Recently this possibility was tested in reverse micelles containing Aerosol OT [sodium *bis*(2-ethylhexyl) sulfosuccinate] and octane (24). The 13-hydroperoxide isomer was the predominant product formed, indicating that the orientation of the substrate on lipoxygenase was not influenced by its orientation in the bulk reaction medium. We have also examined

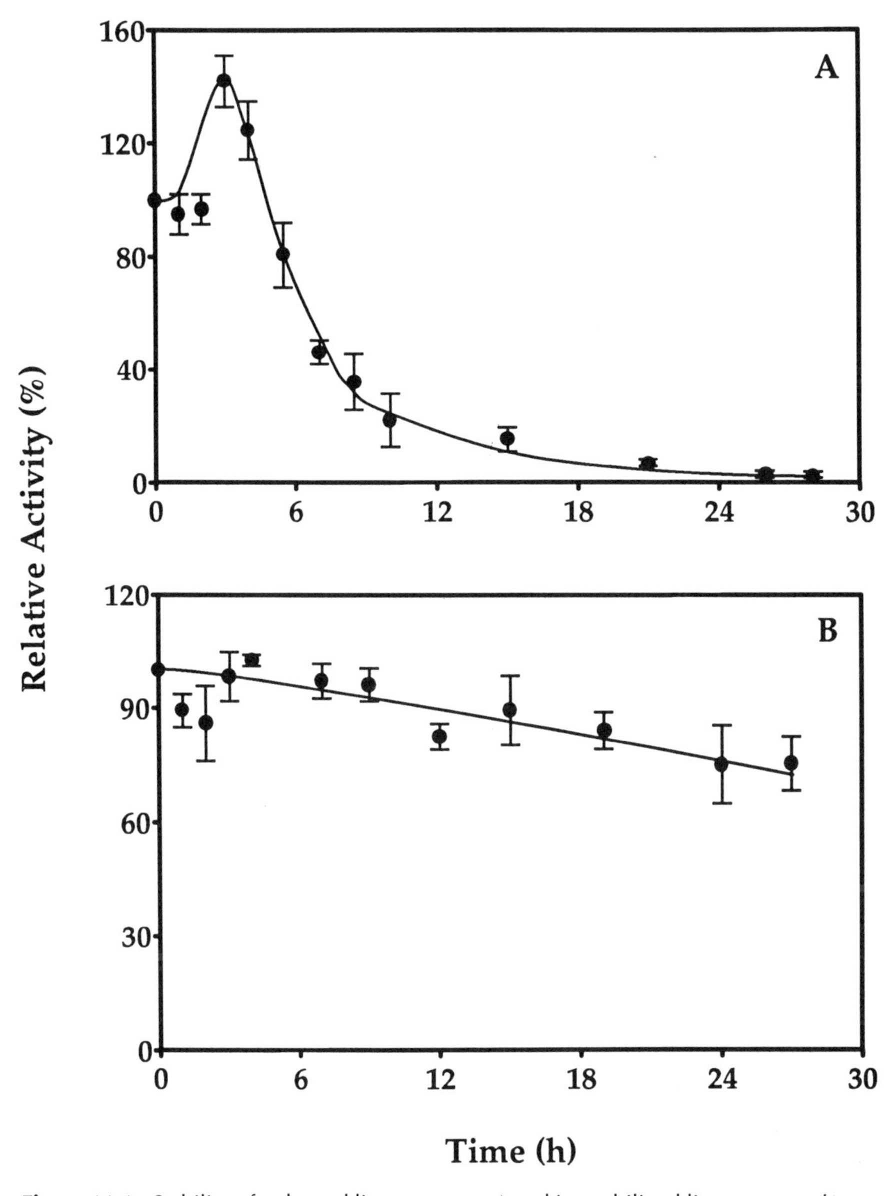

Figure 11.1. Stability of unbound lipoxygenase *a*) and immobilized lipoxygenase *b*) at 15°C. Lipoxygenase (2 mg of powder containing 1 mg of protein) or immobilized lipoxygenase (0.194 g of gel containing 1 mg of protein) was dispersed in 14.5 mL of Tricine buffer, pH 9.0. At the indicated time, 40 mg of LA in 100 μL of ethanol was added. The reaction was allowed to proceed for 15 min. Results are means ± SEM for four determinations.

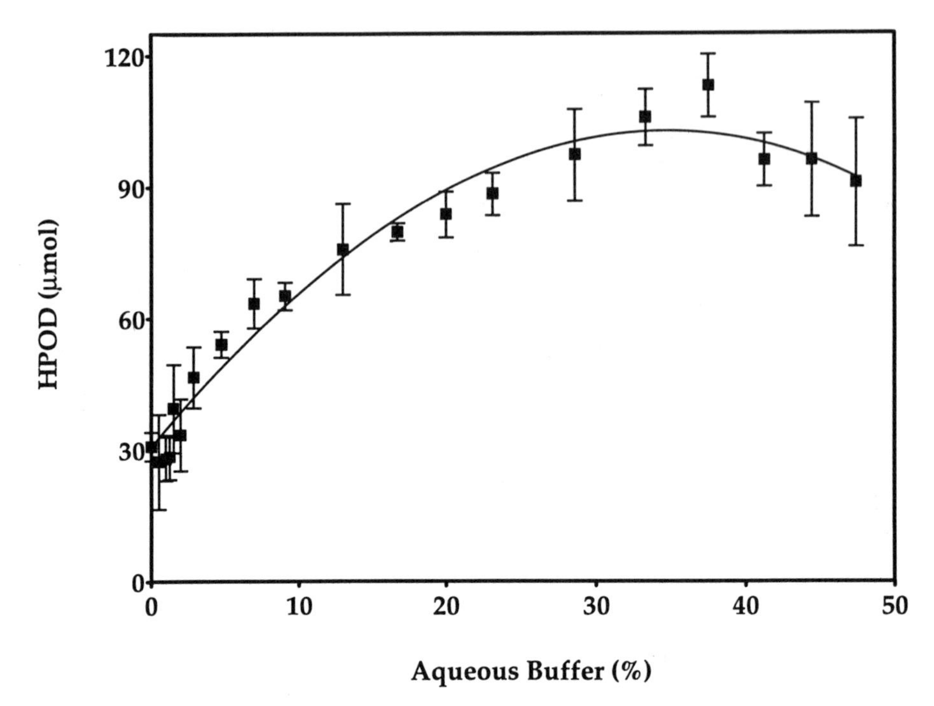

Figure 11.2. Influence of the amount of 0.2 M aqueous borate buffer, pH 9.0, on HPOD formation catalyzed by immobilized lipoxygenase. In addition to the buffer, each assay contained 0.75 g immobilized lipoxygenase containing 3.0 mg bound protein, 40 mg LA, and 15 mL water-saturated *n*-octane. Assays were conducted for 3 h at 15°C. The percentage buffer values were calculated: mL buffer/(mL buffer + mL octane) × 100. The data are the means ± SEM of 10 determinations.

the specificity of immobilized lipoxygenase in media containing both aqueous buffer and hexane (21). Again the 13-hydroperoxide isomer was found to be the predominant product formed.

Action of Lipoxygenase upon Linoleate Esters

Lipoxygenase from different sources and lipoxygenase isozymes from the same source exhibit different degrees of specificity depending on whether the carboxylic acid moiety is free or esterified. The high-pH form of lipoxygenase from soybean, SBL-1, is highly specific for free fatty acid under most conditions, while lipoxygenases from other sources are able to oxidize phospholipids containing the esters of polyunsaturated fatty acids. It has been particularly well documented that lipoxygenases from mammalian sources can oxidize phospholipids (25,26), as well as membranes and lipoproteins primarily containing esterified fatty acids (27–29). However,

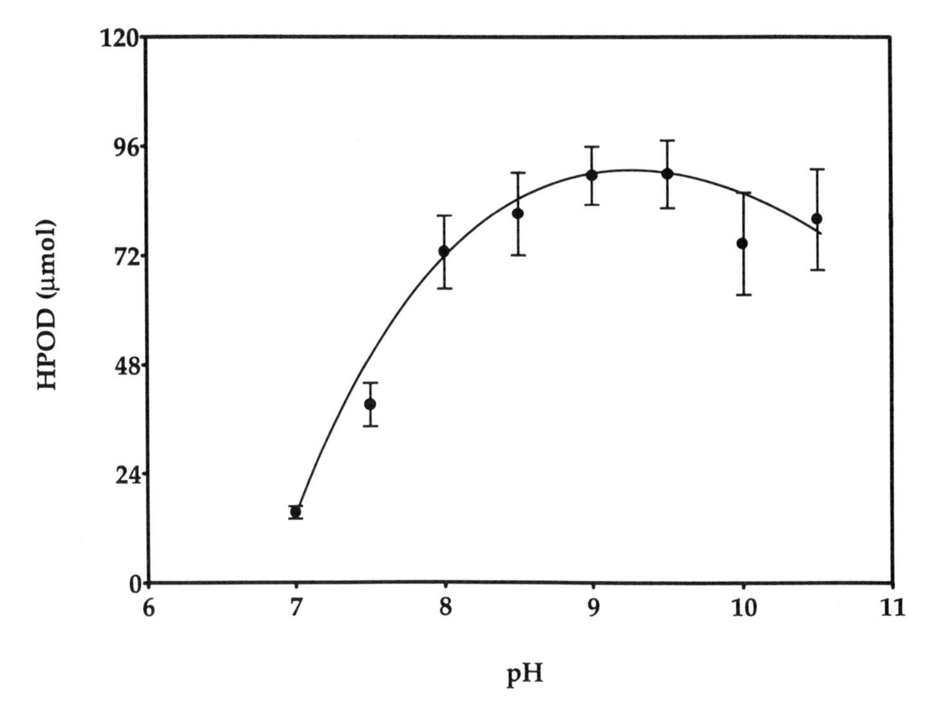

Figure 11.3. Influence of buffer pH on HPOD formation catalyzed by immobilized lipoxygenase. The assays contained 6 mL of buffer containing a mixture of 0.1 M Hepes, Tricine, and 2-amino-2methyl-1-propanol, hydrochloride, 15 mL water-saturated *n*-octane, 0.67 g immobilized lipoxygenase containing 3 mg bound protein, and 40 mg LA. The assays were conducted for 3 h at 15°C. The data are the means ± SEM of six determinations.

some plant lipoxygenase isozymes can oxidize membrane fractions and purified phospholipids (30–32).

Early work on SBL-1 showed that in the presence of Tween 20, free LA was rapidly oxidized, but the LA methyl ester and trilinolein were poor substrates for lipoxygenase (33). It was concluded that lipoxygenase requires the prior action of lipase or phospholipase to release free fatty acid in vivo. Modification of this view was required when it was discovered that soybean lipoxygenase had significant activity with phosphatidylcholine in the presence of bile salt, deoxycholate, cholate, or taurocholate (34). No oxidation occurred in Tween 20, Triton X-100, Tween 80, sodium dodecyl sulfate (SDS), and octyl glucoside. The detergent 3-16 Zwittergent promoted very slow oxidation. Except for SDS, lipoxygenase inactivation by the detergents was eliminated as a reason for the lack of support of oxidation. In addition, it was shown that release of free fatty acid from phosphatidylcholine did not occur prior to oxidation.

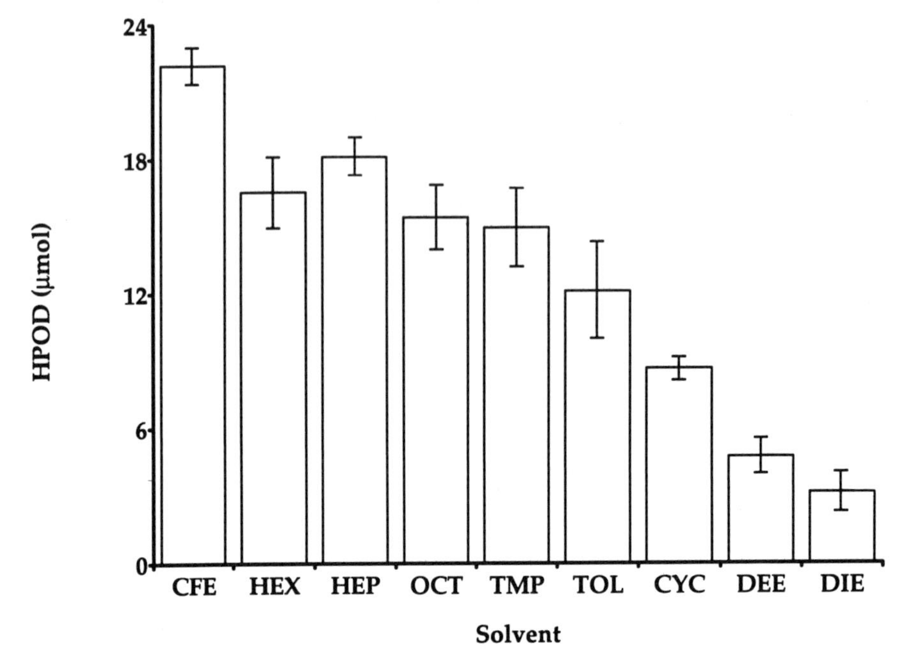

Figure 11.4. Influence of organic solvent on HPOD formation catalyzed by immobilized lipoxygenase. Each assay contained 15 mL water-saturated organic solvent, 6 mL 0.2 M borate buffer, pH 9.0, 0.194 g immobilized lipoxygenase preparation containing 1 mg bound protein, and 40 mg LA. The assays were conducted for 1 h at 15°C. Results are means ± SEM of three repetitions.
Abbreviations: 1,1,2-trichlorotrifluoroethane, CFE; hexane, HEX; heptane, HEP; octane, OCT; 2,2,4-trimethylpentane, TMP; toluene, TOL; cyclohexane, CYC; diethyl ether, DEE; di-isopropyl ether, DIE.

This study was extended by an examination of the structure of the oxidized arachidonyl and linoleoyl moieties in phosphatidylcholine that had been exposed to soybean lipoxygenase in the presence of deoxycholate (35). It was found that regio- and stereospecific hydroperoxide formation took place. Also, when oxidation was conducted in ^{18}O-labeled water, there was no incorporation of ^{18}O in the product, eliminating the possibility of hydrolysis and reesterification during oxidation.

These studies show that phosphatidylcholine is a true substrate for lipoxygenase. Phosphatidylethanolamine was also shown to be a substrate for lipoxygenase (36). When oxidation reactions conducted with soybean-derived phosphatidylethanolamine and phosphatidylcholine were subjected to fractionation by HPLC, it was shown that the oxidized species contained the following fatty acid pairs: 18:3/18:2, 18:2/18:2, and 16:0/18:2.

A single report has shown that lipoxygenase can oxidize fatty acid amides (37). Arachidonylethanolamide was oxidized by porcine leukocyte 12-lipoxygenase and rabbit reticulocyte and soybean 15-lipoxygenases. Human platelet 12-lipoxygenase oxidized the amide only very slowly, while porcine leukocyte 5-lipoxygenase was totally inactive.

Recently the action of soybean lipoxygenase upon neutral glycerides and the LA methyl ester has been investigated (38; Piazza et al., unpublished results). These studies were performed with free lipoxygenase. Presumably the advantages of using immobilized lipoxygenase on LA would also apply with esterified substrates, but this has not yet been tested.

Table 11.1 shows the amounts of oxidation products formed by the action of SBL-1 on several neutral esters in 15-minute assays. Without surfactant, only LA was oxidized at a rapid rate. In Tween 20, LA and monolinolein were oxidized rapidly. Linoleic acid, methyl linoleate, and monolinolein were oxidized rapidly in the presence of deoxycholate. The rate of dilinolein oxidation was about one-half the rate of LA oxidation. Trilinolein was oxidized at a very slow rate. Methyl linoleate and trilinolein have approximately equal solubility in the assay buffer, and thus their highly different rates of oxidation by lipoxygenase must be due to steric constraints in the active site.

The pH profiles of the action of soluble soybean lipoxygenase upon trilinolein and 1,3-dilinolein are shown in Figure 11.5. Both profiles are similar with the optimal pH for oxidation being approximately 8.0. The oxidation rate diminishes at both higher and lower pH values, although action against trilinolein decreases more slowly as the pH is reduced.

TABLE 11.1 Relative Amounts of Oxidized Linoleate Formed by Lipoxygenase in 15 min.

	Relative Amount of Oxidation[a]		
Substrate	Deoxycholate[b]	Without Surfactant	Tween 20[c]
Linoleic acid	100 ± 6	186 ± 1	195 ± 7
Methyl linoleate	90 ± 8	28 ± 1	12 ± 1
1-Monolinolein	126 ± 6	11 ± 1	72 ± 3
1,3-Dilinolein	55 ± 4	NS[d]	NS
Trilinolein	3 ± 1[e]		

[a]Assays were conducted in 10 mL Erlenmeyer flasks, each containing substrate (6 μmol linoleoyl residues), 25 μg lipoxygenase, 0.2 mL surfactant or water, and 1.8 mL aqueous buffer, consisting of an equal 0.1 M mixture of 2-amino-2-methyl-1propanol hydrochloride, *N*-tris(hydroxymethyl)-methylglycine (Tricine), *N*-(2-hydroxyethyl)piperazine-*N'*-(2-ethanesulfonic acid) (Hepes), and 2-(*N*-morpholino)ethanesulfonic acid (Mes). Oxidation was conducted at 15°C with agitation at 250 rpm. The amount of LA oxidized in the presence of deoxycholate was 0.986 μmol (16.4% available LA). Results are means ± SEM for three repetitions.
[b]The deoxycholate concentration was 10 mM.
[c]The Tween 20 concentration was 0.25% (v/v). *Source:* Christopher, Pistorius, and Axelrod (33).
[d]Results were not significantly different from zero.
[e]Data from Piazza and Nuñez (38).

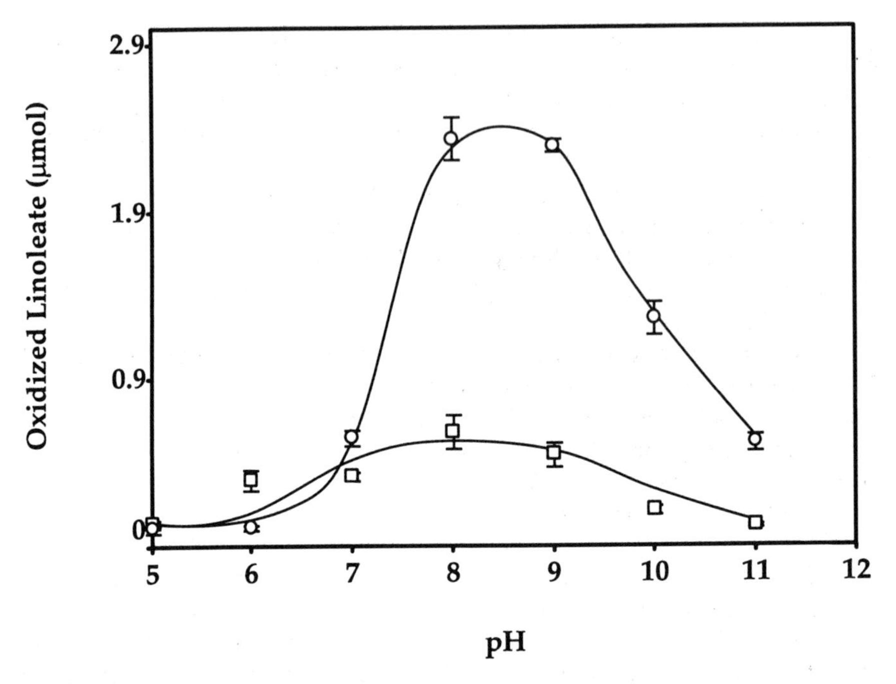

Figure 11.5. Influence of pH on the oxidation of 2 μM trilinolein (□) and 3 μM 1,3-dilinolein (o) by free lipoxygenase in the presence of deoxycholate. Assays were conducted as described in Piazza and Nuñez (38). Results are means ± SEM of three repetitions.

As is clear from Figure 11.5 and Table 11.1, the overall amount of trilinolein oxidized is lower than that of dilinolein within the time period studied. To achieve a better understanding of the potential of lipoxygenase to oxidize these two compounds, complete-reaction-time courses were followed. These showed that only 15% of the available linoleoyl residues were oxidized in trilinolein, and increasing the amount of lipoxygenase did not increase the extent of oxidation. Although direct evidence is lacking, this result may be due to inhibition by the oxidation product. In contrast to the results obtained with trilinolein, 67% of the available linoleate in dilinolein was converted to hydroperoxide by lipoxygenase. When dilinolein oxidation was monitored using HPLC, all of the dilinolein was eventually oxidized, but other unidentified polar materials were formed in addition to mono- and dihydroperoxides. Although exact yields are not available as of this writing, it is clear that both methyl linoleate and monolinolein can also be oxidized by soybean lipoxygenase to a high degree.

Conclusions

In conclusion, soybean lipoxygenase can be immobilized to provide an efficient catalyst for the regio- and stereospecific introduction of hydroperoxide into fatty acids. Recent advances have demonstrated that oxidations can be conducted in reverse micelles, in organic solvent–aqueous mixtures, and that neutral and charged fatty acid esters can be substrates for soybean lipoxygenase provided that proper reaction conditions are utilized. These results indicate that soybean lipoxygenase has the potential to act upon a wide variety of structurally disparate molecules and will eventually find a role in the arsenal of the synthetic chemist.

References

1. Baba, N., Yoneda, K., Tahara, S., Mikami, Y., Iwasa, J., Kaneko, T., and Matsuo, M. (1992) in *Oxygen Radicals,* Elsevier Science Publishers, Amsterdam, pp. 261–264.
2. Luquet, M.P., Pourplanche, C., Podevin, M., Thompson, G., and Larreta-Garde, V. (1993) *Enzyme Microb. Technol. 15,* 842–848.
3. Datcheve, V.K., Kiss, K., Solomon, L., and Kyler, K.S. (1991) *J. Am. Chem. Soc. 113,* 270–274.
4. Yoneda, K., Sasakura, K., Tahara, S., Iwasa, J., Baba, N., Kaneko, T., and Matsuo, J. (1992) *Angew. Chem. Int. Ed. Engl. 31,* 1336–1338.
5. Zhang, P., and Kyler, K.S. (1989) *J. Am. Chem. Soc. 111,* 9241–9242.
6. Hsieh, R.J. (1994) in Lipids in Food Flavors, Ho, C.-T., and Hartman, T.G., *American Chemical Society,* Washington DC, pp. 30–48.
7. Cardillo, R., Fronza, G., Fuganti, C., Grasselli, P., Mele, A., Pizzi, D., Allegrone, G., Barbeni, M., Pisciotta, A. (1991) *J. Org. Chem. 56,* 5237–5239.
8. Piazza, G.J. (1992) *Biotechn. Lett. 14,* 1153–1158.
9. Drouet, P., Thomas, D., and Dominique Legoy, M. (1994) *Tetrahedron Lett. 35,* 3923–3926.
10. Perez-Gilabert, M., Sanchez-Ferrer, A., and Garcia-Carmona, F. (1992) *Biochem. J. 288,* 1011–1015.
11. Messing, R.A. (1975) in *Immobilized Enzymes for Industrial Reactors,* Messing, R.A., Academic Press, New York, pp. 1–10.
12. Grossman, S., Trop, M., Budowski, P., Perl, M., and Pinsky, A. (1972) *Biochem. J. 127,* 909–910.
13. Laakso, S. (1982) *Lipids 17,* 667–671.
14. Yamane, T. (1982) in *Enzyme Engineering,* Chibata, I., Fukui, S., and Wingard, L.B. Jr., Plenum, New York, vol. 6., pp. 141–142.
15. Maguire, N.M., Mahon, M.F., Molloy, K.C., Read, G., Roberts, S.M., and Sik, V. (1991) *J. Chem. Soc. Perkins Trans. 1,* 2054–2055.
16. Battu, S., Cook-Moreau, J., and Beneytout, J.L. (1994) *Biochim. Biophys. Acta 1211,* 270–276.
17. Battu, S., Rabinovitch-Chable, H., and Beneytout J.-L. (1994) *J. Agric. Food Chem 42,* 2115–2119.
18. Parra-Diaz, D., Brower, D.P., Medina, M.B., and Piazza, G.J. (1993) *Biotechnol. Appl. Biochem. 18,* 359–367.

19. Hearn, M.T.W., Harris, E.L., Bethell, G.S., Hancock, W.S., and Ayers, J.A. (1981) *J. Chromatogr. 218,* 509–518.
20. Pitcher, W.H., Jr. (1975) in *Immobilized Enzymes for Industrial Reactors,* Messing, R.A., Academic Press, New York, pp. 151–199.
21. Piazza, G.J., Brower, D.P., and Parra-Diaz, D. (1994) *Biotechnol. Appl. Biochem. 19,* 243–252.
22. Laane, C., Boeren, S., Vos, K., and Veeger, C. (1987) *Biotechnol. Bioeng. 30,* 81–87.
23. Rubio, E., Fernandez-Mayorales, A., and Klibanov, A.M. (1991) *J. Am. Chem. Soc. 113,* 695–696.
24. Shkarina, T.N., Kühn, H., and Schewe, T. (1992) *Lipids 27,* 690–693.
25. Murray, J.J., and Brash, A.R. (1988) *Archives Biochem. Biophys. 265,* 514–523.
26. Takahashi, Y., Glasgow, W.C., Suzuki, H., Taketani, Y., Yamamoto, S., Anton, M., Kühn, H., and Brash, A.R. (1993) *Eur. J. Biochem. 218,* 165–171.
27. Kuhn, H., Belkner, J., Wiesner, R., and Brash, A.R. (1990) *J. Biol. Chem. 265,* 18351–18361.
28. Belkner, J., Wiesner, R., Rathman, J., Barnett, J., Sigal, E., and Kühn, H. (1993) *Eur. J. Biochem. 213,* 251–261.
29. Kühn, H., Belkner J., Suzuki, H., and Yamamoto, S. (1994) *J. Lipid Res. 35,* 1749–1759.
30. Hardy, D.J., Violana Gallegos, M.A., and Gaunt, J.K. (1991) *Phytochemistry 30,* 2889–2894.
31. Kondo, Y., Kawai, Y., Hayashi, T., Ohnishi, M., Miyazawa, T., Seisuki, I., and Mizutani, J. (1993) *Biochim. Biophys. Acta 1170,* 301–306.
32. Maccarrone, M., van Aarle, P.G.M., Veldink, G.A., and Vliegenthart, J.F.G. (1994) *Biochim. Biophys. Acta 1190,* 164–169.
33. Christopher, J., Pistorius, E., and Axelrod, B. (1970) *Biochim. Biophys. Acta 198,* 12–19.
34. Eskola, J., and Laakso, S. (1983) *Biochim. Biophys. Acta 751,* 305–311.
35. Brash, A.R., Ingram, C.D., and Harris, T.M. (1987) *Biochemistry 26,* 5465–5471.
36. Therond, P., Couturier, M., Demelier, J.-F., and Lemonnier, F. (1993) *Lipids 28,* 245–249.
37. Ueda, N., Yamamoto, K., Yamamoto, S., Tokunaga, T., Shirakawa, E., Shinkai, H., Ogawa, M., Sato, T., Kudo, I., Inoue, K., Takizawa, H., Nagano, T., Hirobe, M., Matsuki, N., and Saito, H. (1995) *Biochim. Biophys. Acta 1254,* 127–134.
38. Piazza, G.J., and Nuñez, A. (1995) *J. Am. Oil Chem. Soc. 72,* 463–466.

Chapter 12

Comparison of Lipoxygenase-Null and Lipoxygenase-Containing Soybeans for Foods

Lester A. Wilson

Department of Food Science and Human Nutrition, Iowa State University, Ames, IA

Introduction

With the health concerns about fat (especially saturated fat) in our diets, the development of lower fat foods has greatly increased (1). This is clearly reflected in fast food menus and at the supermarket. Unfortunately, low-fat-formulated products typically suffer from having poor texture, less flavor, and higher cost. For example, replacing the fat with lean meat increases the cost and toughness of the meat products (2–4). One way to improve the quality of low fat products is to formulate them with extenders derived from soybean. In addition, increased use of soy in foods reduces the saturated fatty acid content and adds phytochemicals, such as the isoflavones, which have anticancer properties (5).

The flavor of soybeans and soybean protein ingredients is believed to be one of the major limiting factors for increasing their use in foods (6–11). In 1924 a Hungarian Chemist named Berczeller stated that raw soybeans were "evil tasting" (12). This off or undesirable flavor is currently characterized as "beany," "green," "grassy," "painty", and "bitter" (8,9,10,11,13–18). Many of these flavors can result from the hydroperoxidation of *cis-cis* 1,4-pentadiene-containing fatty acids (linolenic and linoleic acids) by lipoxygenase in the raw soybeans after the beans have been damaged, crushed, ground, or rehydrated during processing and/or storage (17,19,20). The hydroperoxides and their breakdown products, such as hexanal, are known to produce these undesirable flavors (6,8,21,22).

The undesirable flavors generated by lipoxygenase activity are believed to interact with the protein and are often released when the flours, concentrates, and isolates are used as ingredients and extenders, or in fabricated foods (8,9,23–26). Undesirable flavors released upon heating decreases consumer acceptability of these products and the marketability of foods containing these soy ingredients (20). Determining the optimal desirable flavoring of soy foods is still performed by trial and error. Slight changes in the formulation of foods containing soy protein ingredients can cause major changes in the flavor (intensity and balance) of that food. This, in turn, can cause processors to choose other proteins for their product formulations, thus, decreasing the utilization and marketability of soy products.

The flavor problem of soybeans and soybean products has not only limited the use of soy-containing foods, but created a negative image that is applied to all soy-containing foods (30). Approaches used in the past to remove or mask these undesirable flavors are listed in Tables 12.1 and 12.2. The major research thrusts have been

1. Isolate and identify the off-flavors produced from lipoxygenase activity (23,31–34);
2. Use heat treatments, before, during, or after soaking and grinding to inactivate lipoxygenase activity (9,35–41);
3. Attempt to understand how undesirable flavors bind or are adsorbed onto soy protein (24,26–28,42–49); and
4. Use enzymatic improvements of the flavor of soy protein through the use of proteases or aldehyde and alcohol dehydrogenases (23,50–54).

Empirical methods to prevent, remove, or mask off-flavors in soyfoods, soy concentrates, and soy isolates have also been used (Table 12.2 [11,12,17,20,26,50]).

A better approach to solve the undesirable flavor problem in soybeans would be to develop soybeans that lack lipoxygenase. Traditional soybean varieties contain three lipoxygenase isozymes (L-1, L-2, L-3 [55]). In the 1980s and early 1990s, efforts to produce soybeans lacking one or more of the lipoxygenase isozymes were successful. Hildebrand and Hymowitz identified the L-1 null soybean variety (56).

TABLE 12.1 Methods to Prevent or Inhibit Lipoxygenase Activity[a]

Genetic modification of soybeans: lipoxygenase null lines
Wet blanching: water or steam
Dry heating
Grinding at acidic pH
Grinding with H_2O_2 plus calcium chloride
Grinding in hot water (100°C)
Grinding with solvent azeotropes
Grinding and heating in a vacuum
Inhibition by acetylene compounds
Addition of antioxidants
Soaking and heating after alkaline treatment

[a]Modified from Kinsella and Damodaran, 1980 (26); Wilson, 1985 (11).

TABLE 12.2 Methods Used to Minimize Beany Flavors in Soy Proteins[a]

Azeotrope extraction
Enzymatic treatment
Masking the beany flavor with added desirable flavors
Single-solvent extraction
Steam treatment
Vacuum treatment
Supercritical CO_2 extraction
Replacement with another protein source

[a]Modified from Wilson, 1985 (11).

Kitamuri et al. produced an L-3 null soybean variety (57), and Kitamuri et al. and Davies and Nielsen identified an L-2 null soybean variety (58,59). The identification of double-null varieties (L-1,L-3; L-2,L-3) were reported by Kitamuri et al. (58). Hajika et al. successfully produced a triple-null soybean and an L-1,L-2 null soybean (60,61). Nielsen (personal communication) has also identified a putative triple-null soybean from a cross between an L-2,L-3 null Century line and an L-1 null line.

Each of the isozymes has been associated with the production of hexanal, one of the undesirable (beany) flavor compounds produced by traditional raw soybeans. The L-2 isozyme is believed to be responsible for the majority of hexanal production (62,63). The acceptability of these lipoxygenase-null varieties in soyfoods and as soy flours, concentrates, and isolates has only begun to be evaluated, since only a few of these varieties are in commercial production at the present time (Table 12.3). Noncommercial lipoxygenase-null germplasm can be obtained from plant breeders (Hildebrand, personal communication; Nielsen, personal communication).

Complete removal of lipoxygenase isozymes may produce very desirable "bland" soybean and soybean products that can be used in soymilk, tofu, soy yogurt, soy-based ice cream, and as ingredients in bakery, meat, and dairy application (30;

TABLE 12.3 Production Yields and Composition of Commercially Available Lipoxygenase-2 Null Cultivars and Soyfood Varieties

		1994 Field Yield (Bu/A)	P[a]	O[a]	Null
1993 North	IA 1002	34.0	39.2	15.6	2
	IA 1003	36.7	39.7	14.6	2
	IA 2009	36.1	39.3	14.4	2
	IA 2010	34.9	39.9	14.7	2
	IA 2006	33.2	38.6	15.0	2
	IA 2011	40.2	38.3	15.4	2
	Vinton 81	33.3	38.0	15.7	NOT
	Century 84	34.6	38.3	15.4	NOT
1993 Central	IA 1002	35.9	38.1	16.8	2
	IA 1003	37.7	39.0	16.0	2
	IA 2011	38.8	36.5	17.2	2
	IA 2006	33.2	37.7	16.3	2
	Vinton 81	36.6	38.1	16.7	NOT
	LSD (0.05)	1.9	0.5	0.3	
1994 North	IA 1002	49.9	37.7	16.7	2
	IA 2010	48.5	39.8	15.8	2
	IA 103	49.3	38.8	16.1	2
	IA 2009	51.0	38.9	15.9	2
	IA 2011	57.2	35.9	17.7	2
	IA 206	49.4	38.5	16.5	2
	Vinton 81	47.6	38.3	16.7	NOT
	Century 84	51.6	37.9	16.7	NOT
	LSD (0.05)	3.1	0.4	0.2	

(continued)

TABLE 12.3 (continued)

		1994 Field Yield (Bu/A)	P[a]	O[a]	Null
1994 Central	IA 1002	48.2	36.4	17.6	2
	IA 2010	50.0	38.8	16.4	2
	IA 1003	50.1	38.0	16.8	3
	IA 2009	50.0	37.5	16.8	2
	IA 2011	57.2	35.2	18.5	2
	IA 2006	48.6	37.2	17.1	2
	LSD (0.05)	3.1	0.5	0.3	
1995 Northern	IA 1003	50.4	38.6	17.1	2
	IA 2011	55.5	36.0	18.0	2
	P9253[b]	56.0	35.4	19.4	2
	IA 2006	47.5	38.6	17.2	2
	IA 3003	58.4	36.6	18.1	2
	P9305	55.5	34.9	19.1	2
	Vinton 81	49.2	38.8	17.3	NOT
	LSD (0.05)	2.7	0.5	0.3	
1995 Central	IA 1003	46.6	38.6	17.3	2
	IA 2011	56.1	35.6	19.2	2
	IA 2016	47.1	38.2	18.1	2
	IA 2017	54.3	38.0	17.9	2
	P9253	54.5	35.2	19.8	2
	IA 2018	53.0	37.2	17.9	2
	IA 2006	45.6	38.2	17.7	2
	IA 2019	53.6	35.7	18.9	2
	P9305	58.3	35.0	19.5	2
	L960[c]	58.8	35.0	19.0	2
	Vinton 81	46.6	38.2	17.8	NOT
	LSD	2.8	0.5	0.3	
1995 Southern	IA 2020	43.4	38.1	18.3	2
	LSD	3.2			

[a]Protein (P) and Oil (O) at 13% moisture NIR.
[b]P number = Pioneer stock.
[c]L number = Letham stock.
Abbreviation: Least Significant Difference, LSD.
Sources: GoldBook, 1993–1995 (64–66).

Kitamuri, personal communication). However, the blandness of the soybean, or total lipoxygenase removal, may count against whole soybean usage in three ways: enzyme active soy flours are sometimes used in bleaching wheat flours and as dough conditioners in place of chemical bleaching or conditioning agents (10,20); it may be too bland for the older consumers of traditional soyfoods (soymilk and tofu) in other countries; and the blandness may allow the astringency to become more pronounced. The first concern is not a major problem, because the marketplace already differentiates between enzyme active and enzyme inactive (heat-treated) soy flour. Recent reports from Japan support the last two concerns (18,67; Kitamuri, personal communication). Sensory panels composed of first-time soyfood consumers have commented that bland tofu lacked desirable flavor ("Shouldn't this taste like something?—cereal, butter, chicken?" Wilson, unpublished data). The advantages of lipoxygenase-null varieties in soymilk manufacture, tofu production, soy flour, and their use as ingredients will outweigh these few potentially negative attributes.

Economic Significance to Soybean Producers

The removal of the undesirable "beany," "green," and "grassy" flavors by utilizing lipoxygenase-null soybeans would greatly enhance soybean utilization in foods, particularly foods where soy proteins are used for functional or nutritional value. If bland soy flour carried a $0.01/lb premium, an additional $14.3 million/year would result using the existing domestic market (30). Estimates of an additional $0.33/bushel for lipoxygenase removal, with a 20-year lifespan with a net value benefit of $182.4 million to soybean producers was projected in the Beneficiaries of Modern Soybean Traits Special Report (68).

In addition to these existing applications, the utilization of a bland soyflour could be used to replace soy concentrates ($0.15/lb vs. $0.60/lb) in numerous applications where flavor problems have been the limiting factor. Likewise, new applications of bland soy flours and isolates could be used in the meat and bakery industries to reduce the fat content of foods such as breakfast pork sausage for fast-food chains, hot dogs, doughnuts, cookies, muffins, pizza crusts, etc. A conservative estimate for pork sausage and hot dog utilization alone is $3 million.

Review of Lipoxygenase-Null Utilization Research

The commercial availability of the new lipoxygenase-null soybean varieties has limited published research using these varieties. The research has taken three approaches: demonstrations that the null-lines do not produce hexanal; that the soymilk, tofu, or cereal produced from these null-lines have reduce beany aroma and flavor; and that spray-dried soymilk or dried tofu made from null varieties can be used as ingredients in foods to reduce the fat content without producing adverse flavor effects.

Soymilk Results

Using simultaneous distillation and extraction (SDE) combined with gas chromatography-mass spectroscopy, Kobayashi et al. found that the concentration of volatile compounds was greatly reduced in soymilk concentrates produced from soybeans with L-2, L-3 null (Yumeyutaka) and triple-null (Kyushu No. 111) lipoxygenase compared to those produced from normal soybean milk (Suzuyutaka [69]). Kitamuri et al. had previously reported that the L-2,L-3 null variety Yumeyutaka had a better taste and less off-flavor than the parent Suzuyutaka, a leading Japanese cultivar (57,70). Davies et al. found that the L-2 null line produced soymilk and soyflour that had statistically lower rancid aroma and flavor scores, higher cereal flavor scores, and lower beany scores than the Century control (63).

Kitamuri (personal communication) and Lambrecht et al. (personal communication) reported that the L-2,L-3 null lines (Yumeyutaka and Century L_2L_3, respectively) are more resistant to deterioration during adverse storage conditions than their respective controls (Suzuyutaka and Century 84, respectively). This resistance to change during storage was reflected in the superior quality of the deep fried tofu, soymilk, and tofu produced from the stored null lines. Lambrecht et al. also compared storage effects on the quality of soymilk and tofu production from near isogeneic Century 84 and Century L_2,L_3 (lipoxygenase-2, 3 null) soybean varieties. Soybeans stored at 50°C and 70% relative humidity adversely influenced the characteristics (solids, tofu yield, and tofu color) of soymilk and tofu produced from these stored beans. The lipoxygenase-2,3 null soybeans (Century L_2,L_3) were more tolerant to the adverse storage conditions than the control variety.

Lee studied the influence of soymilk-processing methods (traditional/rapid hydration hydrothermally cooking), cultivar (L-2 null/traditional varieties), and the addition of sucrose on the sensory characteristics of the resulting soymilks (71). Table 12.4 shows the influence of cultivars and sucrose level on the sensory attributes as perceived by a trained sensory panel, and the hexanal content of traditional soymilk. The soymilk prepared from L-2 null cultivar (IA 1003) had significantly ($\alpha < 0.05$) less beany flavor and less hexanal content than that prepared from control (HP 204) cultivar. This is consistent with other results (62,63,69). The addition of 3% sucrose statistically lowered the trained panel's perception of beany flavor, which is consistent with a masking effect. Note that the gas chromatographic data did not show a statistical difference in hexanal content, and that the panel was clearly able to detect sweetness (1.88 cm for no added sucrose compared with 11.55 cm for 3% added sucrose in soymilk using a 15-cm line scale)). Thus the panel could detect differences in sweetness, and the 3% sucrose level masked the panelists' ability to perceive the hexanal (beany odor).

TABLE 12.4 Effects of Cultivar and Sucrose Level on the Sensory Attributes and Hexanal Content of Traditionally Processed Soymilk

	Cultivar		Added Sucrose Level	
Attribute	HP 204	IA 1003	0%	3%
Raw beany flavor[a]	7.784A[d]	5.99B	8.37A	5.35B
Sweetness[b]	6.74A	6.70A	1.88A	11.5B
Hexanal content[c]	2.17×10^4A	5.47×10^3B	1.35×10^4A	1.38×10^4A

[a]Mean scores from 15 trained panelists, from 3 replications containing two sucrose levels.
[b]Mean scores from 15 trained panelists, from 3 replications containing two cultivars.
[c]Mean integration units.
[d]Sample evaluation on a 15 cm line scale, where 0 = very little, 15 = very intense; different capital letters in a row are significantly different ($\alpha < 0.05$).
Source: Lee (71).

Methods and Materials

Lipoxygenase null and control soybeans were provided by Walt Fehr (Iowa State University Agronomy Department), Strayer Seed Company (Hudson, IA), and Pioneer Hybrid International (Johnston, IA). The varieties consisted of Vinton 81, a variety commonly used in soy food industries (national and international), Century 84, and lipoxygenase-2 null IA 1002, IA 1003, and near isogeneic Century L_2 and Century L_2L_3 lines.

Traditional Soymilk and Tofu Preparation

The traditional method of soy milk processing was adapted from Johnson (72), Johnson and Wilson (73), and Renata et al. (74), with some modification. For this study, 1.1 kg soybeans were washed and socked in 2.7 L tap water for 12–14 h at 24°C. The beans were drained, rinsed, combined with 6 L tap water and ground twice to a slurry in a Cherry Burrell Vibroreactor (Clinton, IA). The soymilk slurry was transferred to a steam-jacketed kettle (Lee Metal Products Co., Inc., Philipsburg, PA), heated to 95°C, stirred constantly, and held 7 min. The cooked slurry was the poured into a large mesh bag set in a perforated, stainless steel pressing container. The opening of the mesh bag was twisted closed. A cider-press (Day Equipment Corp., Goshen, IN) with a stainless steel press plate was used to provide sufficient pressure (900 psi) to press any remaining soymilk from the water-insoluble residue (okara). After the initial pressing, 500 mL tap water was added to wash the okara. The soymilk slurry was then filtered again through a finer mesh bag and repressed (300–600 psi).

Tofu Production

The slurry was strained through a double-layered medium mesh bag to separate the okara from the soymilk. The okara was pressed by using a hydraulic press (Day Equipment Co., Goshen, IN). The soymilk was adjusted to 5 °Brix (solids) (73) and allowed to cool to 85°C before the correct amount of calcium sulfate dihydrate (slurried in 50–51°C water) was added with rapid stirring. After coagulating, the curds were ladled into stainless steel pressing boxes lined with cheesecloth and pressed for 15 min. (68). The resulting tofu was cut, water packed into plastic containers, and stored at 4°C until evaluated the next day by instrumental and sensory procedures (68).

Traditional Soymilk and Tofu Preparation Using a Commercial Japanese Automated System

Soymilk was produced using a traditional processing procedure utilizing a Takai Automated Soymilk and Tofu System (Ishikawa-ken, Japan [9,72,73]). Prior to tofu production, soymilk was made using 3.0 kg whole soybeans soaked in water for 12–14 h, drained and ground twice in a Stephan Grinder (Stephan Machinery Corp., Columbus, OH), using a 0.5 mm cutter followed by a 0.05 mm cutter. During the grinding operation 40 L water was added (20 L during the first grind, and 20 L during the second pass through the grinder). The resulting slurry was put into the Taki paddle mixer for 10 min. and then injected into the injection steam cooker. The slurry was heated to 95°C and held at this temperature for 7 min. 40 sec. prior to the roller extraction cycle. The roller extractor utilized a 100 mesh and 120 mesh roller and screen, respectively. The soymilk produced was passed through a nylon filter sack into a stainless steel milk can or coagulation vessel where the soymilk was standardized to yield a 5 °Brix (solids) (73) solution for the production of a U.S. firm tofu.

Tofu Production

The soymilk was allowed to cool to 85°C before the correct amount of calcium sulfate dihydrate (slurried in 50–51°C water) was added with stirring. The curds were allowed to stand for 2 min., then broken using a stainless steel perforated plate with a slightly smaller diameter than the coagulating vessel. The broken curds were allowed to stand another 5 min. before they were ladled into the perforated (6 mm perforations spaced 8.25 cm apart) pressing boxes (49 × 43 × 19 cm) and pressed using twin CKD Corporation (Japan) air presses. The tofu was pressed for 5 min. at 2 kg/cm^2, then for 4 min. at 4 kg/cm^2, and finally for 5 min. at 6 kg/cm^2 (4). The temperature of the tofu remained above 60°C prior to cutting, and it was stored at 4°C after cutting.

Sensory Analysis

The tofu samples were served warm, in disposable styrofoam plates coded with random three-digit numbers. Panelists received three samples each time. Twenty trained panelists were used for descriptive and intensity evaluations. Two replications were run with 46 untrained consumer panelists each time. Unsalted crackers and room temperature water were given to cleanse the palate between samples. The samples were evaluated by the panelists utilizing a 15-centimeter, unstructured line scale in the individually divided booths under a white fluorescent light. The consumer panelists evaluated the sample for color, texture, flavor, texture preference, and flavor preference on a scale of 0 for extremely bland, white, soft, undesirable, to 15 for cream, beany, hard, desirable.

Proximate Analysis

Percent moisture, fat, and protein of the tofu were analyzed in triplicate using AOAC and AACC methods (75,76).

Instrumental Analysis

A Hunterlab (Labscan Spectrocolorimeter) L5-5100 Color Difference Meter (Hunter Associates Laboratory, Inc., Reston, VA) was used to measure the color differences among the tofu samples. The tofu samples were measured in small petri dishes wrapped in plastic wrap. The equipment was standardized with a plastic-wrapped white tile. The readings were taken by rotating the samples three times using the 10° Standard Observer and the fluorescent light source. The results were expressed in L (L = 0 for black, L = 100 for white), a (–a = green, +a = red), and b (–b = blue, +b = yellow).

The texture differences among the tofu samples were measured using Texture Profile Analysis on the Instron Universal Testing Machine (Model 1122, Instron, Corp., Canton, MA) equipped with a 5.5 cm diameter compression anvil (68,77). The samples were cut into 20 mm cubes and compressed to 20% of their original height twice. Compression hardness, springiness, cohesiveness, gumminess, and chewiness values were obtained. The crosshead and chart speed were 100 mm/min. and 100 mm/min., respectively. A 50 kg load cell was used (5 kg full scale reading), and the measurements were repeated four times.

Statistical Analysis

Analysis of Variance (ANOVA) and Least Significant Difference (LSD), were used to analyze the data using Statistical Analysis System Package (version 6.03 [78]).

Results

The author has evaluated tofu made from Vinton 81 and L-2 null IA 1003 over the last three years. In addition, tofu made from IA 1002 (L-2 null), and Century L_2L_3 were evaluated for two years and one year, respectively. In general, all of the null lines in each study (Tables 12.5–12.9) were less beany than their respective control cultivars. In the first study IA 1002, IA 1003, and Century L_2L_3 had the lowest beany flavor scores (Tables 12.5–12.6), but there was no statistical flavor preference between these varieties. There were no statistical differences between the two Century null lines. This is consistent with the findings that the L-3 modification is not as significant as the L-2 modification (63).

From a tofu texture standpoint, IA 1003 and Century L_2L_3 had softer texture than the other lines, but there were no textural preferences between the cultivars. There were no significant differences between any of the lines for protein or moisture content. However, IA 1003 was significantly lower in lipids, and it had the highest yield

TABLE 12.5 Composition and Yield of Lipoxygenase-Null and Commercial Soybean Tofu Using a Steam Kettle Pilot Plant Process[a]

Variety	Vinton 81 (STD)	Vinton 81[b]	IA 1002	IA 1003	Century	Century L_2	Century L_2L_3
Moisture	73.7	76.27	73.48	80.26	73.06	73.08	75.94
Protein	43.09	41.57	41.5	42.18	40.69	39.65	44.57
Lipid ($\alpha = 0.0001$)	32.79AB[c]	34.99B	10.85D	26.28C	34.00B	34.05B	—
Yield ($\alpha = 0.02$)	151CD	148CD	137D	224A	1678BCD	169BC	191B

[a]Tofu protein and lipid are % dry weight.
[b]Different capital letters in a row are significantly different.
[c]Different source.

TABLE 12.6 Sensory and Instrumental Characteristics of Tofu Made from Lipoxygenase-Null and Commercial Soybeans

Variety	Vinton 81	Vinton 81[e]	IA 1002	IA 1003	Century	Century L_2	Century L_2L_3
Texture[a]							
Hardness	5.32	3.78	4.08	1.51	4.72	4.25	2.35
Fracturability	2.50	1.76	1.70	0.84	1.92	1.77	1.12
Cohesiveness	0.33	0.31	0.30	0.27	0.32	0.31	0.34
Gumminess	1.73	1.17	1.22	0.41	1.52	1.32	0.80
Color[b]							
L	83.2	86.4	85.8	88.0	82.4	80.5	74.15
a ($\alpha = 0.002$)	0.18AD	0.57C	–0.02AB	–0.01AB	0.13AB	–0.14B	0.12AB
b	12.2	12.777	13.6	15.4	12.9	12.6	10.5
Sensory[cd]							
Color	0.00AC	1.38AD	4.39B	5.99B	–1.26C	1.92D	1.32D
Flavor	0.00AB	–0.453B	–2.15BC	–4.17C	2.01A	–0.88B	–2.01BC
Texture	0.00AB	1.86A	0.62AB	–5.39C	0.21B	–1.19B	–5.10C
Flav-pref	0.00AB	0.08AB	2.70C	2.88C	–1.46A	0.46AB	1.94BC
Tex-pref	0.00A	–0.84A	–0.27A	–2.10A	–0.81A	0.21A	–0.21A

[a]Instron texture units: Hardness, kg; Fracturability, kg; Cohesiveness, No units; Adhesiveness, (–)kg; Springiness, nm; Gumminess, kg; and Chewiness, kg × mm.
[b]Instrumental color (hunter L = 0 (black) – 100 (white); a = (–) green to (+) red; b = (–) blue to (+) yellow.
[c]Different letters indicate statistical significance (color, flavor; texture $\alpha = 0.0001$; flavor preference $\alpha = 0.004$; texture preference $\alpha = 0.05$).
[d]All sensory values were standardized against Vinton 81 (included in each sensory panel). Sensory scales (color: higher scores indicate a more cream-colored sample; flavor: higher scores indicate a more beany flavor; texture: higher scores indicate a harder tofu; flavor and texture preference: higher scores indicate a more preferred sample).
[e]Different source.

TABLE 12.7 Composition and Yield of Tofu Made From IA 1003 (Lipoxygenase-2 Null) and Vinton-81 (1993/1994) Soybeans

Bean Variety	Yield	% Moisture	% Protein[a]	% Oil[a]	% Hardness
IA 1003	222	79.5	55.4	30.4	49
Vinton-81	214	81.4	55.3	28.4	51

[a]Dry weight basis.

TABLE 12.8 Instrumental Color Evaluation of Tofu Made from Commercial IA 1003 (Lipoxygenase-2 Null) and Vinton-81 (1993/1994) Soybeans

Bean Variety	L	a	b
IA 1003	84.4	–0.6	11.1
Vinton-81	76.6	–0.5	11.1

TABLE 12.9 Sensory Evaluation of Tofu Made from Commercial IA 1003 (Lipoxygenase-2 Null) and Vinton-81 (1993/1994) Soybeans

Bean Variety	Color[a]	Flavor[b]	Texture	Flavor Preference[c]	Texture Preference[d]
IA 1003	9.26	6.19	6.82	8.69	8.22
Vinton 81	8.42	6.98	8.81	7.69	8.00

[a]Color: White = 0.00, Cream = 15.00. No preference attached.
[b]Flavor: Weak beany = 0.00, strong beany = 15.00. No preference attached.
[c]Flavor preference: Disliked = 0.00, liked = 15.00.
[d]Texture preference: Disliked = 0.00, liked = 15.00.

of tofu. In the second two-year study using the commercial Takai System (Tables 12.7–12.9), similar results were observed.

The IA 1003 was not different from the Vinton 81 control in composition, yield, texture, or instrumentally measured color (Tables 12.7–12.8). This was supported by the sensory color, texture, and texture preference data (Table 12.9). However, the panelists (Table 12.9) did find significant flavor and flavor preference differences between the two cultivars. The L-2 null line was significantly less beany and preferred over the Vinton 81 control. These studies confirm the decreased beany flavor results from the lipoxygenase-null soymilk and soy flour studies for other null lines (57,63,69,70. They also demonstrate that tofus can be made from lipoxygenase-null lines, without the loss of other compositional, yield, or texture characteristics.

Utilization of Dried Soymilk and Tofu as Ingredients in Meat Products

Over the last 3 years, the author has also investigated the feasibility of reducing fat in meat products utilizing dried soymilk and tofu. Rahardjo et al. reported that the texture of pork sausage was improved with no flavor alterations when spray-dried soymilk, made from Vinton 81 or IA 1003 (lipoxygenase-2 null) soybeans, was added at a 3% level (74). The pork sausage made with the dried soymilk produced a sausage that was 50–60% lower in fat and 41% lower in calories than regular sausage. Ho et al. (submitted for publication) also reported similar results when the soy was added to pork sausage in the form of dried tofu. Unfortunately, the added spice masked differences between the soybean varieties in both of these studies. Rahardjo and Rahardjo et al. found that there were no differences due to the addition of lipoxygenase-2 null (IA 1003) and Vinton 81 spray-dried soymilk to lean frankfurters for flavor preference or yield (79; submitted for publication). The soy addition reduced the fat content of regular frankfurters.

Shen and Shen et al. investigated the influence of dried tofu made from Vinton 81 and IA 1003 (L-2 null line) on the composition and sensory properties of regular (20% fat) and lean (10% fat) hamburgers (80; submitted for publication). As the percentage of dried tofu increased from 0 to 5%, a trained panel observed decreased red color intensity, increased beany flavor, and increased hamburger tenderness and juiciness. This indicated that the L-1 and/or L-3 isozymes contribute to the beany flavor of the hamburger with the increasing levels of L-2 null dried tofu. However, the lipoxygenase-2 null line had less beany flavor than the commercial variety (Vinton 81) in hamburgers (Table 12.10).

Shen et al. found that there were differences in the acceptance of regular and lean beef patties with the addition of 2% dried IA 1003 or Vinton 81 tofu (80). The lipoxygenase-2 null line IA 1003 was preferred over lean, and Vinton 81-containing hamburger patties in a ranking preference test (Fig. 12.1). The IA 1003-added patty was equivalent to the regular hamburger patty in both the ranking test (Fig. 12.1) and in a separate 9-point hedonic scale sensory test (Table 12.11 [80]). The hamburger patties

TABLE 12.10 Mean Sensory Color, Flavor, and Texture Scores of Beef Patties Made with Different Levels of Dried Vinton 81 or Lipoxygenase-2 Null IA 1003 Tofu[a]

Treatments[b]	RBP	LBP	V1BP	V3BP	V5BP
Raw BP color	12.94[c]	12.71[c]	7.11[d]	5.96[d]	5.69[d]
Cooked BP color	13.19[c]	12.39[c]	7.21[d]	6.92[d]	5.94[d]
Beany flavor	3.02[e]	3.23[e]	5.64[d]	6.86[d]	7.23[c]
Beef flavor	8.51[c]	9.60[c]	6.51[d]	5.81[d]	5.16[e]
Tenderness	7.21[cd]	7.03[d]	7.45[cd]	8.21[c]	8.83[c]
Juiciness	6.97[d]	6.05[d]	7.86[cd]	8.58[c]	8.45[c]
Chewiness	7.98[c]	7.54[c]	8.05[c]	7.26[c]	7.40[c]

[a]All attributes were measured on a 15-cm line scale with 0 = bright red, dark, bland, tough, dry, least chewy, and 15 = dark red, light dark, strong beany flavor, beef flavor, extremely tender, juicy, most chewy for raw beef patty (BP) color, cooked beef patty color, beany flavor, beef flavor, tenderness, juiciness, and chewiness.
[b]RBP = regular control, LBP= lean control, V1BP, V3BP, and V5BP were represented as 1%, 3%, and 5% dried Vinton 81 tofu added beef patties, respectively.
[cde]Mean scores with different letters in a row are significantly different ($P < 0.05$).
Source: Chen (80).

containing dried tofu had a redder color, less rancidity, lower cooking losses, lower fat, and higher moisture content than the regular control patties. The lipoxygenase-2 null dried tofu ". . . can be used to formulate lean beef patties (7% fat) that have equivalent or better sensory properties than regular or lean all beef patties." (Shen et al., submitted for publication). These studies (74,79,80; and Ho et al., submitted for publication) demonstrate that:

1. Bland-flavored soybean varieties (low beany-low lipoxygenase activity) can and should be used for these meat applications.
2. Dried soymilk and dried tofu from both null and traditional lines have good water- and fat-holding properties in meats as food ingredients. Similar results have been observed in bakery applications (Wilson, unpublished data).
3. It is possible to make lower fat (lean) pork sausage using dried soymilk or dried tofu that are equal to or better than the control (no soy) and Iota carrageenan-containing pork sausage without compromising the flavor. The soy-containing products had a higher cooking yield and less shrinkage (significant visual difference), and they met composition standards. The soymilk treatments were more tender, juicy, and preferred over the control lean sausage. There was no difference between the control and tofu treatment.
4. It is possible to make lower fat frankfurters using dried soymilk or dried tofu that are equivalent to or better than the control or commercial frankfurters for yield and most sensory characteristics, while still meeting composition standards. Regular and lean frankfurters containing dried tofu were preferred over their controls for texture and overall acceptability.
5. Dried tofu products generally had fewer flavor problems than spray-dried soymilk, and were easier to produce in small scale operations (Wilson, unpublished data).

Conclusions

Food companies, researchers, and nutritionists have been demanding better tasting, less beany flavored soybeans for decades. The plant breeders have done their part by producing lipoxygenase-null soybean varieties. Based upon the soyfood and limited food application research to date, there is no doubt that these null lines have the functional properties of traditional soybeans with a less beany flavor. At this time, it is not known whether the new triple-null lines will be better than single- or double-null lines in food applications. As the triple-null lines become available, they need to be evaluated in traditional soyfoods (soymilk, tofu, soy ice cream, etc.) to determine if these products can become more acceptable in Western diets. The utilization of lipoxygenase-null soybean flour to replace the more costly soy concentrates and isolates as food ingredients also needs study. We now must wait for the farmers, seed companies, and food-processing industry to decide who will take the first step to encourage the production and utilization of these lipoxygenase-null soybean varieties in food.

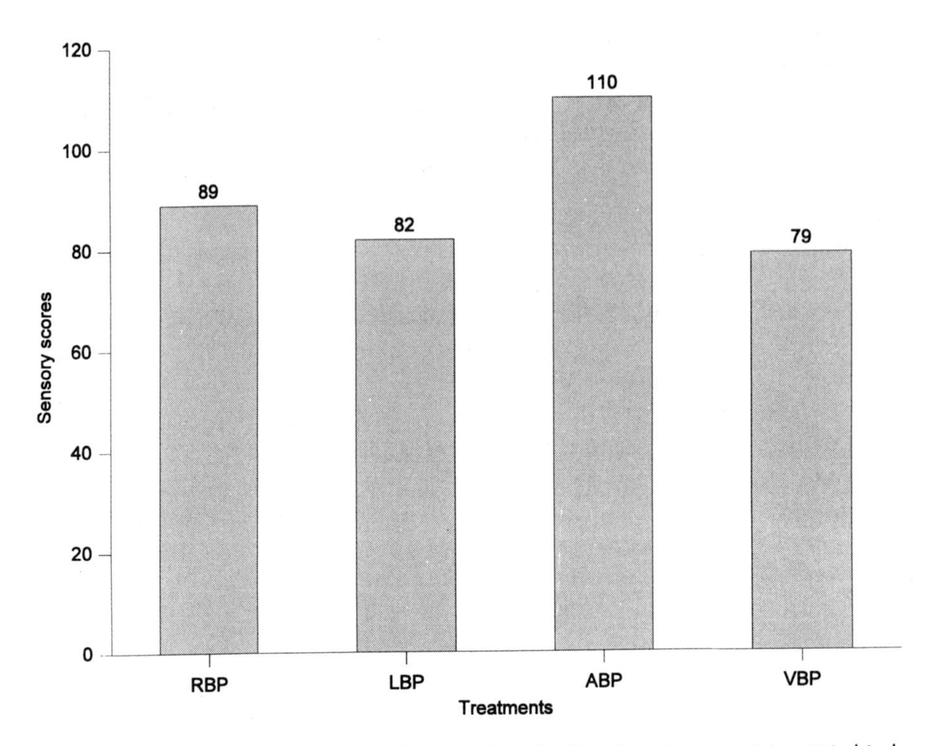

Figure 12.1. Consumer sensory evaluation of cooked beef patties containing 2% dried Vinton 81 or lipoxygenase-2 null IA 1003 tofu. *Source:* Chen (80).

TABLE 12.11 Mean Sensory Scores of Cooked Beef Made with Different Levels of Dried Vinton 81 or Lipoxygenase-2 Null IA 1003 Tofu[a]

Treatment	RBP	LBP	ABP	VBP
Tenderness	5.85[b]	4.95[b]	5.75[b]	5.35[b]
Juiciness	5.25[b]	3.90[c]	4.80[bc]	4.30[bc]
Texture preference	6.05[b]	5.80[b]	6.10[b]	6.05[b]
Flavor	6.15[b]	6.25[b]	5.70[b]	5.35[b]
Overall acceptability	6.15[b]	5.70[b]	6.05[b]	5.95[b]

[a]All attributes were measured on a 9-point scale with 1 = very tough, very dry, least preferable, extreme beef flavor, and extremely undesirable; 9 = extremely tender, extremely juicy, most preferable, extreme beany flavor, and extremely desirable. RBP = Regular beef patties, LBP = Lean beef patties, ABP = Iowa 1003 dried tofu treatment beef patties, and VBP = Vinton 81 dried tofu treatment beef patties.

[bc]Mean scores with different letters in a row are significantly different ($P < 0.05$).

Source: Chen (80).

References

1. American Heart Association (1986) *Circulation 74,* 1465A.
2. Rakosky, J. (1970) *J. Agr. Food Chem. 18,* 1005–1009.
3. Reitmeier, C.A., and Prusa, K.J. (1987) *J. Food Sci. 52,* 916–918.
4. Ahmed, P.O., Miller, M.F., Lyon, C.E., Vaughters, H.M., and Reagan, J.O. (1990) *J. Food Sci. 55,* 625–628.
5. Wang, H., and Murphy, P.A. (1994) *J. Agric. Food Chem. 42,* 1666–1673.
6. Kalbrener, J.E., Moser, H., and Wolf, W.J. (1971) *Cereal Chem. 48,* 595–600.
7. Hammonds, T.M., and Call, D.L. (1972) *Chem. Technol. 2,* 156–162.
8. Rackas, J.J., Sessa, D.J., and Honig, D.H. (1979) *J. Am. Oil Chem Soc. 56,* 262–271.
9. Wilson, L.A. (1995) in *Practical Handbook of Soybean Processing and Utilization,* Erickson, D.R., AOCS Press, Champaign, IL, pp. 428–459.
10. Smith, A.K., and Circle, S.J. (1972) *Soybean Chemistry and Technology, Vol. 1,* AVI Publishing Co., New York, pp. 339–346; 294–338.
11. Wilson, L.A. (1985) *World Soybean Research Conference III: Proceedings,* Shibles, R., ed., Westview Press, Inc., Boulder, CO, pp. 158–165.
12. Wolf, W.J. (1975) *J. Agric. Food Chem. 23,* 126–141.
13. Wolf, W.J., and Cowan, J.D. (1975) *Soybeans as a Food Source,* CRC Press, Cleveland, OH, pp. 65–67.
14. Rackis, J.J., Hong, D.H., Sessa, D.J., and Steggerda, D.F.R. (1970) *J. Agric. Food Chem. 18,* 977–982.
15. Schutte, L., and Van Der Ouweland, G.A.M. (1979) *J. Oil Chem. Soc. 56,* 289–290.
16. Watanabe, T., and Kishi, A. (1984) *The Book of Soybeans,* Japan Publications, Inc., New York, pp. 28–29.
17. Snyder, H.E., and Kwon, T.W. (1987) *Soybean Utilization,* AVI Books, Van Nostrand Reinhold Co., New York, pp. 55–58; 74–144.
18. Wilson, L.A., Murphy, P.A., and Gallagher, P. (1992) *Soyfood Product Markets in Japan: U.S. Export Opportunities.* Matric, Ames, IA, pp. 1–32.
19. Whitaker, J.R. (1994) *Principles of Enzymology,* Marcel Dekker, Inc., New York, pp. 579–593.
20. Lusas, E.W., and Rhee, K.C. (1995) in *Practical Handbook of Soybean Processing and Utilization,* Erickson, D.R., American Oil Chemists' Society Press, Champaign, IL pp. 117–160.
21. Siedow, J.W. (1991) *Ann. Rev. Plant Physiol. Plant Mol. Biol. 42,* 145–188.
22. Zhuang, H., Hildebrand, D.F., Anderson, R.A., and Hamilton-Kempt, T.R. (1991) *J. Agric. Food Chem. 39,* 1357–1364.
23. Arai, S., Noguchi, M., Yamashita, M., Kata, Y., and Fujimaki, M. (1970) *Agric. Biol. Chem. 34,* 1569–1573.
24. Beyeler, M., and Solms, J. (1974) *Lebensm.-Wiss. Technol. 7,* 217–219.
25. Twigg, G., Kotula, A.W., and Young, E.P. (1977) *J. Anim. Sci. 44,* 218–223.
26. Kinsella, J.E., and Damodaran, S. (1980) in *The Analysis and Control of Less Desirable Flavors in Foods and Beverages,* Charalambous, G., ed., Academic Press, New York, pp. 95–131.
27. Damodaran, S., and Kinsella, J.E. (1981) *J. Agric. Food Chem. 29,* 1249–1253.
28. Damodaran, S., and Kinsella, J.E. (1981) *J. Agric. Food Chem. 29,* 1253–1257.
29. Ashraf, H.L., and Snyder, H.E. (1981) *J. Food Sci. 46,* 1201–1204.

30. Wilson, L.A. (1990) *Economic Implications of Modified Soybean Traits,* in Special Report 92, Iowa Agricultural and Home Economics Experiment Station, Iowa State University, Ames, Iowa, pp. 14–15.
31. Fujimaki, M., Arai, S., Kirigaya, N., and Sakurai, Y. (1965) *Agric. Biol. Chem. 29,* 855–863.
32. Goosens, A.E. (1975) *Food Process. Ind. 44,* 29–30.
33. Sessa, D.J. (1979) *J. Agric. Food Chem. 27,* 234–239.
34. Hsieh, O.A.L., Huang, A.S., and Chang, S.S. (1982) *J. Food Chem. 47,* 16–18.
35. Wilkens, W.F., Mattick, L.R., and Hand, D.B. (1967) *Food Tech. 21,* 1630–1633.
36. Mustakas, G.C., Albrecht, W.J., McGhee, J.E., Black, L.T., Brookwalter, G.N., and Griffin, E.L., Jr. (1969) *J. Am. Oil. Chem. Soc. 46,* 623–626.
37. Nelson, A.I., Steinburg, M.P., and Wei, L.S. (1975) U.S. Patent 3,901,978.
38. Nelson, A.I., Steinburg, M.P., and Wei, L.S. (1975) *J. Food Sci. 41,* 57–61.
39. Borhan, M., and Snyder, H.E. (1979) *J. Food Sci. 44,* 586–590.
40. Johnson, L.A., Deyoe, C.W., and Hoover, W.J. (1981) *J. Food Sci. 46,* 239–243.
41. Kwok, K.-C., and Niranjan, K. (1995) *Int. J. Food Sci. Techn. 30,* 263–295.
42. Franzen, K.L., and Kinsella, J.E. (1974) *J. Agric. Food Chem. 22,* 675–678.
43. Gremli, H.A. (1974) *J. Am. Oil Chem. Soc.,* 51, 95A-97A.
44. Aspelund, T.G., and Wilson, L.A. (1980) in *World Soybean Research Conference II: Abstracts,* Corbin, F.T., Westview Press, Boulder, Colorado, pp. 32.
45. Crowther, A.L., Wilson, L.A., and Glatz, C.E. (1980) *J. Food Process Eng. 4,* 99–115.
46. Thissen, J.A. (1982) Interaction of Off-Flavor Compounds with Soy Protein Isolate in Aqueous Systems: Effects of Chain Length, Functional Group, and Temperature. M.S. Thesis, Iowa State University, Ames, Iowa, p. 119.
47. Aspelund, T.G., and Wilson, L.A. (1983) *J. Agric. Food Chem. 31,* 539–545.
48. O'Keefe, S.F., Resurreccion, A.P., Wilson, L.A., and Murphy, P.A. (1991) *J. Food Sci. 56,* 802–806.
49. O'Keefe, S.F., Wilson, L.A., Resurreccion, A.P., and Murphy, P.A. (1991) *J. Agric. Food Chem. 39,* 1022–1028.
50. Maheshwari, P., Ooi, E.T., and Nikolov, Z.L. (1995) *J. Am. Oil Chem. Soc. 72,* 1107–1115.
51. Chiba, H., Takahashi, N., Kitabatake, N., and Sasaki, R. (1979) *Agric. Biol. Chem. 43,* 1891–1897.
52. Chiba, H., Takahashi, N., and Sasaki, R. (1979) *Agric. Biol. Chem. 43,* 1883–1889.
53. Takahashi, N., Sasaki, R., and Chiba (1979) *Agric. Biol. Chem. 43,* 2556–2561.
54. Maheshwari, P. (1995) Characterization and Application of Porcine Liver Aldehyde Oxidase in the Removal of Off-Flavors from Soy Proteins, Ph.D. Thesis, Iowa State University, Ames, Iowa, 129 pp.
55. Axelrod, B., Cheesbrough, T.M., and Laakso, S. (1981) *Methods Enzymol. 71,* 441–451.
56. Hildebrand, D.F., and Hymowitz (1981) *J. Am. Oil Chem. 58,* 583–586.
57. Kitamuri, K., Davies, C.S., Kaizuma, N., and Nielsen, N.C. (1983) *Crop Sci. 23,* 924–927.
58. Kitamuri, K., Kumagai, T., and Kikuchi, A. (1985) *Jpn. J. Breed. 35,* 413–420.
59. Davies, C.S., and Nielsen, N.C. (1986) *Crop Sci. 26,* 460–463.
60. Hajika, M., Igita, K., and Kitamuri, K. (1991) *Jpn. J. Breed. 41,* 507–509.
61. Hajika, M., Kitamuri, K., Igita, K., and Nakazawa, Y. (1992) *Jpn. J. Breed. 42,* 781–792.

62. Matoba, T., Hidaka, H., Kitamuri, K., Kainuma, N., and Kito, M. (1985) *J. Agric. Food Chem. 33,* 852–855.
63. Davies, C.S., Nielsen, S.S., and Nielsen, N.C. (1987) *J. Oil Chem. Soc. 64,* 1428–1433.
64. *Iowa Gold Catalog* (1993) Iowa Department of Agriculture and Land Stewardship Publishers, Des Moines, Iowa, pp. 43–48.
65. *Iowa Gold Catalog* (1994) Iowa Department of Agriculture and Land Stewardship Publishers, Des Moines, Iowa, pp. 33–38.
66. *Iowa Gold Catalog* (1995) Iowa Department of Agriculture and Land Stewardship Publishers, Des Moines, Iowa, pp. 36–41.
67. Wilson, L.A., Murphy, P.A., and Gallagher, P.W. (1991) *Japanese Soyfoods: Products and Markets.* C.P. 1, Center for Crops Utilization Research, Iowa State University, Ames, Iowa, pp. 6–31.
68. McVey, M.J., Pautsch, G.R., and Baumel, C.P. *Beneficiaries of Modified Soybean Traits* (1993) Special Report 93, Iowa Agricultural and Home Economics Experiment Station, Iowa State University, Ames, Iowa, pp. 15–27.
69. Kobayashi, A., Tsuda, Y., Hirata, N., Kubota, K., and Kitamuri, K. (1995) *J. Agric. Food Chem. 43,* 244–245.
70. Kitamuri, K., Ishimoto, M., Kikuchi, A., and Kainuma, N. (1992) *Jpn. J. Breed. 42,* 905–913.
71. Lee, E.B. (1995) Sensory Characteristics of Rapid Hydration Hydrothermally-Cooked Soymilk. M.S. Thesis, Iowa State University, Ames, pp. 84.
72. Johnson, L.D. (1984) Influence of Soybean Variety and Method of Processing on Tofu Manufacturing, Quality, and Consumer Acceptability. M.S. thesis. Iowa State University, Ames, pp. 128.
73. Johnson, L.D., and Wilson, L.A. (1984) *J. Food Sci. 49,* 202–204.
74. Rahardjo, R., Wilson, L.A., and Sebranek, J.G. (1994) *J. Food Sci. 59,* 1286–1290.
75. Association of Official Analytical Chemists (1990) *Official Methods of Analysis, 14th edn.,* Method 925.10, 954.01, 955.04, Association of Official Analytical Chemists, Washington, DC.
76. American Association of Cereal Chemists (1983) *Approved Methods, 8th edn.,* AACC Method 30-25, American Association of Cereal Chemists, Inc., St. Paul, MN.
77. Bourne, M.C. (1968) *J. Food Sci. 33,* 223.
78. SAS Institute, Inc. (1985) *SAS Computer Program, version 6.03,* Trademark of SAS Institute Inc., Cary, NC.
79. Rahardjo, R. (1993) Utilization of Spray Dried Soymilk in Pork Sausage Patties and Frankfurters, M.S. Thesis, Iowa State University, Ames, pp. 118.
80. Shen, C. (1996) Utilization of Dried Tofu to Improve Sensory and Other Properties of Beef Patties. M.S. Thesis, Iowa State University, Ames, pp. 80.

Index